New Perspectives of Colloids for Biological Applications

New Perspectives of Colloids for Biological Applications

Guest Editor

Agnieszka Ewa Wiącek

MDPI

Basel • Beijing • Wuhan • Barcelona • Belgrade • Novi Sad • Cluj • Manchester

Guest Editor
Agnieszka Ewa Wiącek
Department of Interfacial
Phenomena
Maria Curie-Skłodowska
University
Lublin
Poland

Editorial Office
MDPI AG
Grosspeteranlage 5
4052 Basel, Switzerland

This is a reprint of the Special Issue, published open access by the journal *International Journal of Molecular Sciences* (ISSN 1422-0067), freely accessible at: https://www.mdpi.com/journal/ijms/special_issues/Colloids_Biological.

For citation purposes, cite each article independently as indicated on the article page online and as indicated below:

Lastname, A.A.; Lastname, B.B. Article Title. *Journal Name* **Year**, *Volume Number*, Page Range.

ISBN 978-3-7258-4941-3 (Hbk)
ISBN 978-3-7258-4942-0 (PDF)
https://doi.org/10.3390/books978-3-7258-4942-0

Contents

About the Editor

Agnieszka Ewa Wiącek

Agnieszka Ewa Wiącek, a professor at UMCS, received her doctorate from the Faculty of Chemistry at UMCS in 2000, her habilitation in 2013, and has been a professor at the university since 2018. She is the author or co-author of over 100 scientific papers, mostly published in international journals, and nearly 200 conference proceedings. In 2022, Dr. hab. A.E. Wiącek was among the distinguished group of over a dozen UMCS employees among the most frequently cited scientists in the Elsevier ranking (2522 citations according to Google Scholar, IH 25). A.E. Wiącek's research focuses primarily on interfacial phenomena, the stability of dispersed systems, the wetting process, the determination of surface free energy, the properties of biologically active compounds, and their applications in the food industry, pharmacy, and implantology.

International Journal of
Molecular Sciences

MDPI

Editorial

New Perspectives of Colloids for Biological Applications

Agnieszka Ewa Wiącek

Department of Interfacial Phenomena, Faculty of Chemistry, Maria Curie-Skłodowska University, Maria Curie-Skłodowska Sq. 3, 20-031 Lublin, Poland; agnieszka.wiacek@mail.umcs.pl

The field of colloid systems is still a developing scientific area, but a very promising one for many practical applications. This Special Issue of the *International Journal of Molecular Sciences*, entitled New Perspectives of Colloids for Biological Applications, is dedicated to recent advances in the research of colloids and colloid-based systems, focused on highlighting recent, interesting investigations and new perspectives conducted in leading laboratories around the world. The manuscripts included in this Book focus on research data from the field of molecular chemistry; the production, modification, study, and application of colloids and colloidal systems; and the application of highly specialized methods to advance science and, above all, improve health and quality of life. This Reprint covers two main topics. The first group of the collected articles concerns application potential in medical and pharmaceutical sciences; the second covers agricultural, nutritional, and cosmetic applications. All the listed articles (12 in total) represent the latest trends in research on innovative colloidal systems for biological applications.

Aleksander Czogalla et al. contributed a paper entitled "Development of stable, maleimide-functionalized peptidoliposomes against SARS-CoV-2," in which they presented a step-by-step process for the development of maleimide-functionalized liposomes that, when decorated with a SARS-CoV-2-binding peptide, could inhibit the progression of COVID-19 infection. A major emphasis was placed on defining the optimal lipid composition and conditions for the formation of pegylated liposomes. The authors proposed that the developed nanocarrier technology could be used as a universal platform for the construction of many antiviral agents.

Sonia Kudłacik-Kramarczyk and Marcel Krzan et al. contributed the paper "Study on the Effect of Emulsifiers on the Properties of Oleogels Based on Olive Oil Containing Lidocaine", which reports and investigates the synthesis of olive oil-based oleogels, which are among the healthiest and most valuable vegetable fats, rich in unsaturated fatty acids and antioxidants such as vitamin E. Two types of surfactants were used: TWEEN 80, which lowers surface tension and stabilizes emulsions, and SPAN 80, which acts in oil-dominated phases. The oleogels were enriched with lidocaine, an active substance commonly used as a pain reliever and local anesthetic. This research characterized the obtained oleogels through their medical applications, paying particular attention to the influence of surfactant type and amount as well as the active substance on their physicochemical properties. The results indicated that SPAN-stabilized oleogels exhibited better stability and favorable mechanical properties, making them promising candidates for medical applications, particularly in pain relief formulations.

Izabela Nowak et al. contributed a paper entitled "Solid Lipid Nanoparticles Incorporated with Retinol and Pentapeptide-18—Optimization, Characterization, and Cosmetic Application" in which the stability of lipid nanocarriers containing retinol and peptides was investigated. To confirm the effectiveness of method optimization (high shear homogenization, HSH) and proper selection of substrates, SLN dispersions were obtained in three combinations: 1—non-incorporated SLNs; 2—SLNs containing only retinol; and 3—SLNs containing retinol and pentapeptide-18. It was confirmed that the optimized dispersion of lipid nanoparticles incorporated with retinol and oligopeptide led to the desired physicochemical parameters over the required storage period, and the possibility of using the obtained raw material as an ingredient in cosmetic products with anti-aging properties was indicated.

Cheng Peng, Li Guo, and co-workers contributed a paper entitled "Topical Delivery of Dual Loaded Nano-Transfersomes Mediated Chemo-Photodynamic Therapy against Melanoma via Inducing Cell Cycle Arrest and Apoptosis" which presents characteristics of transfersomes (TFS) designed to facilitate transdermal delivery of the chemotherapeutic drug 5-Fluorouracil (5-FU) and the PS Imperatorin (IMP) for combined chemo-photodynamic therapy for melanoma. The cytotoxic and phototoxic effects of TFS-mediated PDT (TFS-UVA) were investigated in A375 cells and mice. The authors also demonstrated that TFS-UVA generated intracellular ROS, induced G2/M phase cell cycle arrest, and promoted cell apoptosis.

Huiqiong Yan et al. contributed a paper entitled "Synthesis, Optimization and Molecular Self-Assembly Behavior of Alginate-g-Oleylamine Derivatives Based on Ugi Reaction for Hydrophobic Drug Delivery" in which, on the basis of previous research, the authors further optimized the synthesis process parameters of alginate-g-oleylamine derivatives (Ugi-FOIT). In addition, they explored the effects of different degrees of substitution (DSs) on the molecular self-assembly properties of Ugi-FOIT and its in vitro cytotoxicity and drug release behavior to achieve an optimal alginate-based oral formulation for the delivery of hydrophobic drugs. The resultant Ugi-FOIT exhibited good amphiphilic properties, and was able to self-assemble into spherical micellar aggregates in an aqueous solution and indicated good cytocompatibility. Ugi-FOIT micellar aggregates with favorable stability also showed a certain sustained and pH-responsive release behavior for the hydrophobic drug ibuprofen.

Hongmei Xia and co-workers contributed a paper entitled "Antioxidant Activity and the Therapeutic Effect of Sinomenine Hydrochloride-Loaded Liposomes-in-Hydrogel on Atopic Dermatitis", which presents a new preparation for drug delivery. In this case, liposomes were dispersed in the colloidal hydrogel on a mesoscopic scale. The results showed that the sinomenine hydrochloride-loaded liposomes-in-hydrogel system could effectively inhibit atopic dermatitis.

Pakorn Kraisit et al. contributed a paper entitled "Fabrication and Characterization of Pectin Films Containing Solid Lipid Nanoparticles for Buccal Delivery of Fluconazole", in which novel FZ-loaded solid lipid nanoparticles (FZ-SLNs) were developed in pectin solutions. To produce pectin films with FZ-SLNs, four formulations were selected based on the small particle size of FZ-SLNs and their suitable polydispersity index, which provided the mucoadhesive matrices. The evaluation of mechanical properties unveiled the influence of particle size variation in FZ-SLNs on the integrity of the film. The Fourier-transform infrared spectra indicated that hydrogen bonds could potentially form between the pectin-based matrix and the constituents of FZ-SLNs. In addition, the films containing FZ-SLNs provided good permeability across the porcine mucosa and antifungal properties, suggesting the potential utility of pectin films incorporating FZ-SLNs for buccal administration.

Wiącek and Furmaniuk contributed a paper entitled "Starch-Based Polysaccharide Systems with Bioactive Substances: Physicochemical and Wettability Characteristics", which presents the physicochemical characteristics of combinations of starch with phospholipids or lysozyme, determining the effects of starch modification (surface hydrophobization or biological additives) and preparation temperature (before and after gelatinization). Surface hydrophobization of starch carried out using an *n*-alkane film does not change its bulk properties and leads to improvements in its mechanical and functional properties. The obtained specific starch-based hybrid systems, characterized in detail by switchable wettability, raises the possibility of determining the energetic state of the starch surface and understanding the strength and specificity of interactions with substances of different polarities in biological processes and their applicability for multidirectional use.

Sujka and Wiącek contributed a paper entitled "Physicochemical Characteristics of Porous Starch Obtained by Combined Physical and Enzymatic Methods, Part 1: Structure, Adsorption, and Functional Properties", analyzing the possibility of obtaining a porous structure for native corn, potato, and pea starches using a combination of ultrasound, enzymatic digestion, and freeze-drying methods. The structure of the native and modified starches was examined using VIS spectroscopy, SEM, ATR-FTIR, and LTNA (low-temperature nitrogen adsorption). The selected properties, such as the water and oil holding capacities, least gelling concentration (LGC), and paste clarity, showed that the corn starch was the most susceptible to the combined modification methods and was therefore best suited for the production of porous starch.

Alejandro Zentella Dehesa and Ashutosh Sharma et al. contributed a paper entitled "A Comparative and Critical Analysis for In Vitro Cytotoxic Evaluation of Magneto-Crystalline Zinc Ferrite Nanoparticles Using MTT, Crystal Violet, LDH, and Apoptosis Assay", which presents the synthesis and characteristics of zinc ferrite nanoparticles (ZFO NPs) as a promising magneto-crystalline platform for nanomedicine-based cancer theranostics. ZFO NPs synthesized using a co-precipitation method we are characterized using UV–visible spectroscopy, FTIR spectra, XRD methods, HR-TEM, XPS, and flow cytometric analysis, confirming that the NPs were more cytotoxic towards triple-negative breast cancer cells (MDA-MB-231) as compared to breast adenocarcinoma cells (MCF-7) and normal cell lines (HEK-293). ZFO can therefore be exploited for cancer therapeutics.

Pastuszak, Jurak, and co-workers contributed a paper entitled "Insight into the Mechanism of Interactions between the LL-37 Peptide and Model Membranes of *Legionella gormanii* Bacteria", in which they presented the properties of the model membranes formed by phospholipids isolated from *L. gormanii* bacteria cultured on both non-supplemented and choline-supplemented media. The effects of the LL-37 peptide on the intermolecular interactions, packing, and ordering under monolayer compression were determined. Penetration tests at a constant surface pressure were carried out, showing that the peptide binds to the anionic bacterial membranes preferentially due to its positive charge and can penetrate into the hydrophobic tails of phospholipids, destabilizing membrane integrity. This process can entail membrane disruption and cell death. The ability to evoke such a great membrane destabilization is dependent on the share of electrostatic, hydrogen bonding, and Lifshitz–van der Waals LL-37−PL interactions. Thus, the LL-37 peptide action depends on the changes in the lipid membrane composition caused by the utilization of exogenous choline by the *L. gormanii*.

Mohammed Ghazwani et al. contributed a paper entitled "QbD-Optimized, Phospholipid-Based Elastic Nanovesicles for the Effective Delivery of 6-Gingerol: A Promising Topical Option for Pain-Related Disorders" which presents elastic nanovesicles, constructed of phospholipids optimized by 6-gingerol (6-G), a natural chemical that may alleviate osteoporosis and musculoskeletal-related pain. A 6-gingerol-loaded transfersome (6-GTF) formulation was optimized and showed more antioxidant activity than the pure 6-G suspension. The formulation was converted into a gel to improve skin retention and efficacy, which demonstrated improved skin absorption, drug release, and antioxidant activity and has the ability to treat pain-related illnesses effectively. Hence, the described study offers a possible topical treatment for conditions connected to pain.

We would like to thank all the authors who submitted manuscripts to our Special Issue for the high quality of their papers. We also express our sincere gratitude to current and future readers. We hope that our Special Issue will be a platform for dialogue between scientists working in the field of colloids and for new perspectives on colloid-based systems for various applications, especially broadly understood biological applications.

Conflicts of Interest: The authors declare no conflict of interest.

List of Contributions:

1. Michel, O.; Kaczorowska, A.; Matusewicz, L.; Piórkowska, K.; Golec, M.; Fus, W.; Kuliczkowski, K.; Sikorski, A.F.; Czogalla, A. Development of Stable, Maleimide-Functionalized Peptidoliposomes Against SARS-CoV-2. *Int. J. Mol. Sci.* **2025**, *26*, 1629. https://doi.org/10.3390/ijms26041629.
2. Kudłacik-Kramarczyk, S.; Przybyłowicz, A.; Drabczyk, A.; Kieres, W.; Socha, R.P.; Krzan, M. Study on the Effect of Emulsifiers on the Properties of Oleogels Based on Olive Oil Containing Lidocaine. *Int. J. Mol. Sci.* **2024**, *25*, 11067. https://doi.org/10.3390/ijms252011067.
3. Pawłowska, M.; Marzec, M.; Jankowiak, W.; Nowak, I. Solid Lipid Nanoparticles Incorporated with Retinol and Pentapeptide-18—Optimization, Characterization, and Cosmetic Application. *Int. J. Mol. Sci.* **2024**, *25*, 10078. https://doi.org/10.3390/ijms251810078.
4. Guo, Y.; Zhong, W.; Peng, C.; Guo, L. Topical Delivery of Dual Loaded Nano-Transfersomes Mediated Chemo-Photodynamic Therapy against Melanoma via Inducing Cell Cycle Arrest and Apoptosis. *Int. J. Mol. Sci.* **2024**, *25*, 9611. https://doi.org/10.3390/ijms25179611.
5. Bu, Y.; Chen, X.; Wu, T.; Zhang, R.; Yan, H.; Lin, Q. Synthesis, Optimization and Molecular Self-Assembly Behavior of Alginate-g- Derivatives Based on Ugi Reaction for Hydrophobic Drug Delivery. *Int. J. Mol. Sci.* **2024**, *25*, 8551. https://doi.org/10.3390/ijms25158551.
6. Chen, X.; Wu, Y.; Jia, R.; Fang, Y.; Cao, K.; Yang, X.; Qu, X.; Xia, H. Antioxidant Activity and the Therapeutic Effect of Sinomenine Hydrochloride-Loaded Liposomes-in-Hydrogel on Atopic Dermatitis. *Int. J. Mol. Sci.* **2024**, *25*, 7676. https://doi.org/10.3390/ijms25147676.
7. Hirun, N.; Mahadlek, J.; Limmatvapirat, S.; Sriamornsak, P.; Yonemochi, E.; Furuishi, T.; Kraisit, P. Fabrication and Characterization of Pectin Films Containing Solid Lipid Nanoparticles for Buccal Delivery of Fluconazole. *Int. J. Mol. Sci.* **2024**, *25*, 5413. https://doi.org/10.3390/ijms25105413.
8. Wiącek, A.E.; Furmaniuk, A. Starch-Based Polysaccharide Systems with Bioactive Substances: Physicochemical and Wettability Characteristics. *Int. J. Mol. Sci.* **2024**, *25*, 4590. https://doi.org/10.3390/ijms25094590.
9. Sujka, M.; Wiącek, A.E. Physicochemical Characteristics of Porous Starch Obtained by Combined Physical and Enzymatic Methods, Part 1: Structure, Adsorption, and Functional Properties. *Int. J. Mol. Sci.* **2024**, *25*, 1662. https://doi.org/10.3390/ijms25031662.
10. de la Fuente-Jiménez, J.L.; Rodríguez-Rivas, C.I.; Mitre-Aguilar, I.B.; Torres-Copado, A.; García-López, E.A.; Herrera-Celis, J.; Arvizu-Espinosa, M.G.; Garza-Navarro, M.A.; Arriaga, L.G.; García, J.L.; et al. A Comparative and Critical Analysis for In Vitro Cytotoxic Evaluation of Magneto-Crystalline Zinc Ferrite Nanoparticles Using MTT, Crystal Violet, LDH, and Apoptosis Assay. *Int. J. Mol. Sci.* **2023**, *24*, 12860. https://doi.org/10.3390/ijms241612860.
11. Pastuszak, K.; Kowalczyk, B.; Tarasiuk, J.; Luchowski, R.; Gruszecki, W.I.; Jurak, M.; Palusinska-Szysz, M. Insight into the Mechanism of Interactions between the LL-37 Peptide and Model Membranes of *Legionella gormanii* Bacteria. *Int. J. Mol. Sci.* **2023**, *24*, 12039. https://doi.org/10.3390/ijms241512039.
12. Ghazwani, M.; Alqarni, M.H.; Hani, U.; Alam, A.QbD-Optimized, Phospholipid-Based Elastic Nanovesicles for the Effective Delivery of 6-Gingerol: A Promising Topical Option for Pain-Related Disorders. *Int. J. Mol. Sci.* **2023**, *24*, 9983. https://doi.org/10.3390/ijms24129983.

International Journal of
Molecular Sciences

MDPI

Article

Development of Stable, Maleimide-Functionalized Peptidoliposomes Against SARS-CoV-2

Olga Michel [1,*], Aleksandra Kaczorowska [1,2], Lucyna Matusewicz [1], Kliwia Piórkowska [3], Marlena Golec [3], Wiktoria Fus [3], Kazimierz Kuliczkowski [3], Aleksander F. Sikorski [4,5] and Aleksander Czogalla [1,*]

1 Department of Cytobiochemistry, Faculty of Biotechnology, University of Wrocław, F. Joliot Curie 14a, 50-383 Wrocław, Poland
2 Department of Biomedical Engineering, Faculty of Fundamental Problems of Technology, Wrocław University of Science and Technology, Wybrzeże S. Wyspiańskiego 27, 50-370 Wrocław, Poland
3 Silesian Park of Medical Technology Kardio-Med Silesia, M. Curie-Skłodowskiej 10C, 41-800 Zabrze, Poland
4 Acellmed Ltd., M. Curie-Skłodowskiej 10C, 41-800 Zabrze, Poland
5 Research and Development Center, Regional Specialist Hospital, Kamienskiego 73a, 51-154 Wrocław, Poland
* Correspondence: olga.michel@umw.edu.pl (O.M.); aleksander.czogalla@uwr.edu.pl (A.C.); Tel.: +48-71-784-06-88 (O.M.); +48-71-375-63-56 (A.C.)

Abstract: Throughout the last 5 years, extensive research has been carried out towards the development of effective treatments for coronavirus disease 2019 (COVID-19). Regardless of the worldwide efforts, only a few drugs have passed clinical trials, and there is still a need to develop therapies, especially for those who are particularly vulnerable to a severe disease course. Maleimide-functionalized liposomes are proposed to serve as a platform for the immobilization, stabilization, and delivery of a short peptide sequence with high affinity towards severe acute respiratory syndrome coronavirus 2 (SARS-CoV-2). However, extensive optimizations should be performed in order to achieve features required for a reliable drug candidate, such as homogeneity of physical parameters and their long-term stability. Here, we present a step-by-step development process for maleimide-functionalized liposomes, which—once decorated with the SARS-CoV-2-binding peptide—could inhibit the infection progress of COVID-19. The main emphasis is placed on defining optimal lipid composition and formation conditions of PEGylated liposomes. We propose that the developed nanocarrier technology can be used as a universal platform for the construction of multiple antiviral agents.

Keywords: decoy receptors; liposomes; extrusion; microfluidization; high-pressure homogenization; SARS-CoV-2; maleimide; peptidoliposomes

1. Introduction

The World Health Organization (WHO) declared the outbreak of the coronavirus disease 2019 (COVID-19) pandemic on 11 March 2020. Since then, over 7,000,000 COVID-19-related deaths have been reported worldwide [1], making causative severe acute respiratory syndrome coronavirus 2 (SARS-CoV-2) one of the deadliest viruses in human history. Given the rapid spread and severity of symptoms, increased efforts have been made to develop drugs against this novel viral infection, resulting in more than 10,000 clinical trials registered for COVID-19 since the pandemic's outbreak [2,3]. However, so far, only several drugs for COVID-19 have been approved or are under Emergency Use Authorization (EUA) by the U.S. Food and Drug Administration (FDA) [4]. These can be divided into four groups: (i) antiviral agents, (ii) immune modulators, (iii) monoclonal antibodies (mAbs) targeting SARS-CoV-2, and (iv) renal replacement therapy solutions. The first group includes drugs

 https://doi.org/10.3390/ijms26041629

that reduce the risk of hospitalization or death—Paxlovid (nirmatrelvir and ritonavir) and Lagevrio (molnupiravir), as well as Veklury (remdesivir), which is also approved to treat or support patients hospitalized due to COVID-19. The primary role of immune modulators is to help suppress the hyperinflammation frequently accompanying SARS-CoV-2 infection. This group currently includes Actemra (tocilizumab), Olumiant (baricitinib), Kineret (anakinra), and Gohibic (vilobelimab). Among the anti-SARS-CoV-2 monoclonal antibodies, Bebtelovimab, Sotrovimab, and REGEN-COV (Casirivimab and Imdevimab) were, at one point, the only mAb-based drugs for the treatment of COVID-19. However, all of them were withdrawn from use due to the high-frequency appearances of non-susceptible SARS-CoV-2 variants. The long-circulating COVID-19-preventive mAb drug, i.e., Evusheld (tixagevimab co-packaged with cilgavimab), met the same fate; however, it was replaced by Pemgarda, which currently remains the only pre-exposure prophylaxis drug for high-risk patients. The last group of drugs is dedicated to patients treated with continuous renal replacement therapy, and it is represented by Regiocit and Fresenius Medical multiFiltrate/multiBic/multiPlus replacement solutions—both authorized under EUA for a certain group of adult patients. The limited number of available drugs and the withdrawal of those released on the market reflect the challenges in the development process of anti-SARS-CoV-2 drugs that are highly effective yet resistant to frequent virus mutations. At the same time, it is commonly believed that the next pandemic is a matter of 'when', not 'if' [5–7]. Developing a universal, antiviral platform is therefore not only an important step towards the enhanced treatment of COVID-19 disease but can also increase our level of preparedness for future pandemics.

Understanding and modifying the immune response to SARS-CoV-2 infection became a matter of utmost importance as the pandemic progressed. The vaccine's development was predominantly focused on the viral spike (S) glycoprotein [8], which plays a key role in virus entry through its interaction with the angiotensin-converting enzyme 2 (ACE2) receptor at host cell surfaces [9,10]. As soon as it was demonstrated that SARS-CoV-2 polyclonal antibodies inhibit SARS-CoV-2 spike-mediated cell penetration [10], the concept emerged to deliver the genetic information to produce the antigen in the form of nucleic acid-based messenger RNA (mRNA) [11]. Unfortunately, the utility of the mRNA vectors is majorly limited by their instability in the human body, as well as their high immunogenicity [12]. The entrapment of mRNA inside the PEGylated liposomes opened new doors for the possibility of adapting to the evolution of the virus [13], but also became a prime example of how mature vesicle technology can be successfully employed to face the SARS-CoV-2 infection. One possible approach to antiviral therapy is multiple applications of decoy receptors using various native or recombinant formulations of ACE2 or peptides derived from the native sequence (for recent reviews see e.g., Li J et al. (2024) [14], Zhang et al. (2023) [15], or Matusewicz et al. (2022) [2]). One of the biggest problems in using recombinant ACE2 or peptides simulating receptor sequences is their formulation, which needs to allow for proper delivery and effective inhibition of the virus–host cell interaction. Our aim is to construct a decoy receptor liposomal carrier that would ensure effective inhibition of virus interaction with cells bearing the ACE2 receptor and be suitable for delivery via nebulization. However, the production of a liposome-based drug requires significant optimization. Besides expressing well-documented functionality, liposomes must be characterized by excellent homogeneity and long-term stability. One way to enhance both the functionality and stability of liposomes is modifying the liposome surface with maleimide. It was demonstrated that employing as little as 0.3 mol% of maleimide into liposomes led to faster internalization and protruded liposomes' presence at the injection site, resulting in the remarkedly improved drug delivery efficiency of liposomes in vitro and in vivo [16]. In-depth studies have revealed that thiol groups facilitate the cellular uptake of maleimide-modified liposomes and that the interaction of maleimide with thiols induces alternative liposome internalization, including endocytosis and energy-independent transport [17]. The therapeutic potential of maleimide-functionalized liposomes has been revealed in several anticancer studies in which enhanced penetration, prolonged retention, and decreased toxicity were achieved in vitro and in vivo [18,19]. These encouraging results indicate that free maleimide groups on liposome surfaces may boost their biological effects. Apart from the entrapment of cytotoxic agents, maleimide-functionalized liposomes have been used in anticancer treatment to carry targeting molecules, such as p53-directed LinTT1 peptide [20], antibodies [21,22] or aptamers [23,24], to name a few. A similar approach is employed in the present study; however, instead of a targeting molecule, liposomes are decorated with a therapeutic agent, a short peptide sequence with high affinity towards the receptor-binding domain (RBD) of the SARS-CoV-2 S protein. The main goal of this study is to prepare and characterize a stable formulation based on small lipid vesicles (liposomes) decorated with human ACE2-derived peptides that can be used for further functionality studies. The greatest emphasis is placed on optimizing the lipid composition and the calibration process (and their impact on the preparation's stability), as well as the process of peptide attachment to vesicles via maleimide residues.

2. Results and Discussion

The results presented here include studies conducted on 51 preparations with 19 different lipid compositions, as shown below (Table 1).

Table 1. Lipid composition of tested liposomal formulations during the development of anti-SARS-CoV-2 maleimide-functionalized therapeutic peptidoliposomes.

	Lipid Concentration in Tested Formulations (mol%)											
Lipid / Formula #	HSPC	CHOL	DSPC	DPPG	DOPG	POPG	EggPG	DSPE-PEG-1K	DSPE-PEG-2K	DSPE-PEG-2K-Mal	LysoPC	*n*
1	90	10	-	-	-	-	-	-	-	-	-	2
2	70	10	10	-	-	5	-	4.5	-	0.5	-	6
3	70	10	10	5	-	-	-	4.5	-	0.5	-	10
4	-	10	80	5	-	-	-	4.5	-	0.5	-	1
5	69	10	10	-	-	5	-	4.5	-	0.5	1	1
6	60	20	10	-	-	5	-	4.5	-	0.5	-	1
7	50	30	10	-	-	5		4.5	-	0.5		1
8	70	10	10	-	-	-	5	4.5	-	0.5	-	4
9	74.5	10	10	-	-	5	-	-	-	0.5	-	1
10	65	10	10	-	-	10	-	4.5	-	0.5	-	1
11	56	39	-	-	-	-	-	-	4.5	0.5	-	3
12	70	10	10	-	5	-	-	4.5	-	0.5	-	1
13	70	10	10	-	-	5	-	4	-	1	-	1
14	70	10	10	-	-	5	-	2.5	-	2.5	-	2
15	70	10	10	-	-	5	-	-	-	5	-	6
16	65	10	10			5	-	-	-	10	-	1
17	55	10	10	-	-	5	-	-	-	20	-	1
18	70	10	10	-	-	5	-	5	-	-	-	1
19	70	10	10	-	-	-	5	-	-	5	-	7
											Σ	51

Abbreviations: CHOL—cholesterol; DSPC—1,2-distearoyl-sn-glycero-3-phosphocholine; DSPE-PEG-1K—1,2-distearoyl-sn-glycero-3-phosphoethanolamine-N-[methoxy(polyethylene glycol)-1000]; DSPE-PEG-2K—1,2-distearoyl-sn-glycero-3-phosphoethanolamine-N-[methoxy(polyethylene glycol)-2000]; DSPE-PEG-2K-Mal—1,2-distearoyl-sn-glycero-3-phosphoethanolamine-N-[maleimide(polyethylene glycol)-2000]; DOPG—1,2-dioleoyl-sn-glycero-3-phospho-(1′rac-glycerol), DPPG—1,2-dipalmitoyl-sn-glycero-3-phospho-(1′rac-glycerol); EggPG—L-α-phosphatidylglycerol from egg; HSPC—hydrogenated soy L-α-phosphatidylcholine; LysoPC—1-stearoyl-sn-glycero-3-phosphocholine; POPG—1-palmitoyl-2-oleoyl-sn-glycero-3-phospho-(1′racglycerol).

Right after formulation, vesicles were calibrated. Next, a portion of the liposomes were retained for further studies (such as phosphorus concentration measurements and, in some cases, dynamic light scattering, DLS, studies) and the rest was conjugated with the peptide. Finally, the excess peptide was removed through dialysis to receive the final product, referred to as "peptidoliposomes". The complete procedure for the preparation of the peptidoliposomes is described and graphically presented in the Section 3.3. The optimization of peptidoliposome production included the selection of the liposome calibration method (pressure extrusion vs. high-pressure homogenization), the selection of the most favorable lipid composition, and the determination of the optimal conditions for peptide attachment to liposomes.

2.1. Selection of Vesicle Calibration Method

2.1.1. The Impact of the High-Pressure Homogenization (HPH) Chamber on the Liposome Calibration Process Performed with the Microfluidics LM20 System

In the project, two techniques were employed for liposome calibration: pressure extrusion and high-pressure homogenization (HPH). Both techniques are widely acknowledged for liposome production [25–30]. For HPH, two Microfluidics LM20 systems were used with different chambers (G10Z vs. F20Y, both provided by Microfluidics). The influence of the microfluidic homogenization chamber on the calibration process was examined in liposomes composed of HSPC/CHOL (Formula #1) and HSPC/CHOL/DSPC/POPG/DSPE-PEG_{1000}/DSPE-PEG_{2000}-Mal (Formula #2). Regardless of the lipid composition, an equilibrium in the hydrodynamic diameter of liposomes was achieved more quickly (with fewer cycles) with a liposome-dedicated chamber, F20Y, than with the universal chamber, G10Z (Figure 1). On the other hand, significant differences were noted only up to the fourth cycle, implying that the impact of the homogenization chamber on liposome calibration diminishes with an increasing number of homogenization cycles. The final average hydrodynamic diameter of liposomes calibrated with the G10Z chamber was 54.16 ($\pm$0.62) nm for Formula #1 (Figure 1A) and 53.81 ($\pm$0.22) nm for Formula #2 (Figure 1B). Correspondingly, the diameter of liposomes calibrated with the F20Y chamber was 52.54 ($\pm$0.32) nm and 56.61 ($\pm$0.74) nm for Formula #1 and #2, respectively. After the last homogenization cycle, liposomes calibrated with the F20Y chamber were characterized by generally lower values of the polydispersity index (PDI) compared to liposomes produced with the G10Z chamber; nevertheless, the differences were not statistically significant ($p > 0.05$).

Previous studies on food nanoemulsions show that the appropriate geometry of the homogenization chamber may support obtaining uniform fluid dynamic conditions, which ensure a narrow droplet size distribution [31]. However, as the microfluidizer is usually equipped with a specific type of chamber, the data on the impact of chamber geometry on nanoemulsions are limited. In a study by Talsma et al. (1989), the M110 HPH apparatus was used with two chambers to produce liposomes (chambers A and B) [25]. The results are consistent with our study, as the impact of the chamber on liposome size was limited and was reduced by the increased number of homogenization cycles. However, chamber geometry was never discussed in the above-mentioned study. In the presented project, two identical LM 20 Microfluidizers were used, each with a different homogenization chamber. G10Z is a Z-type chamber, meaning that the lipid suspension is shot against the wall of the reaction chamber twice, whereas in a Y-type chamber, the liquid flow is split up into two paths, which are then redirected towards each other, thereby doubling the acting shear forces [32]. To our knowledge, our study is the first to compare the outcomes of liposome calibration with Y- and Z-type chambers. For both tested lipid compositions, liposomes subjected from six up to ten homogenization cycles were similar in terms of size and homogeneity. However, the equilibrium was reached with fewer cycles for a setup with an F20Y chamber. Since emulsion over-processing with HPH is possible, leading to phase separation and loss of stability [33], it is favorable to minimize the number of homogenization cycles. Based on those findings, a homogenizer equipped with an F20Y chamber was selected for further studies.

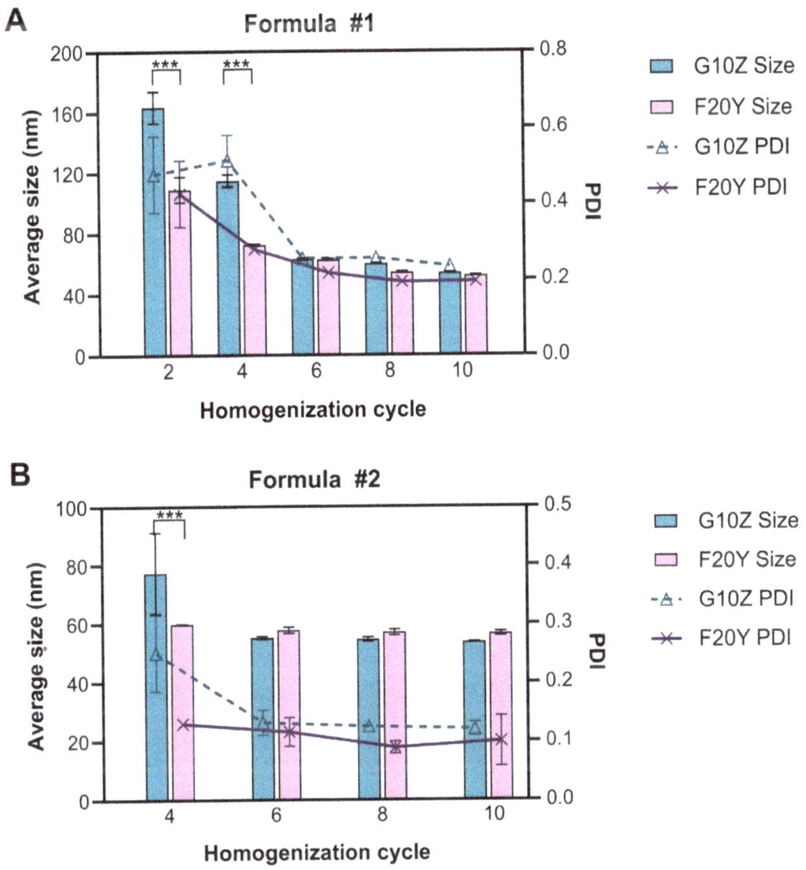

Figure 1. The impact of a high-pressure homogenization chamber (G10Z and F20Y) on the course of the homogenization process at 22,500 psi, as monitored via the size distribution of liposomes composed of HSPC/CHOL (**A**) or HSPC/CHOL/DSPC/POPG/DSPE-PEG$_{1000}$ and DSPE-PEG$_{2000}$-Mal (**B**); *** $p \leq 0.001$.

2.1.2. The Impact of the Calibration on the Long-Term Stability of Peptidoliposomes

As the designed liposomes are intended to serve as a platform for a peptide-based SARS-CoV-2-neutralising system, they must be stable upon storage (most preferably at room temperature or at 4 °C) for at least a couple of months; therefore, long-term stability is one of the most crucial parameters for quality assessment. To determine the optimal calibration conditions in terms of product stability, liposomes consisting of HSPC/CHOL/DSPC/DPPG/DSPE-PEG$_{1000}$/DSPE-PEG$_{2000}$-Mal (Formula #3) were subjected to calibration either by HPH with Microfluidizer LM20 equipped with an F20Y chamber or by the pressure extrusion technique. Both techniques enabled the formation of stable and homogeneous peptidoliposomes (Figure 2).

With the HPH method, smaller peptidoliposomes were produced, with an average hydrodynamic diameter of 59.46 nm (compared to 86.72 nm in their extruded counterparts). On the other hand, the suspension after homogenization was less homogeneous, which was reflected by a higher PDI (0.134 versus 0.065 in the extruded preparation). Homogenized peptidoliposomes remained stable for around 3 months at 4 °C, and afterward, the polydispersity significantly increased (Figure 2A). On the contrary, preparation based on liposomes subjected to pressure extrusion maintained high homogeneity for at least 134 days (Figure 2B), leading to the conclusion that pressure extrusion is superior to HPH in terms of peptidoliposome stability and polydispersity. This observation agrees with other studies with various lipid compositions [25,29]. On the other hand, in the study on adenovirus-containing liposomes, preparations produced with HPH and extrusion had similar performance in terms of transduction efficiencies, physicochemical characterization, and long-term storage stability [34], showing that the effects of calibration may vary depending on liposomes' composition.

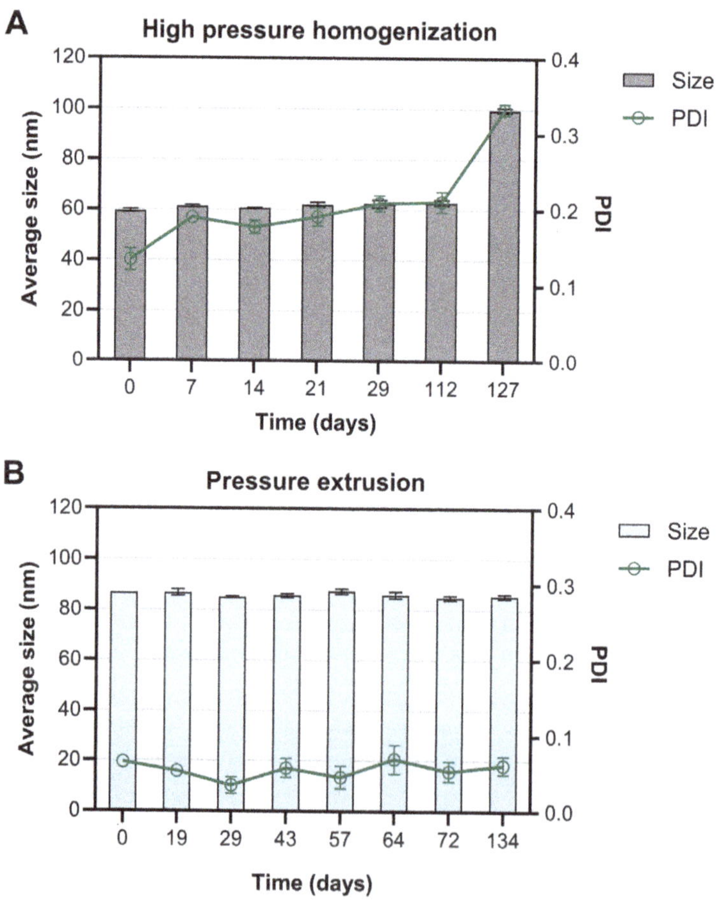

Figure 2. The influence of the calibration method on the long-term stability of peptidoliposomes (Formula #3). (**A**) Calibration with the high-pressure homogenization using the LM20 microfluidizer device; (**B**) calibration with pressure extrusion technique.

2.1.3. The Impact of the Calibration Method on Lipid Loss

Pressure extrusion has been successfully used to produce liposomes for decades [35]. Nevertheless, its use is limited by the lipid concentration, which—once too high—can lead to membrane clogging, slowing down the process and putting the product's sterility at risk when the membrane must be replaced [34–36]. As a consequence, this technique is usually employed for small, lab-scale liposome syntheses. On the contrary, HPH is a high-throughput method designed for the production of nanoemulsions, liposomes, and other nanoparticles on a (semi)-industrial scale in compliance with Good Manufacturing Practice (GMP) standards [34,37–40]. The remaining issue is that homogenization requires more optimizations and may significantly reduce the lipid concentration of processed liposomes [25]. To test the effect of the calibration technique on the effective lipid concentration, liposomes with various lipid compositions (Formula #2 and Formula #3) and the same starting lipid concentration of 10 mg/mL were subjected to calibration with pressure extrusion or HPH (equipped with a F20Y chamber) before the physicochemical characteristics of the liposomes were determined.

Regardless of the lipid composition, liposomes calibrated with pressure extrusion had significantly higher lipid concentrations than their homogenized counterparts (Figure 3). It should be emphasized that the drop in lipid concentration of homogenized liposomes results from liposome dilution with a void volume of the LM20 Microfluidizer device (~14 mL) and not from the material loss in the calibration process. In theory, it would have been possible to retrieve almost the entire sample from the device; however, the last homogenized portions were not used in the study, due to the increased risk of air entering the system, which could have compromised the liposomes' performance. It is worth noting that this is relevant only to low-scale preparations, as the larger the starting volume, the more negligible the dilution effect becomes.

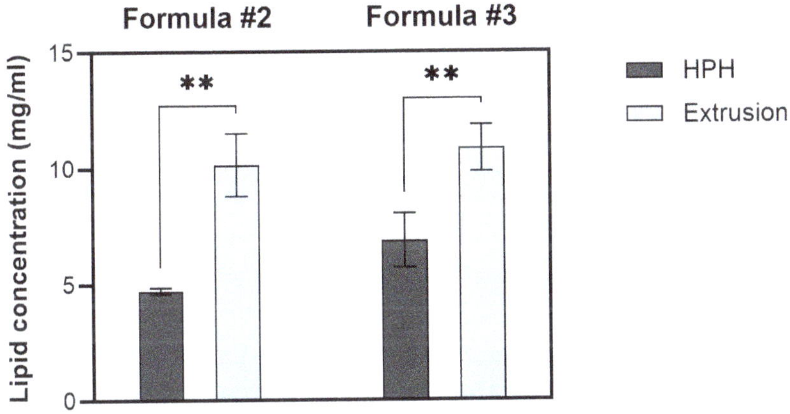

Figure 3. The influence of the calibration method (HPH versus pressure extrusion) on the lipid concentration of liposomes (Formulas #2 and #3); ** $p \leq 0.005$.

2.1.4. The Impact of the Calibration Method on the Functionality of Peptidoliposomes

The interaction of the S protein of the SARS-CoV-2 virus on the surface of infected cells with the neighboring cell receptors may induce the formation of giant structures, referred to as syncytia [41,42]. This phenomenon has further implications for the pathogenesis, viral dissemination, and immune evasion of many viruses, including SARS-CoV-2 [43,44]. The possibility to trigger syncytia in a manner independent of exogenous protease has been explored in the development of various anti-SARS-CoV-2 strategies, including studies on the cross-neutralization activity between convalescent sera from SARS and COVID-19 patients [41], or in the screening of fusion inhibitors [43]. In this study, the syncytia formation assay was employed to compare the functionality of peptidoliposomes consisting of HSPC/CHOL/DSPC/DPPG/DSPE-PEG$_{1000}$/DSPE-PEG$_{2000}$-Mal (Formula #3) calibrated with either HPH or the extrusion method. No significant differences were noted for the tested parameters (Figure 4).

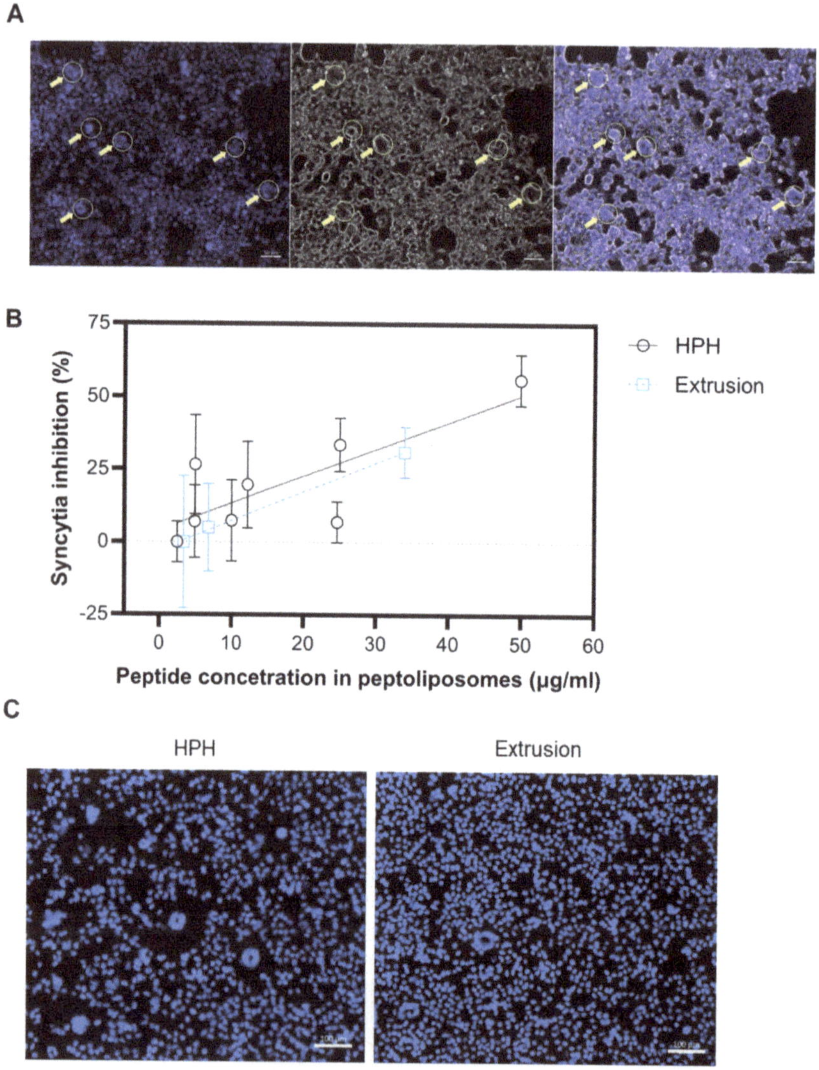

Figure 4. The influence of the calibration method (HPH versus pressure extrusion) on syncytia inhibition. (**A**) Syncytia forming upon the interaction of HEK-hACE2 with 293-SARS2-S, observed under a fluorescent microscope (Hoechst staining) and light microscope. Syncytia are marked with arrows, and the scalebar is 50 µm. (**B**) syncytia inhibition by anti-SARS-CoV-2 peptidoliposomes (Formula #3). Data are presented as single data points (with SD) along with lines from the simple linear regression analysis. (**C**) Representative images of syncytia formation by peptidoliposomes (Formula #3) calibrated with HPH or the extrusion technique. The scalebar is 100 µm.

2.2. Selection of the Optimal Lipid Composition

2.2.1. Modification of Phosphatidylcholine (PC)

Both synthetic and natural phospholipids are used in the pharmaceutical industry. However, the latter are currently preferred as they are derived from renewable sources, are developed in more environmentally friendly production processes, and are available on a larger scale at relatively low costs compared to synthetic phospholipids [45]. On the other hand, synthetic lipids do not require validated purification procedures and have better batch-to-batch reproducibility. Distearoylphosphatidylcholine (DSPC) was found to enhance encapsulation efficiency and liposome stability, allowing for greater effectiveness in bioprocessing [46,47]. In this study, the impact of phosphatidylcholine (PC) modification on the stability of the peptidoliposomal formulation was verified using the DLS and zeta potential (ZP) measurements. The impact of hydro soy phosphatidylcholine (HSPC) replacement with DSPC was examined on two formulations prepared from homogenized liposomes. As confirmed by *t*-test analyses, peptidoliposomes prepared primarily from HSPC (Formula #3) had a smaller average diameter ($p < 0.001$) and PDI ($p < 0.005$) than their DSPC counterparts (Formula #4; Table 2).

Table 2. The average hydrodynamic diameter, polydispersity index (PDI), and zeta potential (ZP) of peptidoliposomes with phosphatidylcholine modifications synthesized with high-pressure homogenization (HPH) or with pressure extrusion techniques. For each formulation, $n = 3$.

	Formula #3	Formula #4	Formula #2	Formula #5
Calibration method	HPH	HPH	Extrusion	Extrusion
Diameter (nm)	59.46 ± 0.56	75.79 ± 1.05	79.07 ± 0.86	75.24 ± 0.11
PDI	0.134 ± 0.015	0.221 ± 0.009	0.067 ± 0.048	0.057 ± 0.007
ZP (mV)	−27.33 ± 0.82	−26.37 ± 0.54	−30.78 ± 1.1	−30.60 ± 0.73

The replacement of HSPC with DSPC impaired the stability of the peptidoliposomal formulation, which was reflected by the increased average hydrodynamic diameter (to 95.08 nm) and elevated PDI (0.276) after 30 days of storage at 4 °C (Figure 5).

DSPC is believed to increase the rigidity of liposomes. However, data from the molecular dynamics studies and flicker noise spectroscopy measurements on HSPC/DSPC bilayers demonstrate that the bending rigidity coefficient was slightly lower for DSPC than HSPC [48]. Nevertheless, HSPC and DSPC bilayers were characterized by similar geometry, with similar outcomes for membrane thickness and area per lipid. To verify if rigidity was a key parameter influencing the stability of tested peptidoliposomes, lysophosphatidylcholine (LysoPC) was introduced into the composition. LysoPCs can occur as minor constituents in biological membranes, where they play an essential role in phospholipid metabolism and signaling [49]. Incorporating lysolipids with a low-value molecular packing parameter ($p < 1$) increases the curvature stress in the bilayer, resulting in more leaky membranes [50–52]. The stability of two peptidoliposomal formulations was compared, each based on extruded liposomes—one composed of HSPC/CHOL/DSPC/POPG/DSPE-PEG_{1000}/DSPE-PEG_{2000}-Mal (Formula #2) and the other with 1 mol% LysoPC in lieu of HSPC (Formula #5). Peptidoliposomes with added LysoPC were slightly smaller than their HSPC counterparts ($p < 0.001$) but did not significantly differ in terms of PDI or ZP (Table 2). Both formulations with or without LysoPC were equally stable for at least 204 days of incubation at 4 °C (Figure 5B) and under the conditions of the accelerated aging assay (72 h at 37 °C in DMEM culture medium; Figure 5C). Although data do not support LysoPC inclusion in the anti-SARS-CoV-2 peptidoliposomes, the knowledge that up to 1 mol% of Lyso PC can be safely included in liposomes can be useful for further pharmacological specification and drug development to increase liposome functionality in vivo [53].

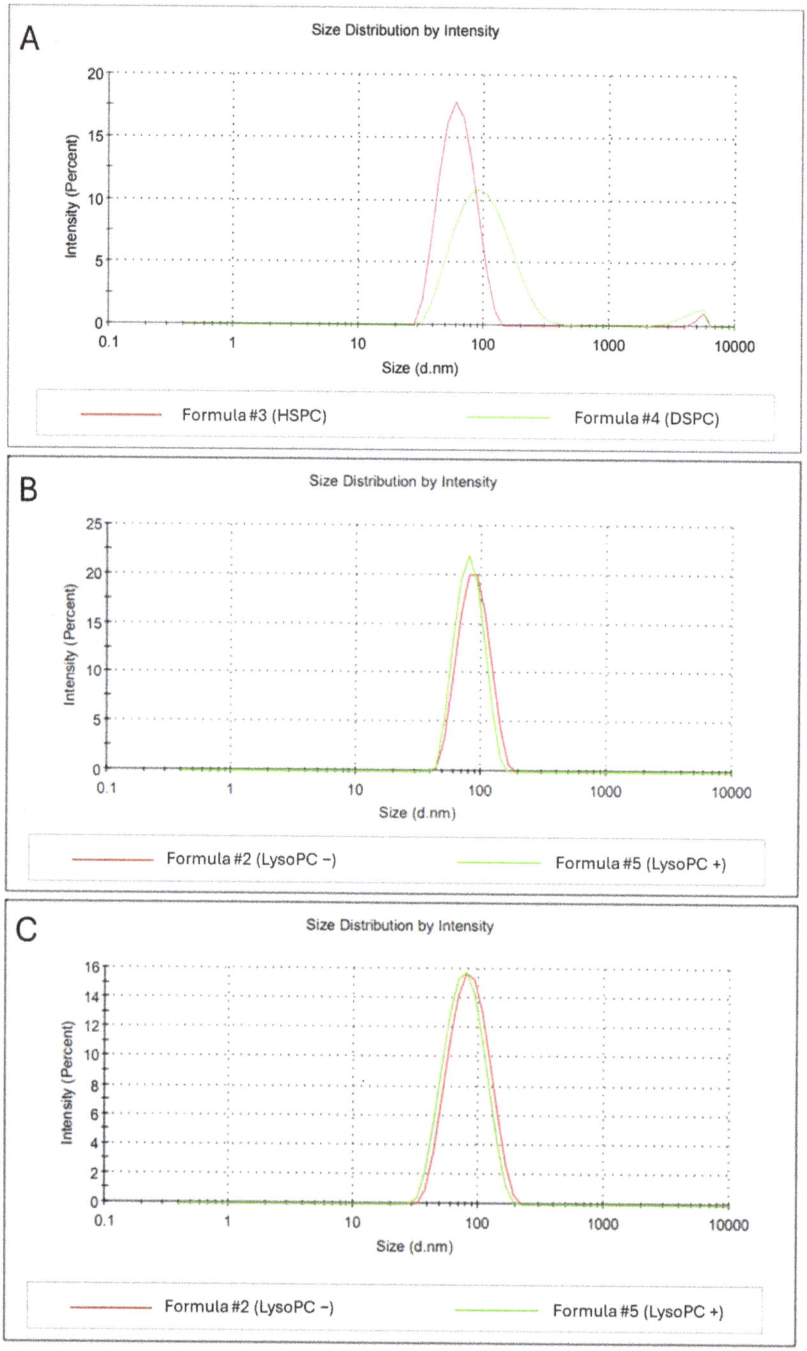

Figure 5. The influence of phosphatidylcholine (PC) modification on the homogeneity of anti-SARS-CoV-2 peptidoliposomes. (**A**) Average hydrodynamic diameters of preparations with the leading component being HSPC (Formula#3) or DSPC (Formula#4) after 30 days at 4 °C. (**B**,**C**) Average hydrodynamic diameters of peptidoliposomes without (Formula #2) and with the addition of LysoPC (Formula #5) after 204 days of incubation at 4 °C and after the accelerated aging assay, respectively.

2.2.2. Modification of the Cholesterol (CHOL) Concentration

Cholesterol (CHOL) is known to exert stabilizing effects when incorporated into the lipid composition of liposomes [54–56]. In the project, we examined the impact of increased CHOL (balanced with HSPC depletion) on the basic characteristics of peptidoliposomes as well as their stability. All preparations were manufactured from liposomes calibrated with the pressure extrusion technique. As confirmed by *t*-test analysis, peptidoliposomes with 10 mol% CHOL were slightly smaller than their counterparts with 20 mol% CHOL ($p < 0.05$) and significantly smaller than peptidoliposomes with 30 mol% CHOL ($p < 0.001$; Table 3). Whereas formulations with 10 and 20 mol% CHOL had similar polydispersity, that of the formulation with CHOL rose to 30 mol% (Formula #7), and it was significantly less homogeneous compared with those with lower cholesterol (Formula #2 and Formula #6; $p \leq 0.01$). Changes in CHOL concentration did not affect the zeta potential of peptidoliposomes (Table 3).

Table 3. The average hydrodynamic diameter (nm), polydispersity index (PDI), and zeta potential (ZP) of peptidoliposomes with increasing concentrations of cholesterol. For each formulation, $n = 3$.

	Formula #2	Formula #6	Formula #7
Cholesterol (mol%)	10	20	30
Diameter (nm)	77.12 ± 1.29	83.02 ± 1.33	107.73 ± 1.68
PDI	0.067 ± 0.017	0.059 ± 0.004	0.177 ± 0.010
ZP (mV)	−28.97 ± 1.26	−31.13 ± 1.25	−30.53 ± 0.69

Under the conditions of the accelerated aging assay, increasing the cholesterol concentration above 10 mol% negatively affected liposome stability (Figure 6).

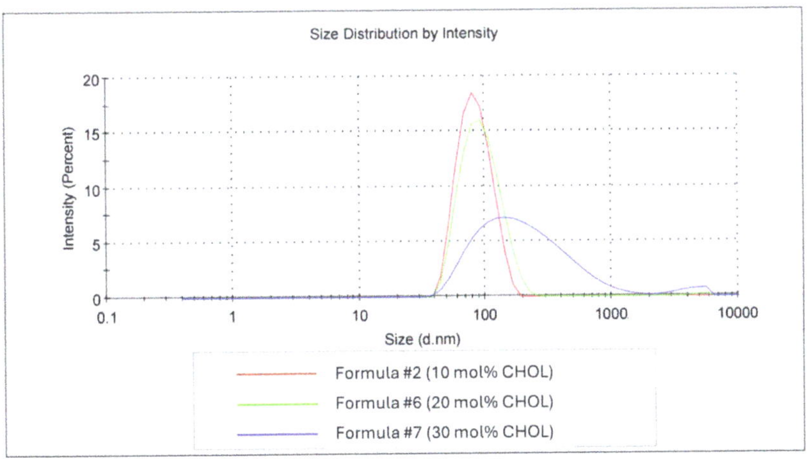

Figure 6. The average hydrodynamic diameters of peptidoliposomes with different cholesterol concentrations following the accelerated aging assay.

The average hydrodynamic diameter of liposomes with 10, 20, and 30 mol% CHOL was 78.19, 86.95, and 158.30 nm, respectively. The average PDI significantly increased with the cholesterol content reaching 0.090, 0.128, and 0.359 for peptidoliposomes with 10, 20, and 30 mol% CHOL, respectively. The increased aggregation was also reflected by the increased precipitation in the culture medium (Supplementary Material, Figure S1). It is known that the impact of cholesterol on liposome behavior may vary, as it depends on the lipid composition [57]. In membranes rich in saturated lipids, cholesterol tends to condense, increasing the order of bilayers, whereas cholesterol's affinity towards unsaturated lipids is much lower [58]. Due to its impact on membrane fluidity, the presence of cholesterol may affect the performance of nanotherapeutics, including the encapsulation efficiency, release profile, and permeability [57,59]. In a study by Briuglia et al. (2015), the most controlled and reproducible release for drugs with different physicochemical characteristics and pharmaceutical applications was achieved for liposomes with phospholipids, with a cholesterol ratio of 2:1 [56]. In our study, such high CHOL concentrations resulted in a loss of stability, and the most favorable properties were characteristic for the preparation which contained only 10 mol% CHOL. Importantly, Matusewicz et al. (2018) demonstrated that cholesterol-free liposomes were toxic to treated cells [60], which indicates that cholesterol may be an indispensable component of the lipid composition and confirms the necessity to optimize cholesterol content during the drug development process.

2.2.3. Effect of Polyethylene Glycol (PEG)-Modified Lipids

Modifications with polyethylene glycol (PEG) are known to shield liposomes from plasma proteins in the bloodstream, significantly increasing their circulation time in vivo [61–63]. The drug candidate developed in this study is intended for inhalation; therefore, these liposomes will be less susceptible to degradation compared to injectable formulations. However, in addition to protection against enzymes, PEG in the right proportions can provide a steric barrier for liposomes, preventing their aggregation in physiological conditions [63]. On the other hand, it has been demonstrated that PEG–lipid hydrolysis products located on the inner surface of liposomes may evoke an increase in membrane permeability; therefore, it seems vital to determine the concentration and role of PEG for each liposomal formulation individually [64]. To assess the impact of PEG on the stability of our formulation, two sets of liposomes were prepared: one containing 5 mol% PEG_{1000} (Formula #2) and the other with no PEG_{1000} (replaced by HSPC; Formula #9). Importantly, all the liposomes still contained 0.5 mol% DSPE-PEG_{2000}-Mal to enable peptide attachment to the liposome surface. Liposomes and peptidoliposomes were stored at 4 °C, and the stability was tracked by DLS measurements. Significant differences were noted after 13 days of storage (Figure 7).

The depletion of PEG_{1000} had a negative impact on the homogeneity of liposomes (Figure 7A), which is reflected by the increased average hydrodynamic diameter (79.68 vs. 89.35 nm) and PDI (0.060 vs. 0.200). Notably, the impact of PEG incorporation into the lipid composition was lower for peptidoliposomes (Figure 7B) than for liposomes not decorated with peptides, suggesting that the presence of the peptide on the liposome surface had a beneficial effect on the formulation's stability. Notwithstanding the latter, peptidoliposomes containing PEG_{1000} were still smaller (80.65 nm vs. 82.24 nm) and more homogeneous than their PEG-deprived counterparts (PDI 0.074 vs. 0.103). This was an expected result, as PEG-conjugated lipids have been known for decades to improve the blood circulation capability of liposomes [65] and have been used in various liposomal drugs, including the first FDA-approved nano-drug Doxil® [66] and mRNA vaccines against SARS-CoV-2 [67]. However, the drawback of applying PEG is that immune responses have been reported after applying PEG-conjugated nanocarriers, including accelerated blood clearance and complement activation-related pseudoallergy [68–70]. Therefore, defining the optimal PEG concentration before functionality, toxicity, and safety studies is crucial to minimize the risk of PEG-related side effects.

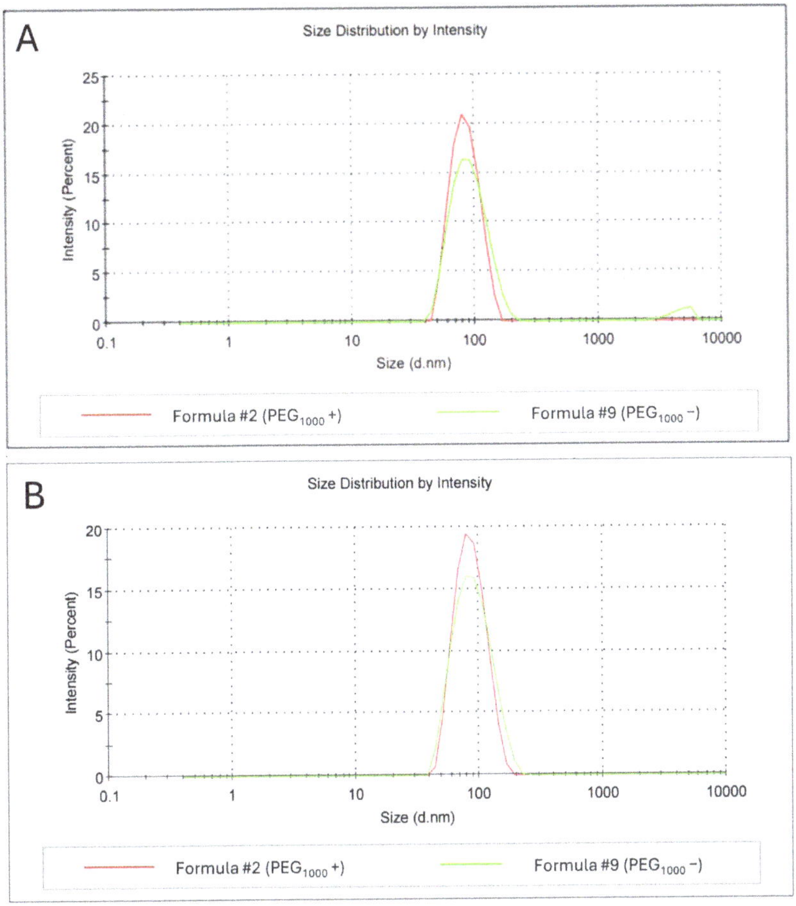

Figure 7. The average hydrodynamic diameters of liposomes (**A**) and peptidoliposomes (**B**) with 5 mol% PEG_{1000} (Formula #2) and without PEG^{1000} (Formula #9) after 13 days of incubation at 4 °C.

2.2.4. Effect of Phosphatidylglycerol (PG) Modifications

Including an anionic phospholipid, such as phosphatidylglycerol (PG), may impart a negative charge to liposomes' surface. By enforcing electrostatic repulsion, a strong charge enhances the colloidal stability of liposomes, preventing their aggregation [71,72] Variations in PG type, such as chain length and saturation, can affect the rigidity and fluidity of liposomes [57,73–75]. The impact of the PG modifications was investigated in terms of the stability and functionality of the final formulation. As demonstrated in Table 4, right after synthesis, peptidoliposomes with 10 mol% dipalmitoylphosphatidylglycerol (DPPG) were of a similar size and polydispersity compared with those with 5 mol% DPPG but had a slightly lower zeta potential (t-test, $p < 0.05$).

Following the two-week incubation at 4 °C, peptidoliposomes with elevated PG concentrations were characterized by slightly bigger average hydrodynamic diameters (83.55 nm vs. 79.29 nm in 5 mol% preparation) and slightly higher PDIs (0.115 vs. 0.090; Figure 8A). Within the tested PG species, the measured average hydrodynamic diameter was similar to that of all the peptidoliposomal preparations (Table 4). The homogeneity was significantly affected only with the dioleoylphosphatidylglycerol (DOPG) inclusion, which was reflected by a significantly higher PDI (t-test, $p < 0.01$). The differences between

DOPG and other PGs intensified over time, and for DOPG peptidoliposomes, an increase in the PDI value to 0.158 was noted after 2 weeks of incubation at 4 °C (Figure 8B). Moreover, after 72 h of incubation in Dulbecco's Modified Eagle Medium (DMEM) in the presence of 10% FBS, at 37 °C, severe precipitation of the DOPG-containing preparation was observed with a light microscope (Supplementary Material, Figure S2). Phosphatidylglycerols (PGs) seem to prevail over other negatively charged phospholipids; therefore, including them in a lipid composition can have a positive influence on liposome stability [57,74]. The ZP of all preparations was similar, indicating that the negative impact of DOPG on liposome stability cannot be attributed to the increased surface potential.

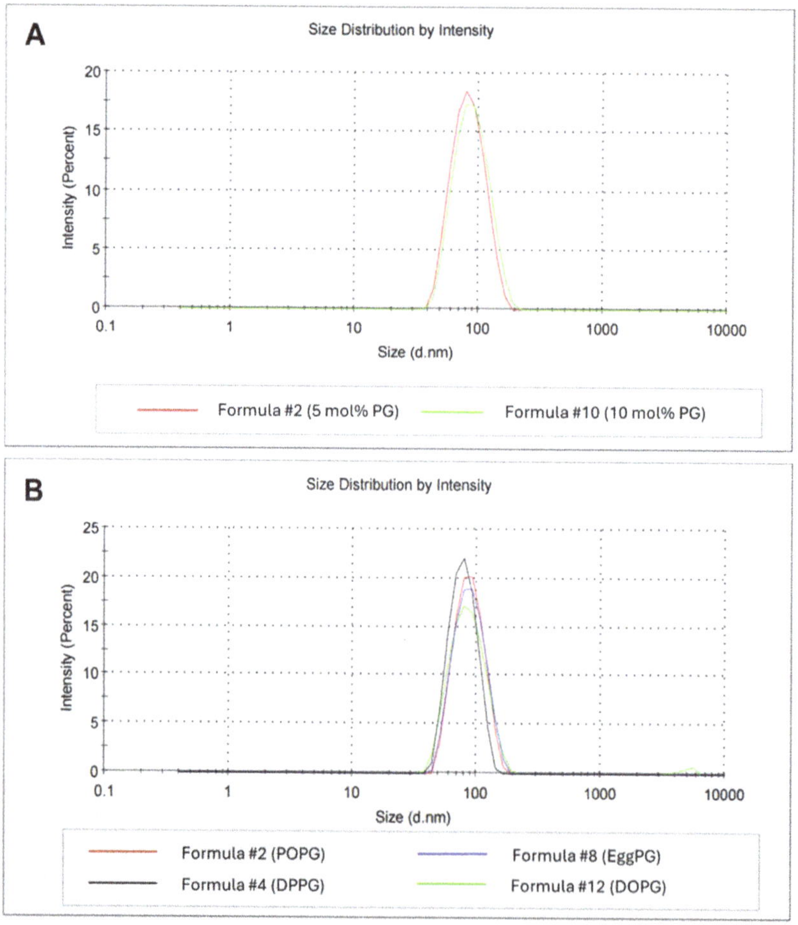

Figure 8. The average hydrodynamic diameters of formulations with different concentrations of POPG (**A**) and different types of PG (**B**) after 2 weeks of incubation at 4 °C. POPG—palmitoyloleoylphosphatidylglycerol; DPPG—dipalmitoylphosphatidylglycerol; EggPG—Egg L-α-phosphatidylglycerol; DOPG—dioleoylphosphatidylglycerol.

As the precipitations formed in DMEM media with DOPG-containing liposomes affected the functionality assays, the following experiments were performed only with peptidoliposomes containing POPG (Formula #2), DPPG (Formula #3), and EggPG (Formula #8). As presented in Figure 9, the inhibition of syncytia formation by peptidoliposomes with various PGs was as follows: EggPG > DPPG > POPG. However, due to the high standard deviation, no statistical significance was achieved ($p > 0.05$).

Table 4. The average hydrodynamic diameter (nm), polydispersity index (PDI), and zeta potential (ZP) of peptidoliposomes with modified PG content. POPG—palmitoyloleoylphosphatidylglycerol; DPPG—dipalmitoylphosphatidylglycerol; EggPG—Egg L-α-phosphatidylglycerol; DOPG—dioleoylphosphatidylglycerol. For each formulation $n = 3$.

	Formula #2	Formula #3	Formula #8	Formula #10	Formula #12
PG (mol%)	5	5	5	10	5
PG type	POPG	DPPG	EggPG	DPPG	DOPG
Diameter (nm)	83.32 ± 0.90	77.12 ± 1.29	83.40 ± 0.86	79.88 ± 0.81	80.21 ± 1.05
PDI	0.075 ± 0.008	0.069 ± 0.017	0.070 ± 0.014	0.071 ± 0.010	0.138 ± 0.004
ZP (mV)	−30.40 ± 0.41	−28.97 ± 1.26	−30.60 ± 0.16	−32.67 ± 0.57	−30.77 ± 1.68

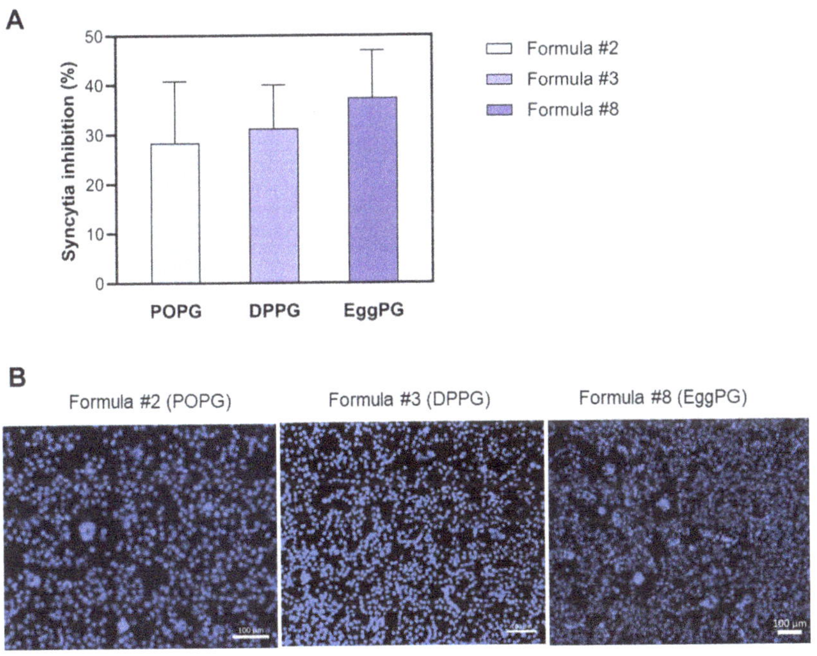

Figure 9. The influence of the type of PG on syncytia formation in HEK-hACE2 and 293-SARS2-S cells by anti-SARS-CoV-2 peptidoliposomes. (**A**) syncytia inhibition by peptidoliposomes containing various types of PG (Formula #2, 3 and 8) at a peptide concentration of ~35 µg/mL. (**B**) Representative images from the assay (cells stained with Hoechst). The scalebar is 100 µm. POPG—palmitoyloleoylphosphatidylglycerol; DPPG—dipalmitoylphosphatidylglycerol; EggPG—Egg L-α-phosphatidylglycerol.

EggPG-containing peptidoliposomes had satisfactory physicochemical properties and slightly better functionality compared to other PG-containing formulations. As the EggPG was also easily accessible in the GMP standard at the time of the study, this lipid was selected as a part of the nanocarrier component.

2.3. Defining the Optimal Conditions for Peptide Conjugation to Maleimide

2.3.1. Determination of Optimal Maleimide Concentration

Maleimide is the most expensive lipid component of the studied liposomes. It was therefore crucial to determine the minimal maleimide concentration that results in satisfactory liposome coverage with the peptide but that is reasonable regarding production costs at the same time. Preparations of peptidoliposomes with various concentrations of PEG-PE maleimide derivative (0.5–20 mol%) were tested for the amount of stably bound peptide through size exclusion chromatography (SEC). The DSPE-PEG_{2000}-Mal concentration was increased in exchange for DSPE-PEG_{1000} and HSPC (see Table 1 for comparison). Apart from the preparation with the highest concentration of maleimide, substitutions did not significantly influence the average hydrodynamic diameter and PDI of peptidoliposomes (Table 5).

Table 5. The average hydrodynamic diameter (nm), polydispersity index (PDI), and zeta potential (ZP) of peptidoliposomes with modified DSPE-PEG_{2000}-Mal content. For each formulation, $n = 3$.

Formula #	DSPE-PEG-2K-Mal (mol%)	Diameter (nm)	PDI	ZP (mV)
2	0.5	81.02 ± 0.43	0.068 ± 0.008	−32.60 ± 0.94
13	1.0	81.04 ± 0.60	0.080 ± 0.006	−23.80 ± 2.25
14	2.5	85.26 ± 0.68	0.077 ± 0.015	−21.17 ± 1.99
15	5.0	88.07 ± 0.052	0.105 ± 0.018	−24.57 ± 1.56
16	10.0	84.02 ± 0.20	0.078 ± 0.009	−25.73 ± 0.50
17	20.0	77.46 ± 0.57	0.183 ± 0.014	−29.53 ± 1.21

Following chromatographic separation, the Brown–Forsythe and Welch ANOVA test revealed that the binding efficacy was positively influenced by the maleimide concentration only in formulations containing up to 5 mol% of the maleimide derivative of PEG-PE (Figure 10).

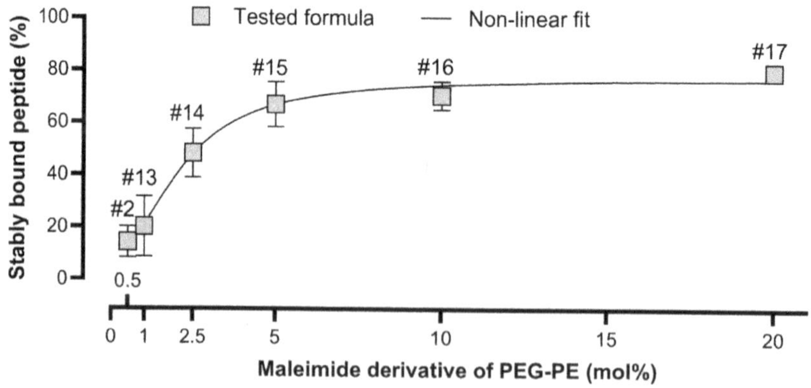

Figure 10. The effect of the maleimide concentration on the amount of peptide stably bound to liposomes in peptidoliposomal preparations following size exclusion chromatography (SEC). The detailed lipid composition is listed in Table 1.

In contrast, no statistically significant differences were noted between formulations containing 5, 10, and 20 mol% of DSPE-PEG_{2000}-Mal ($p > 0.05$). For that reason, 5 mol% DSPE-PEG_{2000}-Mal was selected as the most cost-effective composition for an anti-SARS-CoV-2 liposomal drug candidate. It has to be emphasized that SEC was the only method used to determine the stably bound peptide (SBP) value; therefore, the covalent nature of the binding was not confirmed in the study. However, the peptide and liposomes must have been linked by a strong bond, since a signal from both was obtained in the same fractions. Moreover, gel filtration has been used previously to reveal the covalent nature of bonds between various particles [76,77].

2.3.2. Defining the Optimal Protocol for Peptide Reduction

To maintain free sulfhydryl groups and to ensure the repeatability and reproducibility of the peptide attachment process, the peptide was reduced each time before conjugation to the liposomes. Tris(2-carboxyethyl)phosphine (TCEP) has been widely used to reduce disulfide bonds for over three decades. It is often preferred to other reducing agents due to its stability, wide pH range, odorless nature and easier handling (no need to remove from the reaction mixture) [78]. Here, the SEC method was employed to determine peptide binding efficacy following the reduction process (see Section 3.7). The results of our experiments may indicate that a high TCEP concentration does not favor high binding efficiency between maleimide and peptide (Figure 11A–E). This may result from TCEP's interaction with maleimide, as TCEP had not been removed before the introduction of the ligand to the reduced peptide molecules. Several commercial sources and studies in the literature report the lack of reaction between TCEP and maleimide; therefore, the reducing agent does not have to be removed before the conjugation process [79–82]. Conversely, some studies report that TCEP is likely to react with maleimide-functionalized biomolecules under typical conditions used for bioconjugation [81,82]. Since the results of our SEC experiments favor the latter hypothesis, 1,4-Dithiothreitol (DTT) was also tested as a reducing agent. Due to its reactivity with thiol-alkylating reagents (which reduce yields of alkylation) and high toxicity [83], it is essential to remove DTT from the peptide solution before conjugation with liposomes. Three methods were used for peptide purification: filtration with centrifugation filters, ion exchanging resins, and dialysis to 50 mM HEPES/100 mM NaCl buffer. None of the methods significantly affected binding efficacy ($p > 0.05$; Figure 11F).

Regardless of the purification method, DTT reduction appeared superior to TCEP reduction in peptidoliposomes consisting of the proposed formulation (Formula #19, see Table 1). Due to its better feasibility regarding the production process and higher (but not statistically significant) binding efficacy, centrifugal ultrafiltration is a preferred method for peptide purification before conjugation to liposomes.

2.4. Platform Versatility and Possible Applications

Maleimide–thiol chemistry is widely used in nanocarrier production and has been demonstrated as an effective method for the development of liposomes presenting a variety of particles, including peptides [20,84–86], proteins [87] antibodies [21,22,88], nanobodies [84], and aptamers [23,24]. Here, we propose a platform based on a maleimide–thiol click reaction that has been optimized at different stages of production. In terms of composition, the best physicochemical parameters were achieved for Formula #19, consisting of HSPC:DSPC:CHOL:EggPG:DSPE-PEG_{2000}-MAL in a molar ratio of 70:10:10:5:5. For this formula, stable and homogeneous liposomes were obtained regardless of the calibration method. When adapting the protocol, the features distinguishing liposomes obtained by either of the techniques should be taken into account. High-pressure homogenization allowed us to produce smaller liposomes, but the suspension was less homogeneous compared to the extruded counterparts (Table 6).

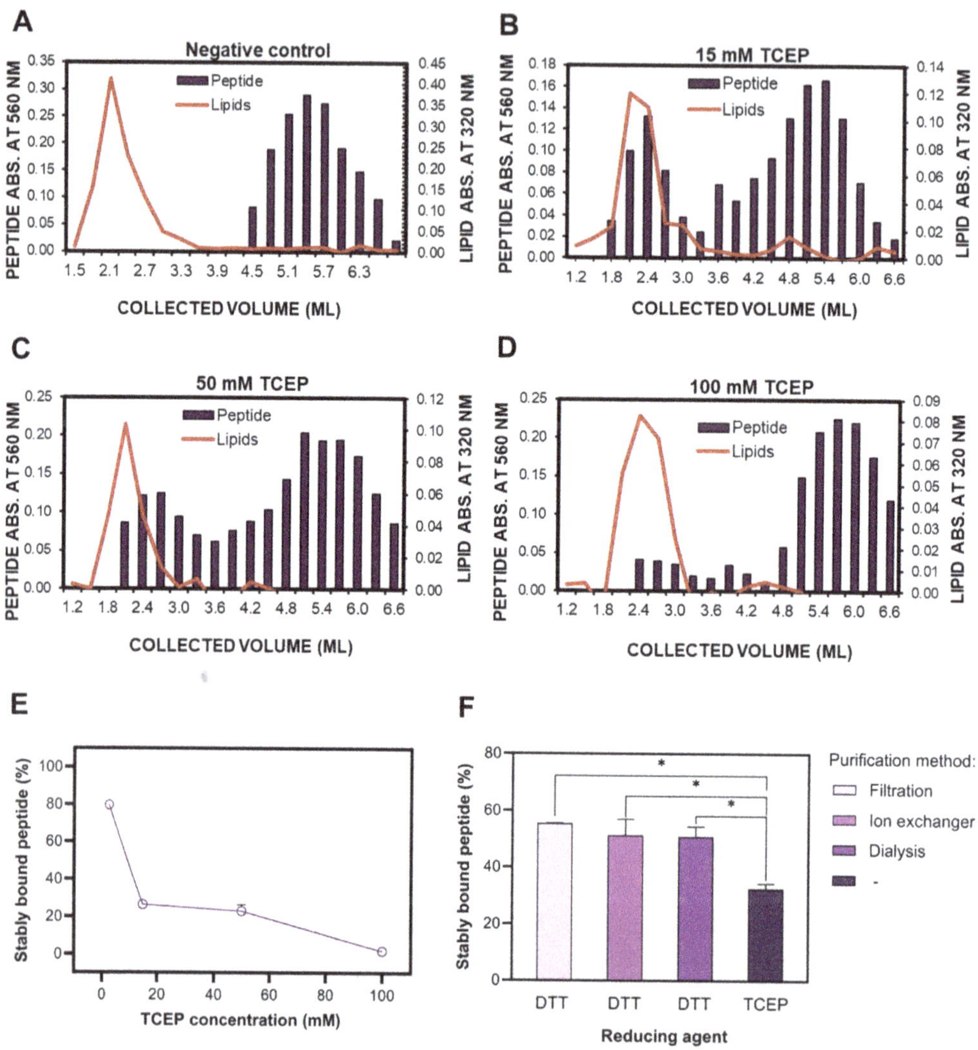

Figure 11. The effect of the peptide reduction on the efficacy of peptide binding to liposomes; (**A**) separation of the maleimide-deprived peptidoliposomes (Formula #18—negative control) from the unbound peptide with size exclusion chromatography (SEC); (**B–D**) separation of peptidoliposomes (Formula #17) from the unbound peptide with SEC following peptide reduction with 15. 50 and 100 mM TCEP; (**E**) the summary data on the TCEP concentration's influence on peptide binding after SEC and the concentration of separated peptidoliposomes (Formula #17); (**F**) the effect of the reducing agent (DTT vs. TCEP) and the purification method (filtration, ion exchanging resin, or dialysis) on the amount of stably bound peptide after SEC and the concentration of separated peptidoliposomes (Formula #19); * $p \leq 0.05$.

After a year of storage at 4 °C, in both cases, the ZP of liposomes was slightly increased. However, it did not have a significant impact on homogeneity, as the PDI remained below 0.2 for homogenized liposomes and below 0.1 for the extruded ones. The extruded peptidoliposomes were subjected to accelerated aging tests and the stability of both liposomes and peptidoliposomes was demonstrated in the presence of cell culture media with the addition of FBS at 37 °C for at least 144 h (Figure S4). The physicochemical parameters achieved for the formulation easily meet the acceptance criteria for drug delivery systems, staying far from the safety margin [89–91]. In addition to the liposome preparation, a conjugation reaction was optimized in terms of the peptide:maleimide ratio and conjugation protocol. In the pilot experiment, the same conditions were used to obtain liposomes conjugated with another peptide, targeting a different virus. The resulting formulations were equally stable and homogeneous, showing that similar conditions can be used to achieve successful conjugation with various molecules. However, it should be noted that the maleimide to thiol molar ratio may require modification for different conjugates to reach the optimal reaction efficiency [84]. Importantly, in the present study, liposomes were not loaded with any agents, but it is possible to use the same formula for the entrapment of various therapeutic agents. The stage at which the agent is added to the formulation should depend on the physicochemical properties and thermal stability of the agent.

Table 6. The average hydrodynamic diameter with the most prominent peak size, width at half height (Peak 1 size (nm), peak intensity (%), HH width (nm), respectively), the polydispersity index (PDI), and the zeta potential (ZP) with the leading peak value, area, and width at half height (Peak 1 ZP (mV), Peak Area (%) and HH Width (mV), respectively), measured in liposomes (Formula #19) manufactured using the high-pressure homogenization method or pressure extrusion method directly after preparation (day 0) and after 365 days of storage at 4 °C.

	Homogenized Liposomes		**Extruded Liposomes**	
Time (days)	0	365	0	365
Diameter (nm)	57.18 ± 0.41	66.96 ± 0.43	80.39 ± 0.71	82.56 ± 1.24
Peak 1 size (nm)	63.69	70.83	85.50	89.13
Peak intensity (%)	99.7	98.2	100.0	100.0
HH Width (nm)	23.00	24.62	21.68	25.33
PDI	0.131 ± 0.015	0.188 ± 0.014	0.55 ± 0.015	0.078 ± 0.021
ZP (mV)	−25.77 ± 0.49	−21.8 ± 1.49	−34.40 ± 1.41	−26.4 ± 0.1
Peak 1 ZP (mV)	−25.8	−29.8	−33.8	−26.4
Peak Area (%)	100.0	86.1	100.0	100.0
HH Width (mV)	9.53	9.63	9.82	8.46

3. Materials and Methods

3.1. Chemicals

The following lipids were used in the study: cholesterol (CHOL, cat. no. PA-03-0100-E), purchased from POL-AURA (Zabrze, Poland) and from Avanti (Alabaster, AL, USA); cat. no. 700000P), hydrogenated soy L-α-phosphatidylcholine (HSPC, cat. no. 840058P), 1-palmitoyl-2-oleoyl-sn-glycero-3-phosphatidylglycerol (POPG, cat. no. 840457P), 1,2-distearoyl-sn-glycero-3-phosphocholine (DSPC, cat. no. 850365C), 1,2-distearoyl-sn-glycero-3-phosphoethanolamine-N-[methoxy(polyethylene glycol)-1000] ammonium salt ($DSPE\text{-}PEG_{1000}$, cat. no. 880720P),1,2-distearoyl-sn-glycero-3-phosphoethanolamine-N-[methoxy(polyethylene glycol)-2000] ammonium salt ($DSPE\text{-}PEG_{2000}$; cat. no. 880120P) and 1,2-distearoyl-sn-glycero-3-phosphoethanolamine-N-[maleimide(polyethylene glycol)-2000] ammonium salt ($DSPE\text{-}PEG_{2000}$-Mal, cat. no. 880126P), purchased from Avanti; HSPC (cat. no. 525600), 1,2-dipalmitoyl-sn-glycero-3-phospho-(1′-rac-glycerol), sodium salt (DPPG, cat. no. 67232-81-9), and L-α-phosphatidylglycerol from egg (EggPG, cat. no. 583500), purchased from Lipoid (Ludwigshafen, Germany). Organic solvents and sodium chloride were purchased from CHEMPUR (Piekary Śląskie, Poland). HEPES and Tris(2-carboxyethyl)phosphine (TCEP) were purchased from ROTH (Karlsruhe, Germany), and 1,4-Dithiothreitol (DTT) was purchased from A&A Biotechnology (Gdańsk, Poland).

3.2. The Anti-SARS-CoV-2 Peptide

Liposomes were designed to expose a short ACE2-based, 32-amino-acid-residue peptide (molecular formula $C_{164}H_{241}N_{43}O_{54}S$), with a molecular weight of 3711.03 Da, a theoretical pI = 4.654 and a net charge of z = −3.608 at pH 7.4, as calculated with the Prot pi calculator (www.protpi.ch/Calculator/ProteinTool; accessed on 8 November 2024). The sequence was taken from Huang et al. (2020) [92]. Each time before conjugation with maleimide, the peptide was reduced with TCEP or DTT. In the latter case, DTT was separated from the peptide either with centrifugation filters (MWCO = 2K, Millipore (Burlington, MA, USA), or Sartorius (Göttingen, Germany), with an ion exchanging resin (DEAE–Sephadex® A-25 chloride form; Sigma (St. Louis, MO, USA), cat. no. A-25-120), or through dialysis to 50 mM HEPES/100 mM NaCl buffer (pH 7.2) using the dialysis membranes (2 kDa, Repligen, Waltham, MA, USA). Following purification from DTT, the peptide concentration was measured before conjugation to liposomes.

3.3. Preparation of Peptidoliposomes

Liposomes were prepared using the thin-film hydration method. Briefly, lipids were dissolved in a ratio of 9:1 in a chloroform: methanol mixture (DSPE-PEG2000-Mal) or in pure chloroform (all other lipids) and transferred in appropriate proportions into a round-bottom flask. The organic solvents were initially slowly evaporated under a nitrogen stream, and then the film was entirely dried overnight with a desiccator. Next, the lipid film was hydrated with 50 mM HEPES/100 mM NaCl buffer (pH 7.2) by gentle mixing with glass beads for 30′ at 64 °C (the number of beads was equal to half of the suspension volume in mL). After the synthesis, liposomes were calibrated through pressure extrusion or HPH. During the extrusion, nitrogen-generated pressure was used to press liposomes through track-etch polycarbonate membranes (Cytiva, Wilmington, DE, USA) in the following order: 0.2 µm (10 cycles), 0.1 µm (10 cycles), and 0.05 µm (5 cycles). Membranes were supported by a PE drain disc (Cytiva). HPH was performed with two LM20 Microfluidizer® devices (Microfluidics, Westwood, MA, USA) equipped with either a F20Y or a G10Z chamber. Each time, 10 homogenization cycles were run at 22,500 psi (1551.3 bar). The liposomes were incubated with a reduced peptide for 24 h at 4 °C, and then the excess peptide was removed through dialysis to the HEPES/NaCl buffer. Dialysis was carried out with dialysis membranes (MWCO 14–50 kDa, Spectra/Por®) at 4 °C in two stages: first for 16 h in sample/buffer v/v 1:100, and then for another 8 h with sample/buffer v/v 1:300. A graphical sketch of the synthesis process is presented in Figure 12.

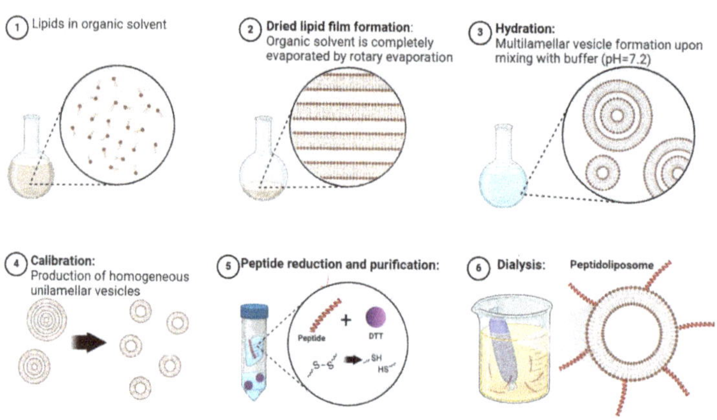

Figure 12. The process of synthesis of maleimide-functionalized liposomes decorated with the ACE2-derived peptide (created with BioRender and adapted and modified from Zhang et al. (2017) [93]).

Each time after synthesis preparations were sterilized. The details on the sterilization process can be found in Appendix A.

3.4. Lipid Concentration Measurements

Lipid concentration was measured with the modified molybdate/ascorbic acid method first described by Murphy and Riley (1962) [94]. In brief, 50 µL of each sample of various dilutions was mixed with 0.5 mL 70% perchloric acid (CHEMPUR) and incubated for 2 h at 200 °C in glass covered vials. Next, 1 mL of ammonium molybdate (Avantor Performance Materials Poland; Gliwice, Poland) at a concentration of 25 mg/mL and 1 mL of ascorbic acid (ROTH) at a concentration of 100 mg/mL were added to each sample, and the samples were incubated for another hour at 37 °C. Next, samples were transferred on the 96-well transparent plate (200 µL per well), and the absorbance of the solutions was measured with a Rayto RT-100 microplate reader (Rayto Life and Analytical Sciences, Shenzhen, China) at λ = 750 nm. At each stage, the absorbance was compared against to the standard curve prepared from the phosphorus standard (Honeywell, Charlotte, NC, USA; cat. no. E1042-100ML).

3.5. Peptide Concentration Measurements

The peptide concentration in peptidoliposomes was measured with the modified bicinchoninic acid (BCA) assay, first described by Smith et al. (1985) [95]. In brief, 135 µL of each sample in various dilutions was mixed with 15 µL of 10% sodium dodecyl sulfate (SDS; Sigma) (for liposome disintegration), and 150 µL of BCA reagent (Sigma) mixed with copper (II) sulfate ((Sigma) in a proportion of 50:1) was added to each sample in a 96-well, transparent plate. Next, samples were incubated for 60 min at 37 °C, and the absorbance was measured with Promega Glomax Discover Microplate Reader (Promega, Madison, WI, USA) at λ = 560 nm. A standard curve was prepared with the freshly dissolved peptide at a concentration of 1 mg/mL.

3.6. Stability Measurements

Stability was assessed primarily based on changes in particle average hydrodynamics and the PDI, measured with the DLS technique. The ZP was tracked using the Dip Cell Kit (Malvern, UK). All measurements were conducted using a ZetaSizer Nano device (Malvern, UK) at 25 °C. Measurements were carried out in samples incubated for a longer duration at 4 °C or following the accelerated aging assay when samples were mixed in a ratio of 1:10 with the culture medium (DMEM. Gibco, Waltham, MA, USA) + 10% FBS and then incubated for at least 72 h at 37 °C.

3.7. Size Exclusion Chromatography (SEC)

Free peptides were separated from peptidoliposomes via SEC using the 5 mL columns with Sepharose® CL-4B (Cytiva; column diameter to volume ratio of 1:25). A sample of 0.3 mL was transferred into a column, and fractions in volumes of 0.3 mL were collected on the 96-well transparent plate; then, the absorbance was measured at 320 nm to distinguish the liposome-containing fractions. Next, a BCA assay was employed to determine the peptide concentration in the collected fractions using two approaches: (i) to assess the separation resolution, the peptide concentration was measured directly in non-diluted fractions, and (ii) for the precise determination of the stably bound peptide, the liposome-containing fractions were concentrated with the Amicon® centrifugation system (100 kDa, Millipore), and the lipid and peptide concentrations were measured in samples before and after chromatographic separation. This approach is independent of the one used for sample dilution with chromatographic columns. The amount of peptide bound to liposomes was calculated according to the following Equation (1):

$$SBP = (m_{PEPTIDE}/m_{LIPID} \text{ after } SEC)/(m_{PEPTIDE}/m_{LIPID}) \text{ before } SEC) \times 100\% \quad (1)$$

where SBP stands for stably bound peptide (%), $m_{PEPTIDE}$ is the mass of peptide per ml (μg), and m_{LIPID} is the mass of lipids (mg) per ml in the peptidoliposomal preparations.

3.8. Syncytia Formation Assay

A syncytia formation test was performed according to the procedure described by Lu et al. (2005) [96]. Giant, multinucleated cells were formed as a result of the fusion of two HEK293 sublines, one exposing the ACE2 protein on the cell membrane surface (HEK-hACE2 (Invivogen, catalog number: hkb-hace2, San Diego, CA, USA)) and the other exposing protein S of the SARS-CoV-2 virus (293-SARS2-S (Invivogen, cat. no. 293-cov2-s)). The test was optimized to be performed on a 24-well plate. Plates were coated 24 h before seeding the cells with a medium composed of albumin BSA (0.01 mg/mL, Sigma, cat. no. A8022), fibronectin (0.01 mg/mL. Sigma, cat. no. F2006), and collagen type I (0.03 mg/mL, Advanced BioMatrix, cat. no. 5006, Carlsbad, CA, USA). Before seeding them, the HEK-hACE2 and 293-SARS2-S cells were mixed in a 1:1 ratio using 20,000 cells of each line, and then incubated without or with the liposome preparation at 37 °C for 120 min. Following seeding, the cultures were incubated on the precoated plates at 37 °C and with 5% CO_2 for 72 h. After this, the cell nuclei were stained with Hoechst 33342 dye (H3570, Invitrogen) and observations were made using the ZEISS Axio Observer fluorescence microscope (ZEISS, Oberkochen, Germany) and processed by ZEN software version 3.4 (blue edition). At each point, the syncytia were counted manually by two independent researchers.

4. Conclusions and Future Perspectives

The presented results suggest the successful development of a liposomal formulation of a decoy receptor intended for therapeutic use in combating early SARS-CoV-2 infection. The aim of this study was to conduct a phase I clinical trial with a novel drug candidate. Therefore, in future research, it is necessary to conduct preclinical in vivo studies with various animal models. We plan to conduct such studies in line with the requirements of the International Council for Harmonisation of Technical Requirements for Pharmaceuticals for Human Use (ICH) and the Organisation for Economic Co-operation and Development (OECD), as well as with recommendations by the European Medicines Agency (EMA) and the Food and Drug Administration (FDA). This in vivo study will include toxicity studies using rat and rabbit models followed by efficacy studies using the hamster model and a pharmacokinetic study on rats, since SARS-CoV-2 has been previously shown to efficiently replicate in the respiratory tract of these animals, invading the central nervous system and generating a cytokine storm resembling the same phenomenon in humans [97,98]. In these non-lethal models, the assessments carried out will be based on parameters such as body weight loss, histopathological changes, drug concentrations in the blood and the bronchoalveolar lavage, etc., in alignment with other animal studies on the development of COVID-19 therapeutics [99]. Similar approaches based on the interaction of a human recombinant soluble ACE2 fragment with S-RBD have been implemented in various anti-COVID-19 therapies. Their effectiveness, however, is limited by low blood concentrations and a short plasma half-life [15,100]. Creating a stable, inhalable carrier for ACE2-derived peptides is a promising approach, with this carrier acting at the site with the highest pathogen concentration. Inhaled liposomal formulations have been used in several studies on lung diseases, particularly on pulmonary fibrosis, lung infections, asthma, and lung cancer [101–104]. The main advantages of introducing liposome-based carriers include a reduction in the toxicity of the therapeutic agent, improvements in its pharmacokinetics and efficacy, and increased delivery, but features that do not affect the functioning of the biologically active compound itself, such as taste masking, may also be beneficial [104–106]. The major advantage of the proposed drug candidate compared to the already existing treatments described in the Section 1 is its very low sensitivity to the mutational variability of the RNA of the viral genome because it is based on the sequence of the human ACE2 derivative with an extremely low mutation rate. Extensive preclinical studies have demonstrated that this approach is successful against SARS-CoV-2 infection. However, half-life extension and effector function enhancement are recommended as directions for further development [107]. The introduction of this liposomal carrier is expected to significantly improve the stability of the peptide and improve its efficacy by allowing for the better exposition and increased visibility of the virus-liposomes conglomerates to the immunological system, which will be the subject of the follow-up functionality studies. The presented work is focused on the development of a carrier for the ACE2 derivative that can serve as a trap to immobilize the SARS-CoV-2 virus on the liposome surface. The same carrier can also be used against other viral infections in future. In fact, the multivalent character of the developed liposomes offers a strategy to enhance the targeting efficiency of various therapeutics [108,109]. The formulation obtained in the project is characterized by its nano-size (diameter below 100 nm), the 8–12-month stability of its physicochemical parameters, and its biological activity, i.e., the effective inhibition of a SARS-CoV-2 protein S–ACE2 interaction in a HEK293 cellular system. The aforementioned protocols can be applied during the development of different drug candidates, and the PEGylated liposomes demonstrated can serve as a platform for the immobilization and stabilization of various therapeutic molecules.

5. Patents

The authors declare pending patents no P.435921 and P.451191 (Patent Office of the Republic of Poland) and no. EP21889696.7 (European Patent Office).

Supplementary Materials: The supporting information can be downloaded at https://www.mdpi.com/article/10.3390/ijms26041629/s1.

Author Contributions: Conceptualization: A.F.S., K.K., A.C., O.M. and L.M.; methodology: A.F.S., A.C., O.M., A.K. and L.M.; validation: O.M., A.K., K.P. and M.G.; formal analysis: A.F.S. and A.C.; investigation: O.M., A.K., K.P., W.F. and M.G.; resources: A.F.S., A.C. and W.F.; data curation: O.M., A.K., K.P. and M.G.; writing—original draft preparation: O.M.; writing—review and editing: O.M., A.F.S., A.C., K.P. and A.K.; visualization: O.M.; supervision: A.F.S., A.C. and K.K.; project administration: A.F.S., A.C. and W.F.; funding acquisition: A.F.S., K.K. and A.C. All authors have read and agreed to the published version of the manuscript.

Funding: This research was funded by the Medical Research Agency (Poland), project number 2021/ABM/05/00002-00, project title "Mutation-resistant nano therapeutics inhibiting infection progress of RNA-virus, SARS-CoV-2".

Institutional Review Board Statement: Not applicable.

Informed Consent Statement: Not applicable.

Data Availability Statement: All data supporting the reported results are available in a repository of project's leaders—Silesian Park of Medical Technology KardioMed Silesia and ACELLMED (contact: biuro@kmptm.pl; biuro@acellmed.pl). Access to the data is subject to approval and data sharing agreement due to the project's data confidentiality (project no 2021/ABM/05/00002-00).

Acknowledgments: The authors would like to express their gratitude to Adam Konka, MBA, and Joanna Zembala-John—the project's consortium partners—for administrative support throughout the project. We are grateful to the Lipoid company for providing free lipid samples for our preliminary studies. We thank Kaviarasi V S and Daid Ahmad Khan for sharing their template on https://BioRender.com. Thanks also go to Devon Stackonis for the linguistic corrections made to the manuscript.

Conflicts of Interest: Author Aleksander F. Sikorski was employed by the company Acellmed Ltd. The remaining authors declare that the research was conducted in the absence of any commercial or financial relationships that could be construed as a potential conflict of interest. The funders had no role in the design of the study; in the collection, analyses, or interpretation of data; in the writing of the manuscript; or in the decision to publish the results.

Appendix A

At each point right after peptidoliposome synthesis and whenever necessary, the preparation was sterilized. The sterilization process was optimized, and two methods were selected for that purpose—extrusion for preparation with small volumes (up to 10 mL) and filtration with the Steriflip vacuum filtration system (Millipore) for larger volumes. Neither of the sterilization methods led to any significant decreases in the lipid and peptide concentrations (Supplementary Material, Figure S3).

References

1. COVID-19 Deaths | WHO COVID-19 Dashboard. Available online: https://data.who.int/dashboards/covid19/deaths (accessed on 14 August 2024).
2. Matusewicz, L.; Golec, M.; Czogalla, A.; Kuliczkowski, K.; Konka, A.; Zembala-John, J.; Sikorski, A.F. COVID-19 Therapies: Do We See Substantial Progress? *Cell. Mol. Biol. Lett.* **2022**, *27*, 42. [CrossRef] [PubMed]
3. Li, G.; Hilgenfeld, R.; Whitley, R.; De Clercq, E. Therapeutic Strategies for COVID-19: Progress and Lessons Learned. *Nat. Rev. Drug Discov.* **2023**, *22*, 449–475. [CrossRef] [PubMed]
4. Coronavirus (COVID-19) | Drugs | FDA. Available online: https://www.fda.gov/drugs/emergency-preparedness-drugs/coronavirus-covid-19-drugs (accessed on 27 August 2024).
5. Frieden, T.R.; Buissonnière, M.; McClelland, A. The World Must Prepare Now for the next Pandemic. *BMJ Glob. Health* **2021**, *6*, e005184. [CrossRef]
6. Neumann, G.; Kawaoka, Y. Which Virus Will Cause the Next Pandemic? *Viruses* **2023**, *15*, 199. [CrossRef] [PubMed]
7. Krammer, F. The Role of Vaccines in the COVID-19 Pandemic: What Have We Learned? *Semin. Immunopathol.* **2024**, *45*, 451–468. [CrossRef] [PubMed]
8. Krammer, F. SARS-CoV-2 Vaccines in Development. *Nature* **2020**, *586*, 516–527. [CrossRef]
9. Yan, R.; Zhang, Y.; Li, Y.; Xia, L.; Guo, Y.; Zhou, Q. Structural Basis for the Recognition of SARS-CoV-2 by Full-Length Human ACE2. *Science* **2020**, *367*, 1444–1448. [CrossRef]
10. Walls, A.C.; Park, Y.J.; Tortorici, M.A.; Wall, A.; McGuire, A.T.; Veesler, D. Structure, Function, and Antigenicity of the SARS-CoV-2 Spike Glycoprotein. *Cell* **2020**, *181*, 281–292.e6. [CrossRef] [PubMed]
11. Corbett, K.S.; Edwards, D.K.; Leist, S.R.; Abiona, O.M.; Boyoglu-Barnum, S.; Gillespie, R.A.; Himansu, S.; Schäfer, A.; Ziwawo, C.T.; DiPiazza, A.T.; et al. SARS-CoV-2 MRNA Vaccine Design Enabled by Prototype Pathogen Preparedness. *Nature* **2020**, *586*, 567. [CrossRef] [PubMed]
12. Bettini, E.; Locci, M. SARS-CoV-2 MRNA Vaccines: Immunological Mechanism and Beyond. *Vaccines* **2021**, *9*, 147. [CrossRef] [PubMed]
13. Auguste, D.T.; Lavik, E.B. What Comes after Liposomes? *Bioconjug Chem.* **2022**, *33*, 1955–1956. [CrossRef] [PubMed]
14. Li, J.; Xiao, H.; Zhang, C.; Liu, G.; Liu, X. From Virus to Immune System: Harnessing Membrane-Derived Vesicles to Fight COVID-19 by Interacting with Biological Molecules. *Eur. J. Immunol.* **2024**, *54*, 2350916. [CrossRef] [PubMed]
15. Zhang, H.; Lv, P.; Jiang, J.; Liu, Y.; Yan, R.; Shu, S.; Hu, B.; Xiao, H.; Cai, K.; Yuan, S.; et al. Advances in Developing ACE2 Derivatives against SARS-CoV-2. *Lancet Microbe* **2023**, *4*, e369–e378. [CrossRef] [PubMed]
16. Li, T.; Takeoka, S. A Novel Application of Maleimide for Advanced Drug Delivery: In Vitro and in Vivo Evaluation of Maleimide-Modified PH-Sensitive Liposomes. *Int. J. Nanomed.* **2013**, *8*, 3855–3866. [CrossRef]
17. Li, T.; Takeoka, S. Enhanced Cellular Uptake of Maleimide-Modified Liposomes via Thiol-Mediated Transport. *Int. J. Nanomed.* **2014**, *9*, 2849–2861. [CrossRef]

18. Li, H.; Tang, C.; Tang, Q.; Yin, D.; He, E.; Li, M.; Tong, Y.; Zhu, Y. Maleimide-Functionalized Liposomes for Tumor Targeting via In Situ Binding of Endogenous Albumin. *J. Biomed. Nanotechnol.* **2021**, *17*, 2382–2390. [CrossRef]
19. Tang, C.; Yin, D.; Liu, T.; Gou, R.; Fu, J.; Tang, Q.; Wang, Y.; Zou, L.; Li, H. Maleimide-Functionalized Liposomes: Prolonged Retention and Enhanced Efficacy of Doxorubicin in Breast Cancer with Low Systemic Toxicity. *Molecules* **2022**, *27*, 4632. [CrossRef]
20. d'Avanzo, N.; Torrieri, G.; Figueiredo, P.; Celia, C.; Paolino, D.; Correia, A.; Moslova, K.; Teesalu, T.; Fresta, M.; Santos, H.A. LinTT1 Peptide-Functionalized Liposomes for Targeted Breast Cancer Therapy. *Int. J. Pharm.* **2021**, *597*, 120346. [CrossRef] [PubMed]
21. Matusewicz, L.; Filip-Psurska, B.; Psurski, M.; Tabaczar, S.; Podkalicka, J.; Wietrzyk, J.; Ziółkowski, P.; Czogalla, A.; Sikorski, A.F. EGFR-Targeted Immunoliposomes as a Selective Delivery System of Simvastatin, with Potential Use in Treatment of Triple-Negative Breast Cancers. *Int. J. Pharm.* **2019**, *569*, 118605. [CrossRef] [PubMed]
22. Meissner, J.M.; Toporkiewicz, M.; Czogalla, A.; Matusewicz, L.; Kuliczkowski, K.; Sikorski, A.F. Novel Antisense Therapeutics Delivery Systems: In Vitro and in Vivo Studies of Liposomes Targeted with Anti-CD20 Antibody. *J. Control. Release* **2015**, *220*, 515–528. [CrossRef] [PubMed]
23. Ara, M.N.; Matsuda, T.; Hyodo, M.; Sakurai, Y.; Hatakeyama, H.; Ohga, N.; Hida, K.; Harashima, H. An Aptamer Ligand Based Liposomal Nanocarrier System That Targets Tumor Endothelial Cells. *Biomaterials* **2014**, *35*, 7110–7120. [CrossRef] [PubMed]
24. Li, L.; Hou, J.; Liu, X.; Guo, Y.; Wu, Y.; Zhang, L.; Yang, Z. Nucleolin-Targeting Liposomes Guided by Aptamer AS1411 for the Delivery of SiRNA for the Treatment of Malignant Melanomas. *Biomaterials* **2014**, *35*, 3840–3850. [CrossRef]
25. Talsma, H.; Özer, A.Y.; van Bloois, L.; Crommelin, D.J.A. The Size Reduction of Liposomes with a High Pressure Homogenizer (Microfluidizer^{TM}). Characterization of Prepared Dispersions and Comparison with Conventional Methods. *Drug Dev. Ind. Pharm.* **1989**, *15*, 197–207. [CrossRef]
26. Barnadas i Rodríguez, R.; Sabés i Xamaní, M. Liposomes Prepared by High-Pressure Homogenizers. *Methods Enzym.* **2003**, *367*, 28–46. [CrossRef]
27. Pupo, E.; Padrón, A.; Santana, E.; Sotolongo, J.; Quintana, D.; Dueñas, S.; Duarte, C.; De La Rosa, M.C.; Hardy, E. Preparation of Plasmid DNA-Containing Liposomes Using a High-Pressure Homogenization–Extrusion Technique. *J. Control. Release* **2005**, *104*, 379–396. [CrossRef] [PubMed]
28. Olson, F.; Hunt, C.A.; Szoka, F.C.; Vail, W.J.; Papahadjopoulos, D. Preparation of Liposomes of Defined Size Distribution by Extrusion through Polycarbonate Membranes. *Biochim. Biophys. Acta (BBA)-Biomembr.* **1979**, *557*, 9–23. [CrossRef]
29. Ong, S.G.M.; Chitneni, M.; Lee, K.S.; Ming, L.C.; Yuen, K.H. Evaluation of Extrusion Technique for Nanosizing Liposomes. *Pharmaceutics* **2016**, *8*, 36. [CrossRef] [PubMed]
30. Mui, B.; Chow, L.; Hope, M.J. Extrusion Technique to Generate Liposomes of Defined Size. *Methods Enzym.* **2003**, *367*, 3–14. [CrossRef]
31. Donsì, F.; Sessa, M.; Ferrari, G. Effect of Emulsifier Type and Disruption Chamber Geometry on the Fabrication of Food Nanoemulsions by High Pressure Homogenization. *Ind. Eng. Chem. Res.* **2012**, *51*, 7606–7618. [CrossRef]
32. Holtzhauer, G.A. Development and Characterization of Parenteral Lipid Emulsions from Vegetable Oil Sources to Reduce Inflammatory Responses. PhD Thesis, ETH Zürich, Zurich, Switzerland, 2022. [CrossRef]
33. Inguva, P.; Grasselli, S.; Heng, P.W.S. High Pressure Homogenization—An Update on Its Usage and Understanding. *Chem. Eng. Res. Des.* **2024**, *202*, 284–302. [CrossRef]
34. Shah, J.R.; Dong, T.; Phung, A.T.; Reid, T.; Larson, C.; Sanchez, A.B.; Oronsky, B.; Blair, S.L.; Aisagbonhi, O.; Trogler, W.C.; et al. Development of Adenovirus Containing Liposomes Produced by Extrusion vs. Homogenization: A Comparison for Scale-Up Purposes. *Bioengineering* **2022**, *9*, 620. [CrossRef] [PubMed]
35. Liposome Extrusion. Method. Patent U.S. 4737323A, 12 April 1988. Available online: https://patents.google.com/patent/US4737323A/en (accessed on 8 October 2024).
36. Maja, L.; Željko, K.; Mateja, P. Sustainable Technologies for Liposome Preparation. *J. Supercrit. Fluids* **2020**, *165*, 104984. [CrossRef]
37. Beltrán, J.D.; Ricaurte, L.; Estrada, K.B.; Quintanilla-Carvajal, M.X. Effect of Homogenization Methods on the Physical Stability of Nutrition Grade Nanoliposomes Used for Encapsulating High Oleic Palm Oil. *LWT* **2020**, *118*, 108801. [CrossRef]
38. Barnadas-Rodríguez, R.; Sabés, M. Factors Involved in the Production of Liposomes with a High-Pressure Homogenizer. *Int. J. Pharm.* **2001**, *213*, 175–186. [CrossRef] [PubMed]
39. Jensen, G.M.; Bunch, T.H.; Hu, N.; Eley, C.G. Process development and quality control of injectable liposomal therapeutics. In *Liposome Technology, Volume I*; CRC Press, Taylor and Francis Group: Boca Raton, FL, USA, 2007; pp. 297–310, ISBN 978-0-8493-9726-4.
40. Lin, H.; Leng, J.; Fan, P.; Xu, Z.; Ruan, G. Scalable Production of Microscopic Particles for Biological Delivery. *Mater. Adv.* **2023**, *4*, 2885–2908. [CrossRef]
41. Ou, X.; Liu, Y.; Lei, X.; Li, P.; Mi, D.; Ren, L.; Guo, L.; Guo, R.; Chen, T.; Hu, J.; et al. Characterization of Spike Glycoprotein of SARS-CoV-2 on Virus Entry and Its Immune Cross-Reactivity with SARS-CoV. *Nat. Commun.* **2020**, *11*, 1620. [CrossRef]
42. Rajah, M.M.; Bernier, A.; Buchrieser, J.; Schwartz, O. The Mechanism and Consequences of SARS-CoV-2 Spike-Mediated Fusion and Syncytia Formation. *J. Mol. Biol.* **2022**, *434*, 167280. [CrossRef]

43. Chan, S.W. Fusion Assays for Screening of Fusion Inhibitors Targeting SARS-CoV-2 Entry and Syncytia Formation. *Front. Pharmacol.* **2022**, *13*, 1007527. [CrossRef]
44. Leroy, H.; Han, M.; Woottum, M.; Bracq, L.; Bouchet, J.; Xie, M.; Benichou, S. Virus-Mediated Cell-Cell Fusion. *Int. J. Mol. Sci.* **2020**, *21*, 9644. [CrossRef]
45. van Hoogevest, P.; Wendel, A. The Use of Natural and Synthetic Phospholipids as Pharmaceutical Excipients. *Eur. J. Lipid Sci. Technol.* **2014**, *116*, 1088–1107. [CrossRef]
46. Anderson, M.; Omri, A. The Effect of Different Lipid Components on the in Vitro Stability and Release Kinetics of Liposome Formulations. *Drug Deliv. J. Deliv. Target. Ther. Agents* **2004**, *11*, 33–39. [CrossRef] [PubMed]
47. Khan, D.R.; Rezler, E.M.; Lauer-Fields, J.; Fields, G.B. Effects of Drug Hydrophobicity on Liposomal Stability. *Chem. Biol. Drug Des.* **2008**, *71*, 3–7. [CrossRef] [PubMed]
48. Drabik, D.; Chodaczek, G.; Kraszewski, S.; Langner, M. Mechanical Properties Determination of DMPC, DPPC, DSPC, and HSPC Solid-Ordered Bilayers. *Langmuir* **2020**, *36*, 3826–3835. [CrossRef]
49. Sasaki, Y.; Asaoka, Y.; Nishizuka, Y. Potentiation of Diacylglycerol-Induced Activation of Protein Kinase C by Lysophospholipids. Subspecies Difference. *FEBS Lett.* **1993**, *320*, 47–51. [CrossRef] [PubMed]
50. Mouritsen, O.G. Lipids, Curvature, and Nano-Medicine. *Eur. J. Lipid Sci. Technol.* **2011**, *113*, 1174–1187. [CrossRef]
51. Van Echteld, C.J.A.; De Kruijff, B.; Mandersloot, J.G.; De Gier, J. Effects of Lysophosphatidylcholines on Phosphatidylcholine and Phosphatidylcholine/Cholesterol Liposome Systems as Revealed by 31P-NMR, Electron Microscopy and Permeability Studies. *Biochim. Biophys. Acta (BBA)-Biomembr.* **1981**, *649*, 211–220. [CrossRef]
52. Lu, J.; Xu, Y.; Chen, J.; Huang, F. Effect of Lysophosphatidylcholine on Behavior and Structure of Phosphatidylcholine Liposomes. *Sci. China C Life Sci.* **1997**, *40*, 622–629. [CrossRef] [PubMed]
53. Lutz, J.; Augustin, A.J.; Jäger, L.J.; Bachmann, D.; Brandl, M. Acute Toxicity and Depression of Phagocytosis in Vivo by Liposomes: Influence of Lysophosphatidylcholine. *Life Sci.* **1994**, *56*, 99–106. [CrossRef]
54. Raffy, S.; Teissié, J. Control of Lipid Membrane Stability by Cholesterol Content. *Biophys. J.* **1999**, *76*, 2072–2080. [CrossRef] [PubMed]
55. Kirby, C.; Clarke, J.; Gregoriadis, G. Effect of the Cholesterol Content of Small Unilamellar Liposomes on Their Stability in Vivo and in Vitro. *Biochem. J.* **1980**, *186*, 591–598. [CrossRef] [PubMed]
56. Briuglia, M.L.; Rotella, C.; McFarlane, A.; Lamprou, D.A. Influence of Cholesterol on Liposome Stability and on in Vitro Drug Release. *Drug Deliv. Transl. Res.* **2015**, *5*, 231–242. [CrossRef]
57. Juszkiewicz, K.; Sikorski, A.F.; Czogalla, A. Building Blocks to Design Liposomal Delivery Systems. *Int. J. Mol. Sci.* **2020**, *21*, 9559. [CrossRef] [PubMed]
58. Ermilova, I.; Lyubartsev, A.P. Cholesterol in Phospholipid Bilayers: Positions and Orientations inside Membranes with Different Unsaturation Degrees. *Soft Matter* **2019**, *15*, 78–93. [CrossRef]
59. Wu, H.; Yu, M.; Miao, Y.; He, S.; Dai, Z.; Song, W.; Liu, Y.; Song, S.; Ahmad, E.; Wang, D.; et al. Cholesterol-Tuned Liposomal Membrane Rigidity Directs Tumor Penetration and Anti-Tumor Effect. *Acta Pharm. Sin. B* **2019**, *9*, 858–870. [CrossRef]
60. Matusewicz, L.; Podkalicka, J.; Sikorski, A.F. Immunoliposomes with Simvastatin as a Potential Therapeutic in Treatment of Breast Cancer Cells Overexpressing Her2—An in Vitro Study. *Cancers* **2018**, *10*, 418. [CrossRef] [PubMed]
61. Milla, P.; Dosio, F.; Cattel, L. PEGylation of Proteins and Liposomes: A Powerful and Flexible Strategy to Improve the Drug Delivery. *Curr. Drug Metab.* **2011**, *13*, 105–119. [CrossRef] [PubMed]
62. Torchilinl, V.P.; Papisov, M.I. Why Do Polyethylene Glycol-Coated Liposomes Circulate so Long?: Molecular Mechanism of Liposome Steric Protection with Polyethylene Glycol: Role of Polymer Chain Flexibility. *J. Liposome Res.* **1994**, *4*, 725–739. [CrossRef]
63. Lee, H. Molecular Simulations of Pegylated Biomolecules, Liposomes, and Nanoparticles for Drug Delivery Applications. *Pharmaceutics* **2020**, *12*, 1–23. [CrossRef] [PubMed]
64. Nakamura, K.; Yamashita, K.; Itoh, Y.; Yoshino, K.; Nozawa, S.; Kasukawa, H. Comparative Studies of Polyethylene Glycol-Modified Liposomes Prepared Using Different PEG-Modification Methods. *Biochim. Biophys. Acta Biomembr.* **2012**, *1818*, 2801–2807. [CrossRef] [PubMed]
65. Allen, T.M.; Hansen, C.; Rutledge, J. Liposomes with Prolonged Circulation Times: Factors Affecting Uptake by Reticuloendothelial and Other Tissues. *Biochim. Biophys. Acta (BBA)-Biomembr.* **1989**, *981*, 27–35. [CrossRef]
66. Barenholz, Y. Doxil®-The First FDA-Approved Nano-Drug: Lessons Learned. *J. Control. Release* **2012**, *160*, 117–134. [CrossRef]
67. Gregoriadis, G. Liposomes and MRNA: Two Technologies Together Create a COVID-19 Vaccine. *Med. Drug Discov.* **2021**, *12*, 100104. [CrossRef]
68. Mohamed, M.; Alaaeldin, E.; Hussein, A.; Sarhan, H.A. Liposomes and PEGylated Liposomes as Drug Delivery System. *J. Adv. Biomed. Pharm. Sci.* **2020**, *3*, 80–88. [CrossRef]

69. Ibrahim, M.; Ramadan, E.; Elsadek, N.E.; Emam, S.E.; Shimizu, T.; Ando, H.; Ishima, Y.; Elgarhy, O.H.; Sarhan, H.A.; Hussein, A.K.; et al. Polyethylene Glycol (PEG): The Nature, Immunogenicity, and Role in the Hypersensitivity of PEGylated Products. *J. Control. Release* **2022**, *351*, 215–230. [CrossRef]
70. Mohamed, M.; Abu Lila, A.S.; Shimizu, T.; Alaaeldin, E.; Hussein, A.; Sarhan, H.A.; Szebeni, J.; Ishida, T. PEGylated Liposomes: Immunological Responses. *Sci. Technol. Adv. Mater.* **2019**, *20*, 710–724. [CrossRef]
71. Smith, M.C.; Crist, R.M.; Clogston, J.D.; McNeil, S.E. Zeta Potential: A Case Study of Cationic, Anionic, and Neutral Liposomes. *Anal. Bioanal. Chem.* **2017**, *409*, 5779–5787. [CrossRef] [PubMed]
72. Németh, Z.; Csóka, I.; Semnani Jazani, R.; Sipos, B.; Haspel, H.; Kozma, G.; Kónya, Z.; Dobó, D.G. Quality by Design-Driven Zeta Potential Optimisation Study of Liposomes with Charge Imparting Membrane Additives. *Pharmaceutics* **2022**, *14*, 1798. [CrossRef]
73. Rahnfeld, L.; Thamm, J.; Steiniger, F.; van Hoogevest, P.; Luciani, P. Study on the in Situ Aggregation of Liposomes with Negatively Charged Phospholipids for Use as Injectable Depot Formulation. *Colloids Surf. B Biointerfaces* **2018**, *168*, 10–17. [CrossRef] [PubMed]
74. Lauraeus, S.; Holopainen, J.M.; Taskinen, M.R.; Kinnunen, P.K.J. Aggregation of Dimyristoylphosphatidylglycerol Liposomes by Human Plasma Low Density Lipoprotein. *Biochim. Biophys. Acta Biomembr.* **1998**, *1373*, 147–162. [CrossRef] [PubMed]
75. Li, J.; Wang, X.; Zhang, T.; Wang, C.; Huang, Z.; Luo, X.; Deng, Y. A Review on Phospholipids and Their Main Applications in Drug Delivery Systems. *Asian J. Pharm. Sci.* **2015**, *10*, 81–98. [CrossRef]
76. Santi, D.V.; Norment, A.; Garrett, C.E. Covalent Bond Formation between a DNA-Cytosine Methyltransferase and DNA Containing 5-Azacytosine. *Proc. Natl. Acad. Sci. USA* **1984**, *81*, 6993–6997. [CrossRef]
77. Jacobsen, C. Covalent Coupling of Bilirubin to Albumin. *Int. J. Pept. Protein Res.* **1976**, *8*, 295–303. [CrossRef] [PubMed]
78. Liu, P.; O'Mara, B.W.; Warrack, B.M.; Wu, W.; Huang, Y.; Zhang, Y.; Zhao, R.; Lin, M.; Ackerman, M.S.; Hocknell, P.K.; et al. A Tris (2-Carboxyethyl) Phosphine (TCEP) Related Cleavage on Cysteine-Containing Proteins. *J. Am. Soc. Mass. Spectrom.* **2010**, *21*, 837–844. [CrossRef] [PubMed]
79. PierceTM TCEP-HCl, No-WeighTM Format. Available online: https://www.thermofisher.com/order/catalog/product/20490 (accessed on 15 October 2024).
80. Espuelas, S.; Roth, A.; Thumann, C.; Frisch, B.; Schuber, F. Effect of Synthetic Lipopeptides Formulated in Liposomes on the Maturation of Human Dendritic Cells. *Mol. Immunol.* **2005**, *42*, 721–729. [CrossRef] [PubMed]
81. Cumnock, K.; Tully, T.; Cornell, C.; Hutchinson, M.; Gorrell, J.; Skidmore, K.; Chen, Y.; Jacobson, F. Trisulfide Modification Impacts the Reduction Step in Antibody-Drug Conjugation Process. *Bioconjug Chem.* **2013**, *24*, 1154–1160. [CrossRef]
82. Kantner, T.; Watts, A.G. Characterization of Reactions between Water-Soluble Trialkylphosphines and Thiol Alkylating Reagents: Implications for Protein-Conjugation Reactions. *Bioconjug Chem.* **2016**, *27*, 2400–2406. [CrossRef]
83. Wang, Y.; Misto, M.; Yang, J.; Gehring, N.; Yu, X.; Moussian, B. Toxicity of Dithiothreitol (DTT) to Drosophila Melanogaster. *Toxicol. Rep.* **2021**, *8*, 124–130. [CrossRef]
84. Martínez-Jothar, L.; Doulkeridou, S.; Schiffelers, R.M.; Sastre Torano, J.; Oliveira, S.; van Nostrum, C.F.; Hennink, W.E. Insights into Maleimide-Thiol Conjugation Chemistry: Conditions for Efficient Surface Functionalization of Nanoparticles for Receptor Targeting. *J. Control. Release* **2018**, *282*, 101–109. [CrossRef] [PubMed]
85. Schelté, P.; Boeckler, C.; Frisch, B.; Schuber, F. Differential Reactivity of Maleimide and Bromoacetyl Functions with Thiols: Application to the Preparation of Liposomal Diepitope Constructs. *Bioconjug Chem.* **2000**, *11*, 118–123. [CrossRef] [PubMed]
86. Oude Blenke, E.; Sleszynska, M.; Evers, M.J.W.; Storm, G.; Martin, N.I.; Mastrobattista, E. Strategies for the Activation and Release of the Membranolytic Peptide Melittin from Liposomes Using Endosomal PH as a Trigger. *Bioconjug Chem.* **2017**, *28*, 574–582. [CrossRef] [PubMed]
87. Reulen, S.W.A.; Brusselaars, W.W.T.; Langereis, S.; Mulder, W.J.M.; Breurken, M.; Merkx, M. Protein-Liposome Conjugates Using Cysteine-Lipids and Native Chemical Ligation. *Bioconjug Chem.* **2007**, *18*, 590–596. [CrossRef] [PubMed]
88. Ansell, S.M.; Harasym, T.O.; Tardi, P.G.; Buchkowsky, S.S.; Bally, M.B.; Cullis, P.R. Antibody Conjugation Methods for Active Targeting of Liposomes. In *Drug Targeting*; Humana Press: Totowa, NJ, USA, 2003; pp. 51–68. [CrossRef]
89. Refai, H.; Hassan, D.; Abdelmonem, R. Development and Characterization of Polymer-Coated Liposomes for Vaginal Delivery of Sildenafil Citrate. *Drug Deliv.* **2017**, *24*, 278–288. [CrossRef] [PubMed]
90. Hasan, M.; Ben Messaoud, G.; Michaux, F.; Tamayol, A.; Kahn, C.J.F.; Belhaj, N.; Linder, M.; Arab-Tehrany, E. Chitosan-Coated Liposomes Encapsulating Curcumin: Study of Lipid-Polysaccharide Interactions and Nanovesicle Behavior. *RSC Adv.* **2016**, *6*, 45290–45304. [CrossRef]
91. Guimarães, D.; Cavaco-Paulo, A.; Nogueira, E. Design of Liposomes as Drug Delivery System for Therapeutic Applications. *Int. J. Pharm.* **2021**, *601*, 120571. [CrossRef]
92. Huang, X.; Pearce, R.; Zhang, Y. De Novo Design of Protein Peptides to Block Association of the SARS-CoV-2 Spike Protein with Human ACE2. *Aging* **2020**, *12*, 11263–11276. [CrossRef] [PubMed]
93. Zhang, H. Thin-Film Hydration Followed by Extrusion Method for Liposome Preparation. In *Liposomes, Methods in Molecular Biology*; D'Souza, G., Ed.; Humana Press: New York, NY, USA, 2017; pp. 17–22. [CrossRef]

94. Murphy, J.; Riley, J.P. A Modified Single Solution Method for the Determination of Phosphate in Natural Waters. *Anal. Chim. Acta* **1962**, *27*, 31–36. [CrossRef]
95. Smith, P.K.; Krohn, R.I.; Hermanson, G.T.; Mallia, A.K.; Gartner, F.H.; Provenzano, M.D.; Fujimoto, E.K.; Goeke, N.M.; Olson, B.J.; Klenk, D.C. Measurement of Protein Using Bicinchoninic Acid. *Anal. Biochem.* **1985**, *150*, 76–85. [CrossRef] [PubMed]
96. Lu, W.; Wu, X.D.; De Shi, M.; Yang, R.F.; He, Y.Y.; Bian, C.; Shi, T.L.; Yang, S.; Zhu, X.L.; Jiang, W.H.; et al. Synthetic Peptides Derived from SARS Coronavirus S Protein with Diagnostic and Therapeutic Potential. *FEBS Lett.* **2005**, *579*, 2130–2136. [CrossRef]
97. Frere, J.J.; Serafini, R.A.; Pryce, K.D.; Zazhytska, M.; Oishi, K.; Golynker, I.; Panis, M.; Zimering, J.; Horiuchi, S.; Hoagland, D.A.; et al. SARS-CoV-2 Infection in Hamsters and Humans Results in Lasting and Unique Systemic Perturbations after Recovery. *Sci. Transl. Med.* **2022**, *14*, eabq3059. [CrossRef]
98. Imai, M.; Iwatsuki-Horimoto, K.; Hatta, M.; Loeber, S.; Halfmann, P.J.; Nakajima, N.; Watanabe, T.; Ujie, M.; Takahashi, K.; Ito, M.; et al. Syrian Hamsters as a Small Animal Model for SARS-CoV-2 Infection and Countermeasure Development. *Proc. Natl. Acad. Sci. USA* **2020**, *117*, 16587–16595. [CrossRef]
99. Muñoz-Fontela, C.; Dowling, W.E.; Funnell, S.G.P.; Gsell, P.S.; Riveros-Balta, A.X.; Albrecht, R.A.; Andersen, H.; Baric, R.S.; Carroll, M.W.; Cavaleri, M.; et al. Animal Models for COVID-19. *Nature* **2020**, *586*, 509–515. [CrossRef]
100. Arimori, T.; Ikemura, N.; Okamoto, T.; Takagi, J.; Standley, D.M.; Hoshino, A. Engineering ACE2 Decoy Receptors to Combat Viral Escapability. *Trends Pharmacol. Sci.* **2022**, *43*, 838–851. [CrossRef] [PubMed]
101. Rudokas, M.; Najlah, M.; Alhnan, M.A.; Elhissi, A. Liposome Delivery Systems for Inhalation: A Critical Review Highlighting Formulation Issues and Anticancer Applications. *Med. Princ. Pract.* **2016**, *25*, 60–72. [CrossRef]
102. Li, Q.; Zhan, S.; Liu, Q.; Su, H.; Dai, X.; Wang, H.; Beng, H.; Tan, W. Preparation of a Sustained-Release Nebulized Aerosol of R-Terbutaline Hydrochloride Liposome and Evaluation of Its Anti-Asthmatic Effects via Pulmonary Delivery in Guinea Pigs. *AAPS PharmSciTech* **2018**, *19*, 232–241. [CrossRef] [PubMed]
103. Bassetti, M.; Vena, A.; Russo, A.; Peghin, M. Inhaled Liposomal Antimicrobial Delivery in Lung Infections. *Drugs* **2020**, *80*, 309–1318. [CrossRef] [PubMed]
104. Cipolla, D.; Gonda, I.; Chan, H.K. Liposomal Formulations for Inhalation. *Ther. Deliv.* **2013**, *4*, 1047–1072. [CrossRef]
105. Ong, H.X.; Traini, D.; Young, P.M. Liposomes for Inhalation. *J. Aerosol Med. Pulm. Drug Deliv.* **2024**, *37*, 100–110. [CrossRef] [PubMed]
106. Yousry, C.; Goyal, M.; Gupta, V. Excipients for Novel Inhaled Dosage Forms: An Overview. *AAPS PharmSciTech* **2024**, *25*, 1–10. [CrossRef]
107. Urano, E.; Itoh, Y.; Suzuki, T.; Sasaki, T.; Kishikawa, J.I.; Akamatsu, K.; Higuchi, Y.; Sakai, Y.; Okamura, T.; Mitoma, S.; et al. An Inhaled ACE2 Decoy Confers Protection against SARS-CoV-2 Infection in Preclinical Models. *Sci. Transl. Med.* **2023**, *15*, eadi2623. [CrossRef]
108. Houck, K.S.; Huang, L. The Role of Multivalency in Antibody Mediated Liposome Targeting. *Biochem. Biophys. Res. Commun.* **1987**, *145*, 1205–1210. [CrossRef]
109. Singla, A.; Simbassa, S.B.; Chirra, B.; Gairola, A.; Southerland, M.R.; Shah, K.N.; Rose, R.E.; Chen, Q.; Basharat, A.; Baeza, J.; et al. Hetero-Multivalent Targeted Liposomal Drug Delivery to Treat Pseudomonas Aeruginosa Infections. *ACS Appl. Mater. Interfaces* **2022**, *14*, 40724–40737. [CrossRef] [PubMed]

International Journal of
Molecular Sciences

MDPI

Article

Study on the Effect of Emulsifiers on the Properties of Oleogels Based on Olive Oil Containing Lidocaine

Sonia Kudłacik-Kramarczyk [1,*], Alicja Przybyłowicz [1,2], Anna Drabczyk [3], Weronika Kieres [1], Robert P. Socha [1,3] and Marcel Krzan [1]

[1] Jerzy Haber Institute of Catalysis and Surface Chemistry, Polish Academy of Sciences, 8 Niezapominajek St., 30-239 Krakow, Poland; alicja.przybylowicz@student.pk.edu.pl (A.P.); weronikakieres1605@gmail.com (W.K.); robert.socha@cbrtp.pl (R.P.S.); marcel.krzan@ikifp.edu.pl (M.K.)

[2] Faculty of Mechanical Engineering, Cracow University of Technology, 37 Jana Pawła II Av., 31-864 Krakow, Poland

[3] CBRTP SA—Research and Development Center of Technology for Industry, Ludwika Waryńskiego 3A St., 00-645 Warsaw, Poland; anna.drabczyk@cbrtp.pl

* Correspondence: sonia.kudlacik-kramarczyk@ikifp.edu.pl

Abstract: Oleogels are semi-solid materials that consist primarily of liquid oil immobilized in a network of organized structural molecules, which provide stability and maintain the oil in the desired shape. Due to their structure, oleogels can stabilize large amounts of liquid, making them excellent carriers for active substances, both lipophilic and hydrophilic. This study presents the synthesis methodology and investigations of olive oil-based oleogels, which are among the healthiest and most valuable vegetable fats, rich in unsaturated fatty acids and antioxidants such as vitamin E. Two types of surfactants were used: TWEEN 80, which lowers surface tension and stabilizes emulsions, and SPAN 80, which acts in oil-dominated phases. The oleogels were enriched with lidocaine, an active substance commonly used as a pain reliever and local anesthetic. This research characterized the obtained oleogels regarding their medical applications, paying particular attention to the influence of surfactant type and amount as well as the active substance on their physicochemical properties. Structural analyses were also conducted using Fourier transform infrared (FTIR) spectroscopy, alongside rheological and sorption studies, and the wettability of the materials was evaluated. The stability of the obtained oleogels was verified using the MultiScan MS20 system, allowing for an assessment of their potential suitability for long-term pharmaceutical applications. The results indicated that SPAN-stabilized oleogels exhibited better stability and favorable mechanical properties, making them promising candidates for medical applications, particularly in pain relief formulations.

Keywords: oleogels; lidocaine; long-term stability; rheology; emulsions; physicochemical properties

1. Introduction

Oleogels are semi-solid materials that consist mainly of liquid oil immobilized in a network of organized structural molecules that provide stability and hold the oil in the desired shape [1–3]. The properties of oleogels are due to their structure, which allows them to stabilize large amounts of liquid oil. These materials function as excellent carriers for lipophilic and hydrophilic active substances [4]. Due to their texture, they are easily adapted to a variety of applications, such as skin hydration and protection from external agents or controlled drug release [3,5]. Oleogels are also used extensively in the food industry, including, among others, as replacements for common fats in food. In addition, they enable the manufacture of healthier products without losing flavor and textural properties [6–8].

In this paper, the synthesis methodology and investigations on olive oil-based oleogels are presented. Olive oil is one of the healthiest and most valuable vegetable fats. It is rich in unsaturated fatty acids and contains antioxidants, including vitamin E. It also demonstrates anti-inflammatory properties [9–11]. In cosmetics, this oil is widely used as a moisturizer. Due to the presence of antioxidants, it helps fight skin aging [12,13].

In the synthesis, two types of surfactants, namely, TWEEN 80 and SPAN 80, were used. Span 80 (sorbitan monooleate) and Tween 80 (polysorbate 80) are widely used emulsifiers in oleogel formulations [14–16]. Span 80, an emulsifier with a low HLB (Hydrophilic–Lipophilic Balance) value of 4.3, favors the formation of w/o (water-in-oil) emulsions, while Tween 80, due to its high HLB value of 15.0, prefers to stabilize o/w (oil-in-water) emulsions [17,18]. The key parameter that determines their properties is HLB (Hydrophilic–Lipophilic Balance), which describes the balance between the hydrophilic and lipophilic properties of an emulsifier and allows an assessment of how much a given surfactant prefers to dissolve in water or oil. The HLB value determines an emulsifier's ability to stabilize a particular type of emulsion—low HLB values indicate a more hydrophobic character, which favors the stabilization of w/o (water-in-oil) emulsions, while high HLB values indicate greater hydrophilic properties, which supports the formation of o/w (oil-in-water) emulsions. This makes the HLB parameter a key indicator for selecting the right emulsifier for specific applications and conditions, such as oil-in-oil or water-in-water phase stability in oleogels [19–21]. HLB values for hydrophilic surfactants typically range from 8 to 18, while those for hydrophobic surfactants are below [22–24]. This selection of emulsifiers allows for the precise control of the structure and rheological properties of oleogels by adjusting phase distribution and intermolecular interactions. In the present study, these two surfactants were selected to verify the impact of emulsifiers with different HLB values on selected physicochemical properties of oleogels including, among others, wettability, stability, and sorption ability. The use of surfactants with different HLB values makes it possible to characterize their interactions with the oil and aqueous phases, which is important for understanding the mechanisms shaping the structure and stability of the final oleogels.

Importantly, the oleogels were enriched with lidocaine. This active substance is widely used as an arrhythmic drug and a local anesthetic [25]. Its action is based on blocking sodium channels found in the membranes of nerve cells, resulting in rapid local anesthesia. It is prepared in the form of injections, sprays, gels, ointments, and patches [26,27].

In the literature, several investigations focused on the characteristics of systems containing both olive oil and lidocaine have been so far presented. For example, in the work of Dogrul et al. [28], olive oil was used as one of the oil phases in microemulsions containing lidocaine hydrochloride. It was observed that the use of olive oil as an oil phase led to an increase in the droplet size of microemulsions in which aqueous solutions of lidocaine hydrochloride were dispersed. The increase in the size of these aqueous droplets in the emulsion was due to the interaction of oleic acid present in the oil with lecithin and ethanol (used as surfactant and co-surfactant, respectively), which affected the structure and stability of the film at the water–oil interface. In addition, a reduction in the rate of drug release (only 12.6% of the drug was released within 6 h) was also observed for the systems obtained with olive oil compared to those obtained with Miglyol. In turn, the main goal of the research of Zhang et al. [29] was to develop microemulsions without co-emulsifiers to minimize toxicity and irritation associated with their use. To achieve this, olive oil was mixed with α-linolenic acid and linoleic acid, and vitamin E succinate was used as an auxiliary oil. The final optimal formulation contained 3.23% olive oil to improve the skin permeability of lidocaine and increase its retention. Key findings of the study demonstrated that the use of olive oil in combination with fatty acids and vitamin E significantly improved the skin permeation of lidocaine and prolonged the duration of anesthesia compared to a commercial cream. The microemulsion showed good stability, minimal irritation, and significant anesthetic efficacy. Next, the study of Daryab et al. [30] evaluated gel-based microemulsions as vehicles for the transdermal delivery of lidocaine, including the use of olive oil and various surfactants and co-emulsifiers. A key goal was to investigate how olive oil, in combination with other reagents, affects the solubility and permeability of lidocaine through the skin. The use of olive oil in microemulsions was shown to increase the solubility of lidocaine. In turn, studies performed by Qiao et al. [31] evaluated various formulations of microemulsions containing lidocaine, including those using olive oil as the oil phase. The main objective of the research was to optimize the microemulsions for better transdermal penetration and retention of lidocaine in the skin while reducing the amount of surfactants, which should reduce toxicity and the risk of skin irritation. The study used multi-object optimization algorithms to achieve the best pharmaceutical properties. The key findings of the study indicate that a microemulsion with olive oil (1.11%) formulated with 0.64% α-linolenic acid, 2.54% linoleic acid, 0.71% vitamin E succinate, 15% surfactant, 5% lidocaine, and 75% water provided the highest skin permeation of lidocaine (0.17 μg/cm^2·s) and retention in the skin (0.74 mg/cm^2). Another optimal formulation with a higher ratio of olive oil to fatty acids (1.42) and less surfactant content (13.6%) improved anesthesia duration, achieving a lidocaine retention of 0.80 mg/cm^2. These formulations showed better penetration and retention of lidocaine compared to other investigated systems.

However, no studies focusing on investigations of olive oil-based oleogels containing lidocaine and discussing the application of two types of surfactants in the synthesis of these oleogels have been presented so far. Hence, the aim of this research was to characterize the obtained oleogels in terms of their application for medical purposes, with particular attention paid to the influence of the surfactant type (and amount) and active substance used on their physicochemical properties. In the course of the conducted research, the structure of the oleogels was also characterized using FT-IR spectroscopy to verify potential interactions between the oleogel and the introduced active substance. Performed studies also included the analysis of rheological and sorption properties of oleogels, as well as their wettability. In addition, the stability of the obtained materials over time was verified using the MultiScan system to determine their potential suitability for long-term pharmaceutical applications.

2. Results

2.1. Stability of Obtained Oleogels within the Period of Two Months

Figure 1 presents the appearance of the obtained samples directly after their preparation and after two months, while in Figure 2, the results of the conducted analysis of oleogels' stability via the MultiScan system are compiled.

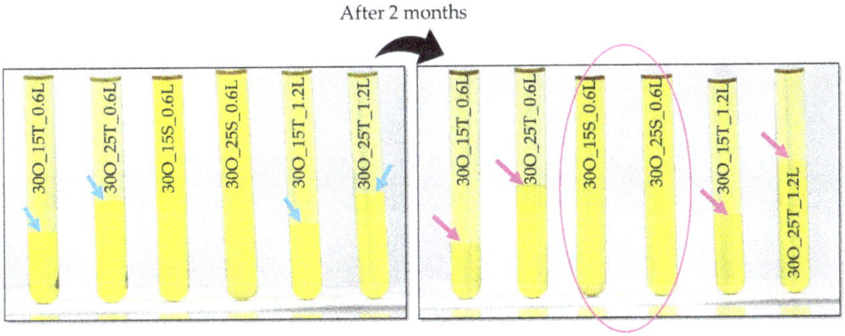

Figure 1. Images presenting the appearance of oleogel samples directly after synthesis (**left**) and after two months (**right**) (blue arrows demonstrate the level of oleogel in the test tubes after the synthesis; pink arrows demonstrate the level of oleogel in the test tubes after 2 months storage; pink circle shows samples in the case of which any difference has been observed).

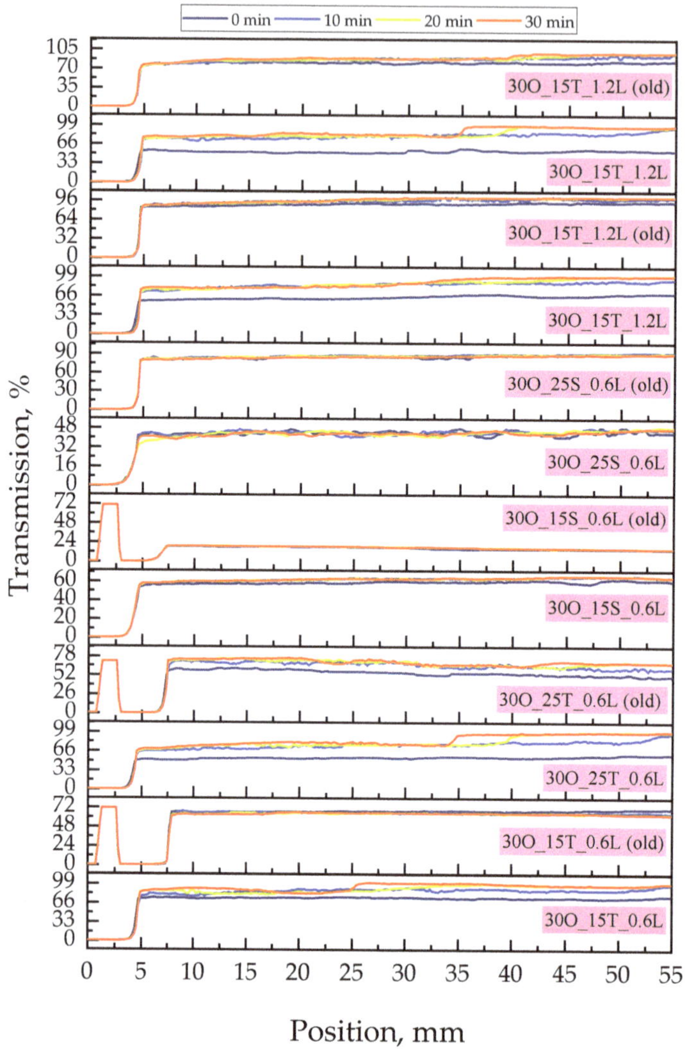

Figure 2. Results of the stability analysis via the MultiScan system for the oleogel samples; analysis after synthesis and after two months (old).

The principle of operation of this MultiScan system is illustrated in the diagram below (Figure 3).

The MultiScan system operates by analyzing the transmission and absorption of light through the oleogel samples. It utilizes a range of wavelengths to assess the structural integrity and stability of the emulsions over time. By comparing the transmission spectra of freshly synthesized samples with those stored for two months, the MultiScan system can identify changes in phase separation, which may indicate destabilization. Additionally, the device helps evaluate the interactions between the oleogels and the active substance, lidocaine, providing insights into how these factors influence the overall physicochemical properties of the materials. Overall, this method is crucial for assessing the long-term suitability of the oleogels for pharmaceutical applications.

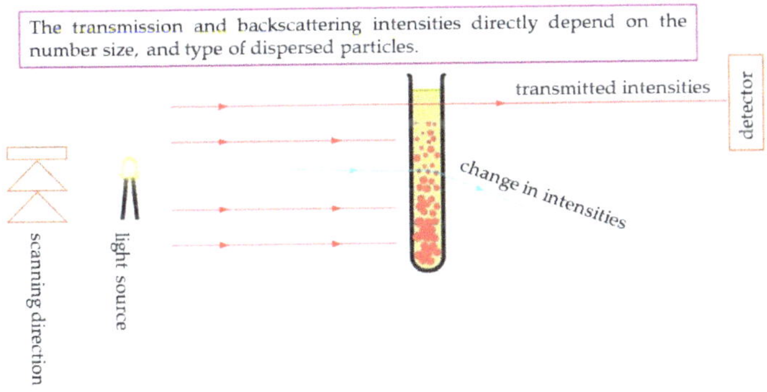

Figure 3. Schematic of the transmission measurement process using the MultiScan MS 20 device (The DataPhysics Instruments GmbH, Filderstadt, Germany) for the tested materials.

2.2. Results of FT-IR ATR Spectroscopy

The results of spectroscopic analysis are shown below. Figure 4a shows the FT-IR spectra of the obtained oleogels while Figure 4b presents the FT-IR spectrum of lidocaine.

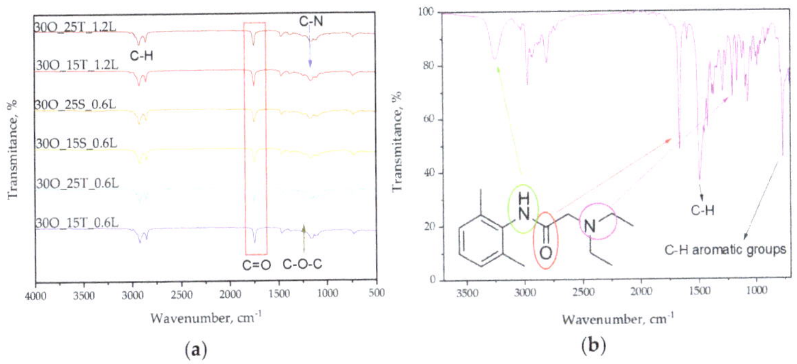

Figure 4. FT-IR spectra of oleogel samples (**a**) and lidocaine (**b**).

The above graphs present a comparison of FT-IR spectra for all studied samples—both immediately after synthesis and the spectrum for pure lidocaine for the purpose of identifying characteristic groups. Table 1 provides the relevant compilation for the analysis of the obtained spectra. Figure 5 presents a zoomed-in view, highlighting the changes in the intensity of bands for hydroxyl groups present in the structure of the emulsifiers used.

Table 1. Summary of all vibrations with their corresponding chemical compounds involved in the obtained materials.

Wavenumber [cm^{-1}]	Type of Vibration	Assigned Group	Component
3400–3200	Stretching	O-H (hydroxyl)	SPAN, TWEEN
2950–2850	Symmetric and asymmetric stretching	C-H (alkanes)	Olive oil, TWEEN 80, SPAN 80
1740–1735	Stretching	C=O (ester)	Olive oil, TWEEN 80, SPAN 80
1650–1620	Stretching	C=O (amide)	Lidocaine
1200–1100	Stretching	C-O-C (ether)	TWEEN 80, SPAN 80
1200–1350	Stretching	C-N (amine)	Lidocaine
1480	Bending	C-H (aromatic groups)	Lidocaine
758	Bending	C-H (aromatic groups)	Lidocaine

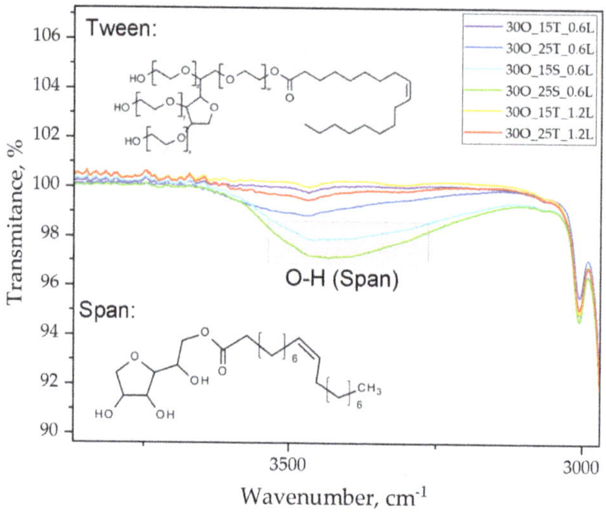

Figure 5. Approximation of the FT-IR spectrum from Figure 4a after overlaying all spectra, revealing the precise depth of absorption bands originating from the surfactants TWEEN 80 and SPAN 80.

This study confirmed the presence of characteristic groups derived from the substrates used in the obtained materials and allowed for the observation of differences in band intensity depending on the amount of substrate applied during the synthesis process.

2.3. Sorption Properties of Oleogels

The results of the performed experiments aimed at determining the sorption ability of oleogels are presented in Figure 6.

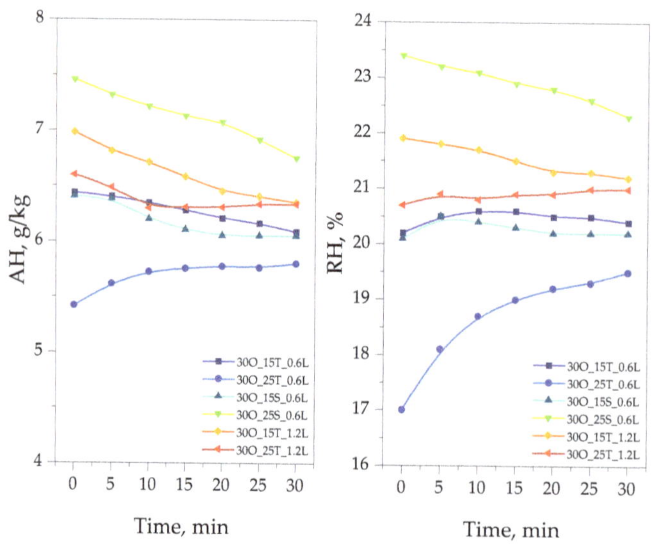

Figure 6. Changes in the absolute (**AH**) and relative (**RH**) humidity of oleogels over time.

The study of the material's reaction to moisture present in the environment allows for the assessment of the material's interaction with the environment, which may simulate its potential applications. A detailed discussion of the obtained results is presented in the following subsection.

2.4. Wettability of Oleogels

Figure 7 compiles wetting angles for oleogels determined for three types of materials, i.e., block mica, Teflon plate, and glassy carbon plate. In Figure 8, the images showing the behavior of a drop of oleogel after initial contact with the test material are presented.

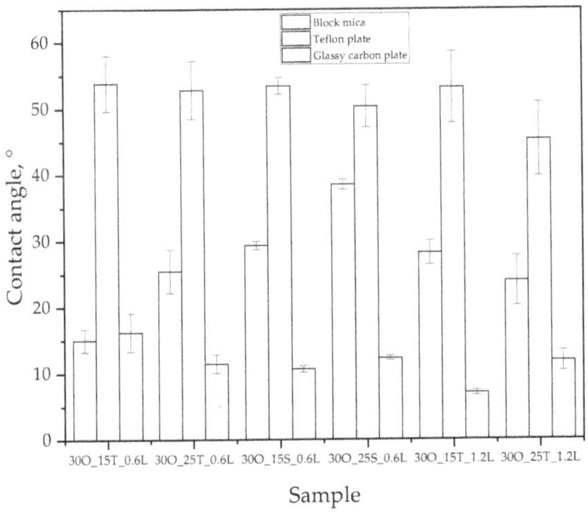

Figure 7. Contact angles for the obtained oleogel materials with standard deviations calculated from five repetitions.

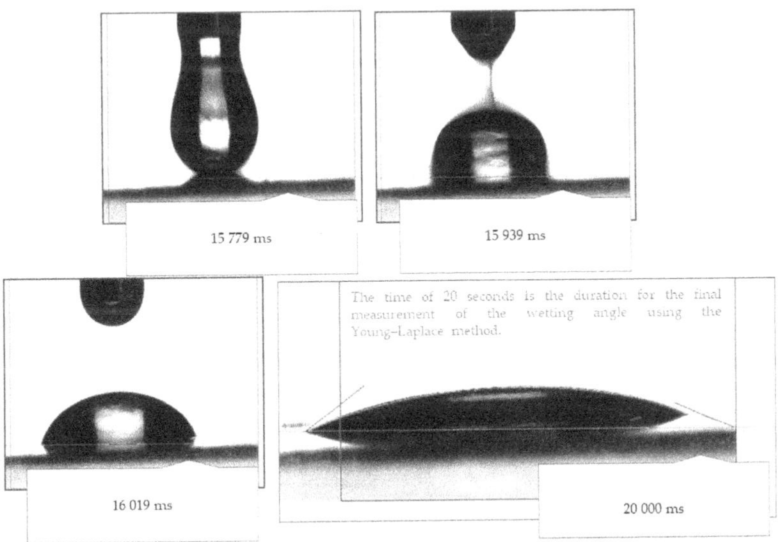

Figure 8. Images of the oleogel drop behavior during the initial contact with the test material.

The study of surface wettability allows for the characterization of the hydrophilic or hydrophobic nature of the obtained materials. The diagram presented in Figure 8 illustrates the measurement principle applied during all measurements conducted for the obtained materials. Collecting measurements after a specified time for the tested materials is essential and enables the acquisition of more reliable research results.

2.5. Results of Studies on Rheological Properties of Oleogels

In Figure 9, the results of the performed rheological analyses are presented.

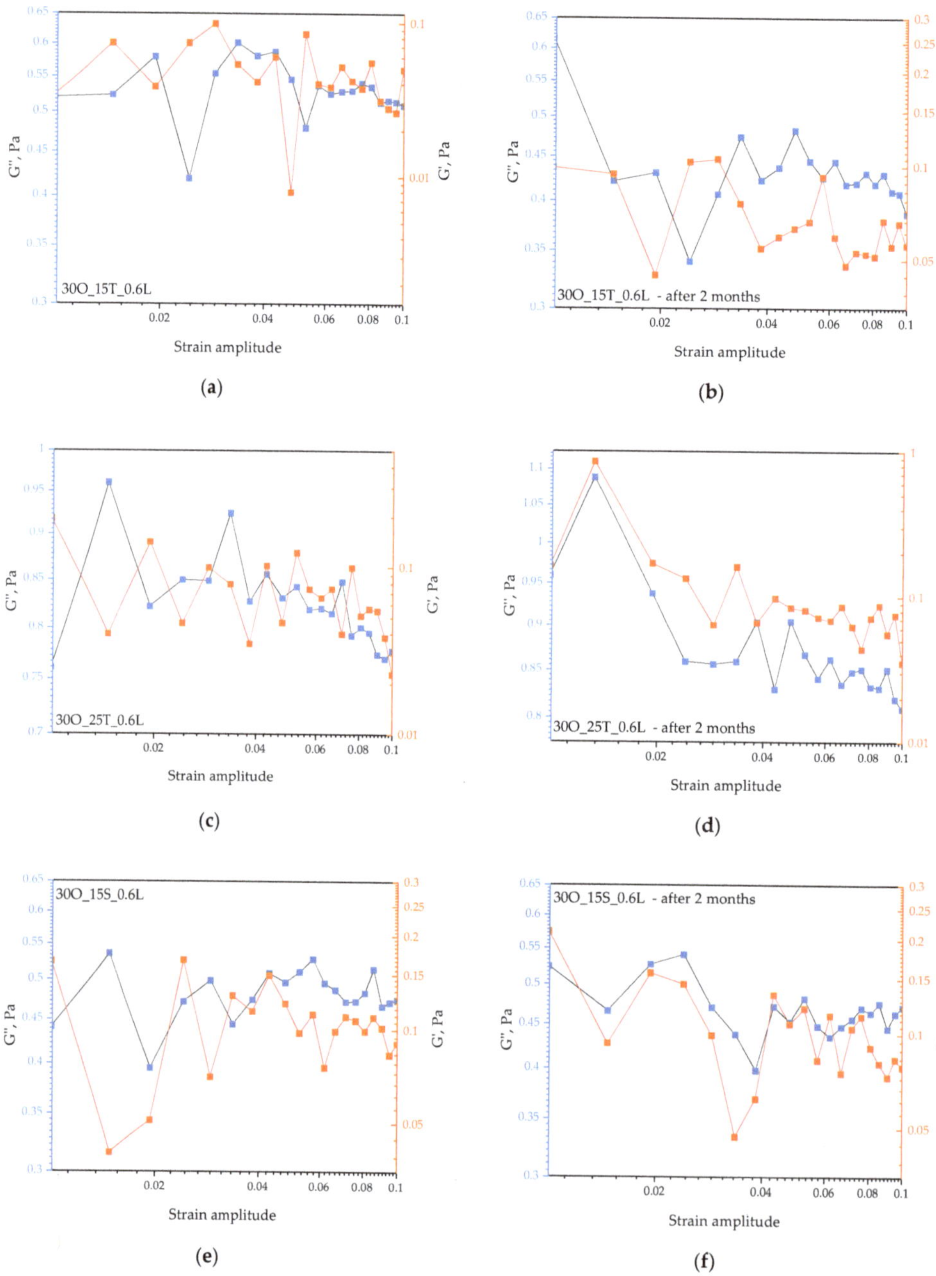

Figure 9. *Cont.*

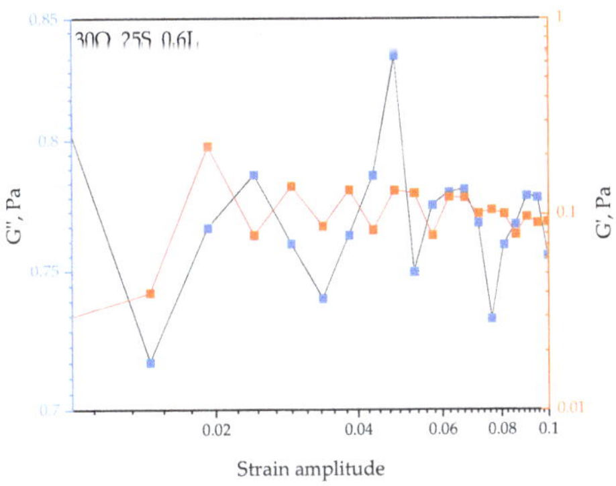

(g)

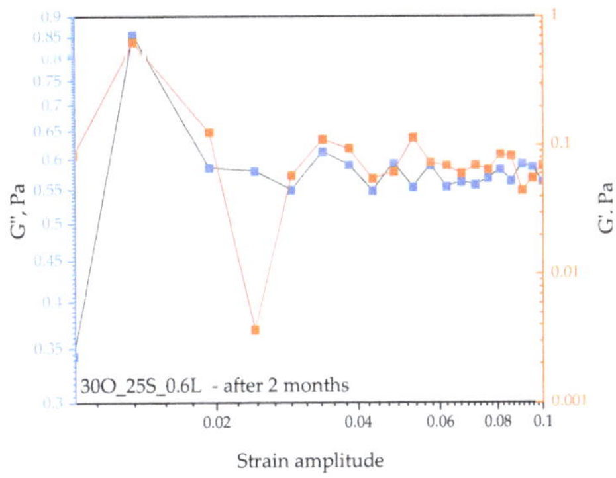

(h)

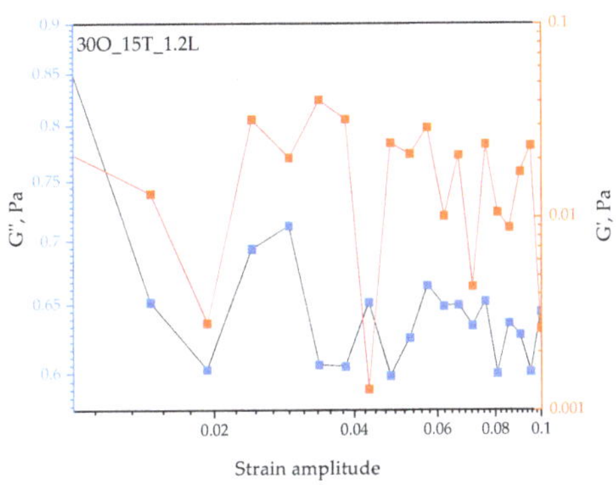

(i)

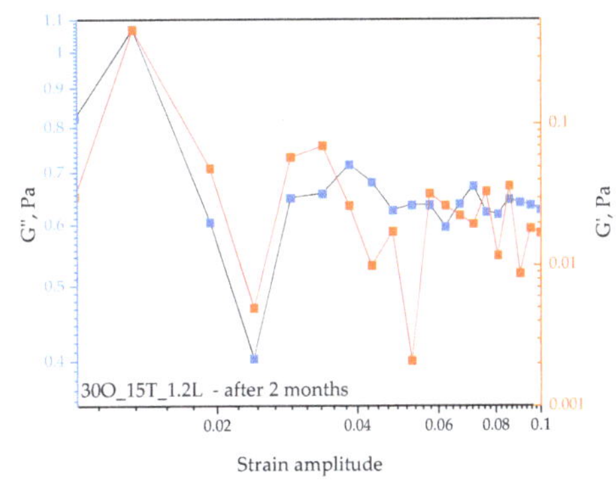

(j)

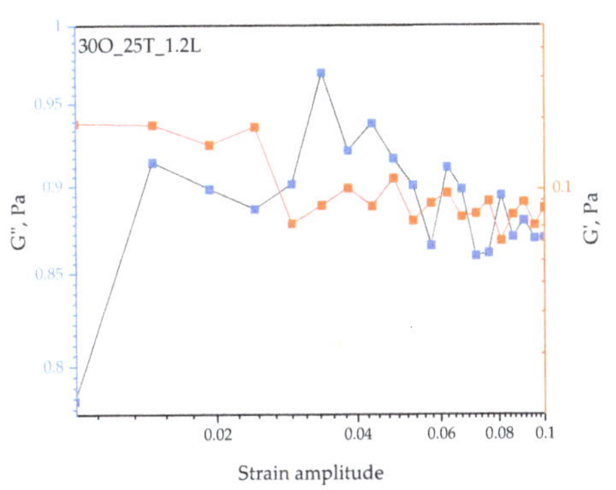

(k)

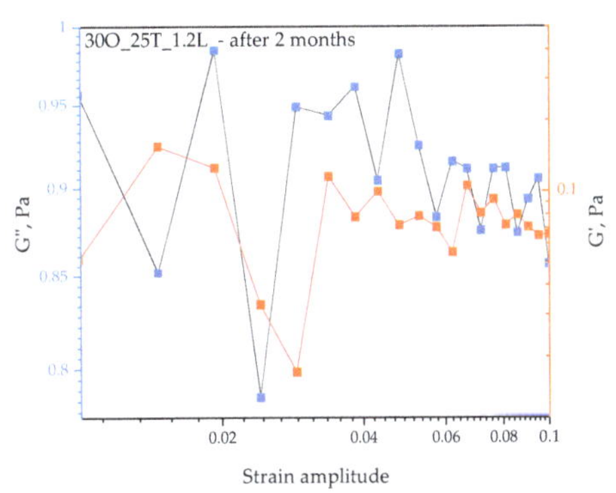

(l)

Figure 9. Summary of G′ and G″ for materials tested immediately after synthesis (**a**,**c**,**e**,**g**,**i**,**k**) and 2 months after synthesis (**b**,**d**,**f**,**h**,**j**,**l**).

3. Discussion

Based on the results obtained from the MultiScan analysis, we can assess the stability of the emulsion after two months of storage at 6 °C and attempt to interpret the type of emulsion present in our materials. Analyzing Figure 2, we can begin by comparing the transmission spectrum for sample 30O_15T_0.6L and the same sample stored for two months (30O_15T_0.6L (old)). It is evident that the oleogel examined immediately after synthesis shows a significantly different spectrum compared to the material stored for two months, which may indicate changes occurring in the material over time.

Focusing on the detailed interpretation of the spectrum for the 30O_15T_0.6L sample, we observe that up to the 4 mm position, the light does not pass through the sample. This is most likely due to significant scattering, as nothing suggests that absorption would occur here, considering the nature of the components in the sample. Next, we observe an increase in transmission to almost maximum levels, indicating that the material is almost fully transparent to light in this range.

However, the 30O_15T_0.6L (old) sample exhibits a different behavior. From the start, we observe a transmission increase to 72% compared to the freshly synthesized sample. It is likely that prolonged storage caused the partial separation of the emulsion, leading to reduced light scattering and thus higher transmission relative to the initial sample. A subsequent drop in transmission can be observed in the spectrum, which may suggest the formation of heterogeneities or aggregates as a result of prolonged storage. Following this, transmission levels rise again, but only to around 72%, compared to about 97% in the "fresh" sample. This may suggest the continued presence of an emulsion, albeit with an altered structure, indicating partial destabilization caused by the storage process.

Comparing the MultiScan results with the photograph presented in Figure 1, we can initiate an interesting discussion. Figure 1 shows the appearance of the systems immediately after synthesis and after two months of storage. In the photograph, we can observe the separation of phases in the sample with the naked eye. For the phase comparison between 30O_15T_0.6L and 30O_15T_0.6L (old), we see that the phase separation visible in the photo is also reflected in the transmission spectrum. At 25 mm, after 30 min, we observe a slight increase in transmission, which most likely corresponds to the phase separation seen in the photograph. The increase in transmission suggests that above the visible phase boundary, we are most likely observing just oil, which has a lower capacity for light scattering, thereby increasing transmission. This is, of course, an undesirable phenomenon as it indicates the destabilization of the final product.

The destabilization is observed in all samples except those obtained using SPAN as the emulsifier. Based on this, we can conclude that lidocaine does not affect the stability of the emulsion, but the choice of emulsifier does. Additionally, a long storage period allows for the creation of a stable emulsion throughout that time, provided that the appropriate amount of emulsifier is used. The samples marked with a pink circle in Figure 1 show no visible phase separation, even after two months of storage. However, when analyzing the transmission graph for these materials, we see that the sample containing 15 g SPAN exhibited changes in transmission after two months, indicating that only the sample containing 25 g SPAN remained stable for the entire two-month period.

The overall conclusions from this study are that lidocaine does not influence emulsion stability; rather, it is the emulsifier that plays a key role. The difference between emulsions stabilized with SPAN and those with TWEEN is due to the fact that TWEEN is a hydrophilic surfactant (with a high HLB, around 15), meaning it stabilizes oil-in-water (o/w) emulsions, where the oil phase is dispersed in the water phase. TWEEN 80 prefers interactions with water and stabilizes emulsions with a more "aqueous" structure. SPAN, on the other hand, is a lipophilic surfactant (with a low HLB, around 4.3), which is better at stabilizing water-in-oil (w/o) emulsions, where the water phase is dispersed in the oil phase. SPAN 80 prefers interactions with oil and stabilizes emulsions with a more "oily" structure. The role of this surfactant in stabilizing w/o emulsions has also been described in [32,33].

To investigate the structural differences between the emulsifiers used and to assess their impact on the material's structure, FT-IR analysis was conducted, as presented in Figures 4 and 5. Figure 4a shows the spectrum comparison for all samples, while Figure 4b presents the spectrum for lidocaine alone. Analyzing the full set of spectra, we can observe bands corresponding to C-H (alkanes) and C=O (esters), which originate from components such as olive oil, TWEEN, and SPAN. Due to its structure, TWEEN contains more ester groups compared to SPAN, and this can be observed in the reduced intensity of the C-O-C band, marked in Figure 4a for samples containing SPAN, i.e., 30O_15S_0.6L and 30O_25S_0.6L.

Furthermore, the spectra show C-N groups originating from lidocaine, but we do not observe a significant deepening of this band depending on the amount of lidocaine in the sample. This is likely due to the limitations in the accuracy of this particular method. In addition, Figure 5 provides a zoomed-in view of the 4000 to 3000 cm^{-1} range from Figure 4a. The aim was to examine the intensity of bands originating from the emulsifiers and to check for any potential differences in this region (since spectral separation often leads to the flattening of the spectra, making it difficult to observe differences, we decided to isolate the section that exhibited the most prominent differences).

In Figure 5, we can see that the bands from SPAN are significantly more intense than those from TWEEN. Moreover, a higher quantity of SPAN in the sample further confirms this relationship, as the band for the sample with 25 g of SPAN is more intense than the one containing 15 g of SPAN. In TWEEN 80, many hydroxyl groups may be more "hidden" within the structure, especially when interacting with other components such as olive oil or lidocaine. These groups may be less capable of forming strong hydrogen bonds, which could result in a lower intensity of the O-H band. Additionally, the structure of SPAN 80 is less "congested" on the hydrophilic side, which may increase the availability of O-H groups for detection.

Staying within the scope of comparing the properties of emulsifiers and the impact of hydroxyl groups in their structures on material characteristics, a moisture measurement study was conducted. The samples were sealed in an airtight container with temperature monitoring. As shown in the results presented in Figure 6, the sample containing 15 g of TWEEN and 0.6 g of lidocaine quickly reached moisture equilibrium—within just 5 min. This indicates that this system responded rapidly to environmental conditions and stabilized in a very short time.

The next sample, which contained 25 g of TWEEN and the same amount of lidocaine, released much more moisture into the environment throughout the experiment. Interestingly, in the case of the sample with 25 g of TWEEN but 1.2 g of lidocaine, this effect was not observed, likely because the higher lidocaine content blocked this effect and allowed the material to reach moisture equilibrium.

The sample containing less SPAN and lidocaine behaved similarly to the sample with less TWEEN, suggesting that 15 g of SPAN provides a similar ability to retain moisture as 15 g of TWEEN, although through different mechanisms of water interaction. For the sample with a higher SPAN content, the decrease in humidity over time indicates that the sample started absorbing moisture from the environment, lowering the RH level. This may confirm the more intense hydroxyl group band observed in the FT-IR spectrum, where a greater number of hydroxyl groups enhances their ability to interact with moisture, resulting in the observed decrease in RH levels.

Next, a surface wettability study was conducted on different materials, and the results are presented in Figure 7. The general principle for interpreting this analysis is that a small contact angle indicates good surface wettability, i.e., better adhesion of the sample to the surface. In contrast, a larger contact angle suggests weaker interactions with the surface [34].

Analyzing the results for the mica surface, we observe that samples containing SPAN exhibit larger contact angles, which is attributed to the hydrophobic nature of this emulsifier [35]. Consequently, the larger the amount of SPAN (e.g., 25 g), the greater the contact angle, compared to the sample containing 15 g of SPAN. Mica, being a hydrophilic material [36], highlights how the addition of a hydrophobic substance in the samples decreases surface wettability.

Regarding the effect of lidocaine on surface wettability, when there is a smaller amount of TWEEN in the sample, we notice reduced wettability, as evidenced by the larger contact angle. However, in the sample with a higher TWEEN concentration, the increased amount of lidocaine does not significantly impact the contact angle. This suggests that an appropriate ratio of emulsifier to lidocaine can result in a product that maintains consistent wettability, independent of minor variations in lidocaine content.

Focusing on the results for the Teflon surface, we observe no significant differences between the samples. The results are quite similar, and no clear dependencies emerge. This is due to the hydrophobic nature of both the oleogels and the Teflon surface itself [37], which limits the ability to draw meaningful conclusions, as the differences in contact angle values fall within the range of the standard deviations.

The final surface tested was glassy carbon, which has relatively balanced properties—neither distinctly hydrophobic nor hydrophilic [38]. Based on the results, we can observe that samples containing SPAN show the most consistent results. This consistency also appears in the mica test, which is important as it suggests that these oleogels likely have the most homogeneous structure and the most stable emulsion. This conclusion was also confirmed through MultiScan tests and by examining photographs taken with a highly sensitive camera.

The final study was a rheological analysis, where materials were tested immediately after synthesis and also after two months of storage (Figure 9). The rheological study also confirmed the results from previous analyses, especially those obtained using the MultiScanning technique. Based on the results presented in Figure S1a, we can observe that the sample with a higher amount of SPAN and 0.6 g of lidocaine exhibits a modulus with slight fluctuations, suggesting that the material becomes more "solid-like" and well maintains its structure under stress. Consequently, we can conclude that this sample immediately after synthesis has the most stable and increasing elastic modulus, indicating that it is a material with a strong structure, resistant to deformation. Regarding the impact of storing materials for 2 months, we can confirm that prolonged storage at 6 °C has positively affected the structural stability of most samples, improving their mechanical properties, except for Sample 30O_15T_1.2L, which still shows signs of instability.

Analyzing the graphs presented in Figure S2, we can observe the absence of intersection points ($G' = G''$) for the samples tested immediately after synthesis, suggesting that all samples after synthesis have dominant viscous properties. The energy loss associated with viscous deformation (G'') is greater than the energy recovered by the material during elastic deformations (G'). The most balanced viscoelastic properties (the smallest differences between G' and G'') are exhibited by samples 30O_25S_0.6L and 30O_25T_1.2L. Although there are no clear intersection points, the differences between G' and G'' are smaller, indicating the greater elasticity of these samples compared to others. Intersection points ($G' = G''$) usually indicate a change in the dominant properties of the material—where elasticity dominates, $G' > G''$, and where viscosity dominates, $G'' > G'$. The absence of an intersection point suggests that the material consistently maintains the dominance of one property over the other.

If we focus on comparing these properties for materials tested after the 2-month storage period at 6 °C, we can observe a lack of intersection points in each of the samples. Sample 30O_25S_0.6L shows a favorable change in terms of elasticity after 2 months, suggesting that it may be more suitable for applications requiring better mechanical properties.

4. Materials and Methods

4.1. Materials

The main components of the oleogel were highly refined olive oil (low acidity), TWEEN 80, SPAN 80, and lidocaine. All materials were sourced from Sigma Aldrich (Saint Louis, MO, USA). TWEEN 80 (Polysorbate 80) is an amphiphilic surfactant that reduces surface tension, aiding in the formation of stable emulsions, while SPAN 80 (sorbitan monooleate) is a lipophilic emulsifier effective in stabilizing water-in-oil emulsions. Lidocaine, a neuroprotective agent with a purity of ≥98% (GC), was also included in powder form. During the synthesis and all investigations a double deionized distilled water was applied. All reagents were used as received without further purification.

4.2. Synthesis of Olive Oil-Based Oleogels Containing Lidocaine

The first step involved dissolving the appropriate amount of lidocaine in olive oil at 60 °C, while stirring for 30 min (8500 rpm) using a Unidrive X1000 CAT (Ingenieurbüro CAT, M. Zipperer GmbH, Ballrechten-Dottingen, Germany) homogenizer to obtain a homogeneous mixture. In the second stage, the appropriate amount of emulsifier was added to the mixture (still at 60 °C) and vigorously stirred using the same homogenizer (20,000 rpm). The material was then transferred into sterile glass cells, which were placed at 6 °C for 48 h. Afterward, the studies described in this article were conducted. The tested series was divided into several vials, with part of each series stored for 2 months at 6 °C before being subjected to the tests.

In Table 2, the compositions of all prepared samples are compiled.

Table 2. Compositions of prepared oleogels.

<table>
<tr><th>Sample Name</th><th>Olive Oil, g</th><th>TWEEN 80, g</th><th>SPAN 80, g</th><th>Lidocaine, g</th></tr>
<tr><td>30O_15T_0.6L</td><td rowspan="6">30</td><td>15</td><td>-</td><td rowspan="4">0.6</td></tr>
<tr><td>30O_25T_0.6L</td><td>25</td><td>-</td></tr>
<tr><td>30O_15S_0.6L</td><td>-</td><td>15</td></tr>
<tr><td>30O_25S_0.6L</td><td>-</td><td>25</td></tr>
<tr><td>30O_15T_1.2L</td><td>15</td><td>-</td><td rowspan="2">1.2</td></tr>
<tr><td>30O_25T_1.2L</td><td>25</td><td>-</td></tr>
</table>

After synthesis, the obtained samples were subjected to numerous investigations aimed at characterizing their stability, chemical structure, sorption, and rheological properties as well as wettability.

4.3. Verification of Stability of Obtained Oleogels

The stability of oleogels was measured via a stability MultiScan MS20 system. The first step of the research included measuring samples immediately after synthesis. Next, the oleogels were stored at 6 °C for two months and investigated again. Measurements were performed at room temperature.

4.4. Characterization of Oleogels Using FT-IR ATR Spectroscopy

The presence of individual functional groups within the structure of developed materials was verified via Fourier-transform infrared (FT-IR) spectroscopy. The study was conducted using a Thermo Scientific Nicolett iS5 (Thermo Fisher Scientific Inc., Waltham, MA, USA) spectrometer equipped with an attenuated total reflection (ATR) attachment containing a diamond crystal. The measurement range was 4000–500 cm^{-1} (32 scans at 0.4 cm^{-1} resolution). Analysis was conducted at room temperature.

4.5. Sorption Ability of Developed Systems

In order to characterize the sorption properties of the obtained oleogels, experiments were conducted involving determining the relative humidity (RH) changes within the tightly closed Duran glass bottle containing two Petri dishes: one with a tested sample (weighing 1.0 g) and a second one with a double-deionized distilled water. The experiments were conducted for 30 min while the relative humidity was read every 5 min. In addition, the temperature changes in the test vessel were also monitored.

The determined relative humidity values allowed the researchers to calculate absolute humidity (AH) using the following equation (Equation (1)):

$$AH = \left[\left(\frac{1320.65}{T + 273.15} \right) \cdot 10 \left(\frac{7.4475 \cdot T}{T + 233.71} \right) \right] \cdot RH \quad (1)$$

where:

- AH is absolute humidity, g/kg;
- RH is relative humidity, %.

Measurements were carried out using a thermo/hydrometer Elmetron PWT-401 (Elmetron GP, Zabrze, Poland) which enables simultaneous measurements of temperature and relative humidity (range 0–100%; resolution 0.1% RH).

4.6. Analysis of Wetting Properties of Oleogels

The wetting properties of the obtained oleogel samples—directly after synthesis and after two months of the storage—were determined using the Drop Shape Analyzer Kruss DSA100M optical contact angle measuring instrument (A. KRÜSS Optronic GmbH, Hamburg, Germany) equipped with a digital camera (200 fps; the device applies a digital image processing algorithm to determine the contact angles by means of the Laplace–Young approximations or tangents) and an optical microscope. Every measurement was performed using sterile syringe needles (NE 44, Kruss Gmbh, Hamburg, Germany). The oleogel samples were dispensed onto the test substrate (three types of substrates were analyzed, i.e., block mica, Teflon plate, and glassy carbon plate) using a syringe positioned inside the environmental chamber. The needle's position was adjusted accordingly. As each sample spread across the surface, forming a layer, it was necessary to wait 10 s for the sample to stabilize. Once the oleogel had settled, the contact angles were recorded on video and measured after setting the baseline. After the liquid on each substrate stabilized, the device displayed the calculated contact angle values. Such analysis was performed in triplicates for each sample. The procedure applied was analogous to the procedure described in detail in our previous paper [39].

4.7. Studies on Rheological Properties

A significant aspect of the research was to verify the rheological properties of the obtained oleogels as well as the potential change in these properties after two months of storage. For this purpose, the Malvern Bohlin Gemini II (Malvern Panalytical Ltd., Cambridge, UK) rheometer, additionally equipped with sample temperature monitoring via a Peltier system and the software Bohlin R6.51.0.3 (Malvern Panalytical Ltd., Cambridge, UK), was employed. During the experiments, 1.0 g of each sample was used, while the gap between the plate probe (angle = 4°; diameter: 40 mm) and the cone was set to 1 mm. The complex shear modulus (denoted $G^* = G' + iG''$) was determined as a function of increasing applied strain amplitude at a fixed frequency (equal to 1 Hz) and received based on the amplitudes and phases of the first harmonic strain signals. A frequency sweep test (within the range 0–10 Hz) was utilized to obtain the elastic modulus (G') and loss modulus (G'') of the test samples in the linear viscoelastic region (LVR). All measurements were applied at 25 ± 1 °C.

5. Conclusions

Based on the conducted studies on emulsions containing lidocaine, the following conclusions can be drawn regarding their potential application in medicine. Emulsions stabilized with SPAN showed better stability and lower tendencies for phase separation compared to those stabilized with TWEEN. The proper choice of emulsifier is crucial for obtaining stable oleogels, which is important in terms of their storage and clinical application. Lidocaine does not significantly affect the stability of the emulsions, suggesting that its presence can be well tolerated in formulations without disrupting the physicochemical properties of the final product. Samples containing higher amounts of SPAN exhibited moisture absorption capacity, which may be relevant for their use in therapies where moisture control and moisturizing properties are essential. Oleogels with lidocaine stabilized with SPAN demonstrated favorable mechanical properties, which could translate to better application and adherence to the skin in therapeutic uses. Due to their stability, rheological properties, and ability to retain moisture, these emulsions have significant potential in medicine, particularly in pain relief and dermatological formulations where long-lasting effects and application comfort are important.

In summary, oleogels with lidocaine, especially those stabilized with SPAN, may be promising candidates for medical applications.

Supplementary Materials: The following supporting information can be downloaded at: https://www.mdpi.com/article/10.3390/ijms252011067/s1.

Author Contributions: Conceptualization, S.K.-K.; methodology, S.K.-K. and A.D.; software, S.K.-K., A.P. and W.K.; validation, S.K.-K., A.D. and A.P.; formal analysis, S.K.-K.; investigation, S.K.-K. and A.P.; resources, M.K.; data curation, S.K.-K., A.P., W.K. and R.P.S.; writing—original draft preparation, S.K.-K.; writing—review and editing, S.K.-K.; visualization, A.P., W.K. and M.K.; supervision, S.K.-K.; project administration, R.P.S. and M.K.; funding acquisition, S.K.-K. and A.D. All authors have read and agreed to the published version of the manuscript.

Funding: This paper received funding from the National Science Centre of Poland (grant number 2024/08/X/ST8/00054).

Institutional Review Board Statement: Not applicable.

Informed Consent Statement: Not applicable.

Data Availability Statement: The original contributions presented in the study are included in the article and Supplementary Materials, further inquiries can be directed to the corresponding authors.

Conflicts of Interest: The authors declare no conflicts of interest.

References

1. Stender, S.; Astrup, A.; Dyerberg, J. A Trans European Union Difference in the Decline in Trans Fatty Acids in Popular Foods: A Market Basket Investigation. *BMJ Open* **2012**, *2*, e000859. [CrossRef] [PubMed]
2. Patel, A.R.; Dewettinck, K. Edible Oil Structuring: An Overview and Recent Updates. *Food Funct.* **2016**, *7*, 20–29. [CrossRef] [PubMed]
3. Chowdhury, R.; Warnakula, S.; Kunutsor, S.; Crowe, F.; Ward, H.A.; Johnson, L.; Franco, O.H.; Butterworth, A.S.; Forouhi, N.G.; Thompson, S.G.; et al. Association of Dietary, Circulating, and Supplement Fatty Acids with Coronary Risk: A Systematic Review and Meta-Analysis. *Ann. Intern. Med.* **2014**, *160*, 398–406. [CrossRef] [PubMed]
4. Pinto, T.C.; Martins, A.J.; Pastrana, L.; Pereira, M.C.; Cerqueira, M.A. Oleogel-Based Systems for the Delivery of Bioactive Compounds in Foods. *Gels* **2021**, *7*, 86. [CrossRef] [PubMed]
5. Pawar, V.U.; Dessai, A.D.; Nayak, U.Y. Oleogels: Versatile Novel Semi-Solid System for Pharmaceuticals. *AAPS PharmSciTech* **2024**, *25*, 146. [CrossRef]
6. Chowdhury, B.; Sharma, A.; Akshit, F.N.U.; Mohan, M.S.; Salunke, P.; Anand, S. A Review of Oleogels Applications in Dairy Foods. *Crit. Rev. Food Sci. Nutr.* **2023**, *1*, 1–19. [CrossRef]
7. Puscas, A.; Muresan, V.; Socaciu, C.; Muste, S. Oleogels in Food: A Review of Current and Potential Applications. *Foods* **2020**, *9*, 70. [CrossRef]
8. Pehlivanoğlu, H.; Demirci, M.; Toker, O.S.; Konar, N.; Karasu, S.; Sagdic, O. Oleogels, a Promising Structured Oil for Decreasing Saturated Fatty Acid Concentrations: Production and Food-Based Applications. *Crit. Rev. Food Sci. Nutr.* **2018**, *58*, 1330–1341. [CrossRef]
9. Millman, J.F.; Okamoto, S.; Teruya, T.; Uema, T.; Ikematsu, S.; Shimabukuro, M.; Masuzaki, H. Extra-Virgin Olive Oil and the Gut-Brain Axis: Influence on Gut Microbiota, Mucosal Immunity, and Cardiometabolic and Cognitive Health. *Nutr. Rev.* **2021**, *79*, 1362–1374. [CrossRef]
10. Abenavoli, L.; Milanović, M.; Milić, N.; Luzza, F.; Giuffrè, A.M. Olive Oil Antioxidants and Non-Alcoholic Fatty Liver Disease. *Expert Rev. Gastroenterol. Hepatol.* **2019**, *13*, 739–749. [CrossRef]
11. Gorzynik-Debicka, M.; Przychodzen, P.; Cappello, F.; Kuban-Jankowska, A.; Gammazza, A.M.; Knap, N.; Wozniak, M.; Gorska-Ponikowska, M. Potential Health Benefits of Olive Oil and Plant Polyphenols. *Int. J. Mol. Sci.* **2018**, *19*, 686. [CrossRef] [PubMed]

12. La, T.; Rohde, J.; Phillips, N.D.; Hampson, D.J. Comparison of Brachyspira Hyodysenteriae Isolates Recovered from Pigs in Apparently Healthy Multiplier Herds with Isolates from Herds with Swine Dysentery. *PLoS ONE* **2016**, *11*, e0160362. [CrossRef] [PubMed]
13. Berglund, G.; Petersson, L.M.; Eriksson, K.R.N.; Häggman, M. "Between Men": Patient Perceptions and Priorities in a Rehabilitation Program for Men with Prostate Cancer. *Patient Educ. Couns.* **2003**, *49*, 285–292. [CrossRef] [PubMed]
14. Chu, Y.; Xu, T.; Gao, C.C.; Liu, X.; Zhang, N.; Feng, X.; Liu, X.; Shen, X.; Tang, X. Evaluations of Physicochemical and Biological Properties of Pullulan-Based Films Incorporated with Cinnamon Essential Oil and Tween 80. *Int. J. Biol. Macromol.* **2019**, *122*, 388–394. [CrossRef] [PubMed]
15. Rashid, R.; Kim, D.W.; Din, F.U.; Mustapha, O.; Yousaf, A.M.; Park, J.H.; Kim, J.O.; Yong, C.S.; Choi, H.G. Effect of Hydroxypropylcellulose and Tween 80 on Physicochemical Properties and Bioavailability of Ezetimibe-Loaded Solid Dispersion. *Carbohydr. Polym.* **2015**, *130*, 26–31. [CrossRef] [PubMed]
16. Wu, C.; Wang, H.; Cao, J. Tween-80 Improves Single/Coaxial Electrospinning of Three-Layered Bioartificial Blood Vessel. *J. Mater. Sci. Mater. Med.* **2023**, *34*, 6. [CrossRef]
17. Kassem, M.G.A.; Ahmed, A.-M.M.; Abdel-Rahman, H.H.; Moustafa, A.H.E. Use of Span 80 and Tween 80 for Blending Gasoline and Alcohol in Spark Ignition Engines. *Energy Rep.* **2019**, *5*, 221–230. [CrossRef]
18. Farooq, A.; Shafaghat, H.; Jae, J.; Jung, S.-C.; Park, Y.-K. Enhanced Stability of Bio-Oil and Diesel Fuel Emulsion Using Span 80 and Tween 60 Emulsifiers. *J. Environ. Manag.* **2019**, *231*, 694–700. [CrossRef]
19. Yamashita, Y.; Sakamoto, K. Hydrophilic–Lipophilic Balance (HLB): Classical Indexation and Novel Indexation of Surfactant. In *Encyclopedia of Biocolloid and Biointerface Science 2V Set*; Wiley: Hoboken, NJ, USA, 2016; pp. 570–574.
20. Smejkal, G.; Gross, V.; Lazarev, A. Theoretical and Experimental Determinations of the Hydrophilic–Lipophilic Balance (HLB) of Representative Oils and Lecithins. *Colloids Interfaces* **2024**, *8*, 21. [CrossRef]
21. Wu, J.; Yan, F.; Jia, Q.; Wang, Q. QSPR for Predicting the Hydrophile-Lipophile Balance (HLB) of Non-Ionic Surfactants. *Colloids Surfaces A Physicochem. Eng. Asp.* **2021**, *611*, 125812. [CrossRef]
22. Zheng, Y.; Zheng, M.; Ma, Z.; Xin, B.; Guo, R.; Xu, X. Sugar Fatty Acid Esters. In *Polar Lipids*; Elsevier: Amsterdam, The Netherlands, 2015; pp. 215–243.
23. Rousseau, D.; Rafanan, R.R.; Yada, R. Microemulsions as Nanoscale Delivery Systems. In *Comprehensive Biotechnology*; Elsevier: Amsterdam, The Netherlands, 2011; pp. 675–682.
24. Available online: https://www.sciencedirect.com/science/article/abs/pii/B9780857094285500059 (accessed on 12 October 2024).
25. Bursi, R.; Piana, C.; Grevel, J.; Huntjens, D.; Boesl, I. Evaluation of the Population Pharmacokinetic Properties of Lidocaine and Its Metabolites After Long-Term Multiple Applications of a Lidocaine Plaster in Post-Herpetic Neuralgia Patients. *Eur. J. Drug Metab. Pharmacokinet.* **2017**, *42*, 801–814. [CrossRef] [PubMed]
26. Dunn, L.K.; Durieux, M.E. Perioperative Use of Intravenous Lidocaine. *Anesthesiology* **2017**, *126*, 729–737. [CrossRef] [PubMed]
27. Hermanns, H.; Hollmann, M.W.; Stevens, M.F.; Lirk, P.; Brandenburger, T.; Piegeler, T.; Werdehausen, R. Molecular Mechanisms of Action of Systemic Lidocaine in Acute and Chronic Pain: A Narrative Review. *Br. J. Anaesth.* **2019**, *123*, 335–349. [CrossRef] [PubMed]
28. Dogrul, A.; Arslan, S.A.; Tirnaksiz, F. Water/Oil Type Microemulsion Systems Containing Lidocaine Hydrochloride: In Vitro and in Vivo Evaluation. *J. Microencapsul.* **2014**, *31*, 448–460. [CrossRef]
29. Zhang, D.; Ye, D.; Jing, P.; Tan, X.; Qiu, L.; Li, T.; Shen, L.; Sun, Y.; Hou, H.; Zhang, Y.; et al. Design, Optimization and Evaluation of Co-Surfactant Free Microemulsion-Based Hydrogel with Low Surfactant for Enhanced Transdermal Delivery of Lidocaine. *Int. J. Pharm.* **2020**, *586*, 119415. [CrossRef]
30. Daryab, M.; Faizi, M.; Mahboubi, A.; Aboofazeli, R. Preparation and Characterization of Lidocaine-Loaded, Microemulsion-Based Topical Gels. *Iran. J. Pharm. Res.* **2022**, *21*, e123787. [CrossRef]
31. Qiao, Y.; Wang, X.; Ren, H.; Cui, Y.; Ren, J.; Hao, C.; Zhao, Z.; Liu, J.; Zhao, R.; Li, Y.; et al. A Study on the Preparation Conditions of Lidocaine Microemulsion Based on Multi-Objective Genetic Algorithm. *Front. Pharmacol.* **2023**, *14*, 1272454. [CrossRef]
32. Zembyla, M.; Murray, B.S.; Sarkar, A. Water-in-Oil Emulsions Stabilized by Surfactants, Biopolymers and/or Particles: A Review. *Trends Food Sci. Technol.* **2020**, *104*, 49–59. [CrossRef]
33. Almeida, M.L.; Charin, R.M.; Nele, M.; Tavares, F.W. Stability Studies of High-Stable Water-in-Oil Model Emulsions. *J. Dispers. Sci. Technol.* **2017**, *38*, 82–88. [CrossRef]
34. Huhtamäki, T.; Tian, X.; Korhonen, J.T.; Ras, R.H.A. Surface-Wetting Characterization Using Contact-Angle Measurements. *Nat. Protoc.* **2018**, *13*, 1521–1538. [CrossRef]
35. Li, P.; Huang, H.; Fang, Y.; Wang, Y.; No, D.S.; Bhatnagar, R.S.; Abbaspourrad, A. Interfacial Engineering of Clear Emulsions: Surfactant Hydrophobicity and the Hidden Role of Chain Structure. *Colloids Surfaces A Physicochem. Eng. Asp.* **2023**, *676*, 132242. [CrossRef]
36. Koishi, A.; Lee, S.S.; Fenter, P.; Fernandez-Martinez, A.; Bourg, I.C. Water Adsorption on Mica Surfaces with Hydrophilicity Tuned by Counterion Types (Na, K, and Cs) and Structural Fluorination. *J. Phys. Chem. C* **2022**, *126*, 16447–16460. [CrossRef] [PubMed]
37. Asrafali, S.P.; Periyasamy, T.; Kim, S.-C. Hydrophilic Nature of Polytetrafluoroethylene through Modification with Perfluorosulfonic Acid-Based Polymers. *Sustainability* **2023**, *15*, 16479. [CrossRef]

38. Uskoković, V. A Historical Review of Glassy Carbon: Synthesis, Structure, Properties and Applications. *Carbon Trends* **2021**, *5*, 100116. [CrossRef]
39. Kudłacik-Kramarczyk, S.; Drabczyk, A.; Przybyłowicz, A.; Krzan, M. Linseed Oil-Based Oleogel Vehicles for Hydrophobic Drug Delivery—Physicochemical and Applicative Properties. *Pharmaceutics* **2024**, *16*, 600. [CrossRef]

International Journal of
Molecular Sciences

MDPI

Article

Solid Lipid Nanoparticles Incorporated with Retinol and Pentapeptide-18—Optimization, Characterization, and Cosmetic Application

Małgorzata Pawłowska [1,2], Marta Marzec [1,*], Waldemar Jankowiak [2] and Izabela Nowak [1,*]

1 Department of Applied Chemistry, Faculty of Chemistry, Adam Mickiewicz University, Poznań, Uniwersytetu Poznańskiego 8, 61-614 Poznan, Poland; malgorzata.pawlowska@amu.edu.pl
2 Dottore Polska Sp. z o.o., Margonińska 22, 60-425 Poznan, Poland; klinika@kec.pl
* Correspondence: marta.marzec@amu.edu.pl (M.M.); nowakiza@amu.edu.pl (I.N.); Tel.: +48-618-291580 (I.N.)

Abstract: Solid lipid nanoparticles (SLNs) incorporated with retinol and oligopeptide can have a full spectrum of effects on the skin as a compatible combination of ingredients with broad anti-aging properties. The research's main objective was to ensure the stability of lipid nanocarriers containing retinol and peptide due to the planned use of this dispersion as a cosmetic raw material. To confirm the effectiveness of method optimization (high shear homogenization, HSH) and proper selection of substrates, SLN dispersions were obtained in three combinations: 1—non-incorporated SLNs; 2—SLNs containing only retinol; 3—SLNs containing retinol and pentapeptide-18; these were then stored at different temperatures (4, 25, 45 °C) for 4 weeks. The desired values of the physicochemical parameters of the optimized dispersion of lipid nanoparticles incorporated with retinol and oligopeptide over the required storage period were confirmed: mean particle size (Z-Ave) = 134.7 ± 0.3 nm; polydispersity index (PDI) = 0.269 ± 0.017 [−]; zeta potential (ZP) = 42.7 ± 1.2 mV (after 4 weeks at 25 °C). The results confirmed the proper selection of the SLN production method and the effectiveness of the optimization performed. The possibility of using the obtained raw material as an ingredient in cosmetic products with anti-aging properties was indicated.

Keywords: solid lipid nanoparticles; retinol; oligopeptide; high shear homogenization; optimization; stability

1. Introduction

Systems for the delivery of active substances to the skin are a subject under constant development and improvement processes. Numerous studies carried out in the pharmaceutical and cosmetic industries have contributed to significant advances in the formulation of pharmaceutical and cosmetic products containing encapsulated bioactive substances. However, the physiology and characteristic structure of the skin, which acts as a natural barrier against the external environment, still limit the effectiveness of therapeutics and skin care, causing difficulties in the penetration and absorption of active substances [1]. In the process of overcoming the epidermal barrier and facilitating the penetration of active substances, different types of mechanisms are involved, such as intercellular diffusion, active transport, and carrier-mediated transport [2–4]. The transport of bioactive substances can take place through three routes: through the structures of the epidermis, through the outlets of sweat glands, and through hair follicles and sebaceous glands [5]. The effectiveness of the last three ways is less significant due to the small total area that these structures occupy on the skin's surface (0.1%) [3]. However, in the case of large polar molecules and ionic compounds, which have difficulty penetrating the epidermal layers, these routes play an important role [3]. The route through the appendages of the skin can be used to transport active substances with the participation of appropriately selected lipid carriers, such as solid lipid nanoparticles (SLNs).

In the case of retinol, transport using skin appendages is the primary route of penetration into the deeper layers of the skin [6]. Vitamin A is the first vitamin approved by the U.S. Food and Drug Administration (FDA) as an anti-aging agent that affects the appearance and condition of the skin and exhibits anti-aging activity [7]. Retinoids are an attractive group of compounds with a broad spectrum of effects on the skin. They regulate numerous physiological processes, such as epidermal cell growth, the differentiation of the keratinocytes, and sebaceous gland function. In addition, they exhibit brightening properties, contributing to the proper distribution of melanocytes within the epidermis, and most importantly, they affect the synthesis of collagen and elastin, so they have anti-aging and anti-wrinkle effects on the skin [8,9]. Retinol, as a fat-soluble substance, can penetrate through the epidermis and, to a small extent, into the dermis. For effective operation, it is important to increase its penetration rate [7]. The incorporation of retinoids into the SLN significantly influences the controlled and prolonged release of the substance in the skin and achieves a high penetration rate [8]. Ferreira et al. [6] also point to this process minimizing the side effects of retinoids (in an encapsulated form) including erythema reaction due to irritation or the degree of epidermal desquamation.

On the other hand, peptides—compounds of hydrophilic nature—do not exhibit the possibility of passive absorption through the skin [2]; hence, the availability of delivery methods for peptide compounds is quite limited and constantly a topic of scientific interest. The history of the use of peptides in skin care dates back to the 1980s [10], but there was a particular surge of interest in these compounds at the beginning of the 21st century when the activity of peptides and their involvement in numerous skin processes were recognized [11]. The safety of the substances has been confirmed by the Scientific Committee on Consumer Safety (SCCS) [12] and the FDA in its Cosmetic Ingredient Review (CIR). Peptides, as bioactive compounds, have a multidirectional effect on skin conditions. They exhibit anti-wrinkle and anti-aging effects by inhibiting the enzymatic activities of hyaluronidase, collagenase, and elastase. In addition, they regulate the activity of tyrosinase responsible for the formation of hyperpigmentation [13,14]. Currently, special attention is paid to peptide compounds of higher molecular mass [2]. The use of solid lipid nanoparticles as transdermal carriers for peptide compounds provides them with protection from enzymatic degradation, increases the bioavailability of this active substance, and reduces the likelihood of side effects [15].

Pentapeptide-18 (Tyr-D-Ala-Gly-Phe-Leu) belongs to the group of neurotransmitter peptides. This neuropeptide is associated with the SNAP-25 receptor protein (SyNaptosome-Associated Protein, 25 kDa), which is involved in the release of acetylcholine and the formation of the SNARE (SNAP receptors—Soluble N-ethylmaleimide sensitive factor Attachment protein REceptor) complex. Acetylcholine, responsible for causing muscle contractions and the formation of facial wrinkles, may negatively affect the condition of the skin. Pentapeptide-18 affects receptors at neuromuscular synapses acting like enkephalins. By binding to the receptors of enkephalins, it closes calcium channels and reduces the release of acetylcholine. Clinical studies have unequivocally confirmed the anti-aging and anti-wrinkle effects of the said oligopeptide on skin [16,17].

The use of SLN is a promising way to deliver active ingredients in a cosmetic product due to several advantages that they show over traditional methods [18]: (i) the protection of the incorporated active ingredients (e.g., retinol) from chemical degradation [19]; (ii) the possibility of the controlled and prolonged release of the active ingredient [5,20]; (iii) increasing the stability of the introduced bioactive substances [20]; (iv) the increased absorption of substances across the epidermal barrier and improved depth of penetration through the skin [5]; (v) the occlusive properties of the lipid carriers themselves, which result in reduced transepidermal water loss (TEWL) and have a beneficial effect on skin hydration levels [5,19]. The lack of toxicity and biodegradability of solid lipid nanoparticles are perfectly in line with the trend of sustainable cosmetics and green chemistry. In addition, due to its efficiency, the carrier-based method allows for the use of fewer active ingredients while maintaining effective therapeutic effects.

The purpose of the presented research was to develop a method for obtaining solid lipid nanoparticles containing retinol and an oligopeptide (Tyr-D-Ala-Gly-Phe-Leu). A detailed optimization of the qualitative composition of lipid nanoparticle dispersions was performed, taking into account, among other things, the type of solid lipid, the surfactant, and the presence of liquid lipid in the lipid matrix. The stability of the obtained SLNs was a basic assumption for the optimization of the method since the designed dispersion was ultimately one of the key components of the cosmetic formulation. Moreover, the additive effect of the oligopeptide and retinol (contained in the cosmetic product formulation) created the possibility of providing effective anti-aging therapy in line with the trend of skin minimalism.

2. Results and Discussion

2.1. Optimization of the Qualitative Composition of Lipid Nanoparticle Dispersions

2.1.1. Lipid Screening—Selection of the Solid Lipid

Lipids, as non-polar compounds, are components of lipid nanoparticle dispersions at levels of 0.1–30.0 wt.% [21]. Moreover, lipids such as mono-, di- or triglycerides of medium- or long-chain fatty acids are, in fact, components of most available emulsion-like cosmetic formulations [22]. The selection of a solid lipid began with a choice of four lipid compounds most commonly used in the formulation of cosmetic products, namely mono- (glyceryl monostearate—Imwitor® 900 K) and diglycerides (glyceryl dibehenate—Compritol® 888 ATO, glyceryl distearate—Precirol® ATO 5) of long-chain fatty acids and mixtures thereof (Witepsol® H15). The selected solid lipid had to be compatible with the incorporated lipophilic active compounds—in this case, retinol—since it is the properly formed lipid matrix that is crucial for the stability of solid lipid nanoparticles [23]. What is more, the solid lipids selected for optimization were safe and commonly used lipid compounds, which, at the same time, ensured production continuity for further commercial implementation. In the course of the experiment conducted on the selection of the solid lipid, Witepsol® H15 and Precirol® ATO 5 were excluded from further testing after the first hour of observation following the solidification of the system (Table 1). These two lipids showed incompatibility with retinol after a preliminary organoleptic evaluation—the resulting mixtures were heterogeneous with irregularities on the surface of the system, which was interpreted as the immiscibility of retinol with the above-mentioned two solid lipids. After an assumed time of 72 h, the quality and compatibility of the remaining systems were evaluated. Two solid lipids, Imwitor® 900 K and Compritol® 888 ATO, were qualified for further optimization of the composition of solid lipid nanoparticles.

Table 1. Results of lipid screening—observation of homogeneity of mixtures (solid lipids combined with retinol) over time.

Solid Lipid	Solubility					
	0 min	15 min	30 min	1 h	24 h	72 h
Compritol® 888 ATO	+	+	+	+	+	+
Imwitor® 900 K	+	+	+	+	+	+
Precirol® ATO 5	−	−	−	−	−	−
Witepsol® H15	+	+/−	+/−	+/−	−	−

"+" soluble; "−" insoluble.

The selected lipids Imwitor® 900 K (1.0 wt.%) and Compritol® 888 ATO (1.0 wt.%) were used to produce lipid nanoparticle dispersions. Physicochemical parameters were determined for the SLN samples, differing in the type of solid lipid used. The test was carried out on the day that the dispersion was obtained (D0). Due to the unsatisfactory parameters of the sample containing Compritol® (Z-Ave = 560.2 ± 7.4 nm; PDI = 0.563 ± 0.040; ZP = 45.7 mV ± 0.4), testing after 14 days (D14) was repeated for the formulation containing Imwitor® only. The purpose of the retest was to confirm the validity of the choice of solid lipid. The dispersion in which Imwitor® was used in the formulation was characterized by the following values for the determined parameters: Z-Ave = 207.4 ± 1.1 nm; PDl = 0.446 ± 0.015; ZP = 39.7 ± 0.2 mV (on day 0). The test repeated on day 14 indicated a decrease in the values of individual parameters as follows: by 16% for Z-Ave = 174.4 ± 0.8 nm; by 17% for PDl = 0.369 ± 0.026; and by 3% if ZP = 38.7 ± 0.3 mV was considered.

For the SLN samples whose matrixes were based on Imwitor®, the most favorable results were observed, given the assumption made regarding the expected size of the lipid structures tested (Z-Ave < 300 nm). This lipid was selected as a solid lipid suitable for the formation of the lipid matrixes of SLNs. The stability of the optimized nanoparticles was confirmed by test results after 14 days of storage at 25 °C. According to Almeida [15], the selection of a suitable solid lipid in the lipid matrix can affect the release efficiency of the active substance, especially protein substances, which are in the research area of this paper. The presented comparison of the results obtained for samples based on Compritol® and Imwitor® indicated that more favorable effects of delivering the active compound deep into the skin using lipid nanoparticles would be obtained by just using glyceryl monostearate. Thus, monoglycerides have been identified as more useful in terms of epidermal use. These observations corresponded with the results obtained by Abdelbar's team, who also based their study on LNs containing Imwitor®, Compritol® and Witepsol® —analogous sizes of the resulting carrier structures were noted [24]. Satisfactory results for SLN production using glyceryl monostearate were also generated by Prof. Müller's research group, which confirmed the effectiveness of the synthesis of SLNs occurring within the limits of 60 °C [25]. Moreover, the reason for this can be attributed to the fact that Imwitor® 900 K has a lower melting point [15], which will translate into the compatibility of the obtained dispersion in terms of cosmetic/skin surface application. It should also be said that despite Compritol®'s compatibility with retinol, this solid lipid was rejected due to its too high melting point, which may affect retinol's stability [26]. Nevertheless, it should be taken into account that the nature of the lipid matrix also depends on the amounts and types of lipids and supporting emulsifiers selected [27].

2.1.2. Addition of Phosphatidylcholine as a Stabilizing Agent

The parameters obtained for the dispersion containing no phosphatidylcholine were compared with those of the sample containing this substance (0.25 wt.%) and the solid lipid Imwitor® 900 K. The comparison was carried out due to the analogous composition and the identical production method. The only element differentiating the compared formulations was the presence or absence of the aforementioned raw material (Figure 1). Testing of the physicochemical parameters was carried out on the day that the sample was made (D0) and after 14 days of storage (D14).

For the sample containing no phosphatidylcholine, the following values of physicochemical parameters (D0) were obtained: Z-Ave = 232.9 ± 2.8 nm; PDl = 0.309 ± 0.016 [−]; ZP = 43.3 ± 0.2 mV. After 14 days, there was a noticeable increase in the values of the studied parameters to the following: Z-Ave = 326.9 ± 7.4 nm (+40%); PDl = 0.634 ± 0.087 [−] (+105%); ZP = 35.7 ± 0.1 mV (+6%). It was noted that for the dispersion containing phosphatidylcholine, much more satisfactory parameter values were obtained compared to the results observed for formulations without this raw material. Moreover, a stabilization of the values of the determined parameters of the dispersion over time could be observed for the sample in which phosphatidylcholine was included. Similar observations were reported in the Molet-Rodriguez study—the increased stability of W/O/W emulsions was noted when lecithin was used in their formulation [28]. In contrast, in the case of a sample without phosphatidylcholine, a significant increase in the values of individual parameters and a decrease in the quality of the dispersion over time were noticeable. Despite comparable initial parameters, the measurement made after 14 days verified the necessity of using phosphatidylcholine in an amount of 0.25 wt.% in the final formulation. Phosphatidylcholine, acting as a stabilizing agent, does not exhibit irritant properties, which is important for the use of optimized dispersion in cosmetic formulation [29]. It excels in processes where lower temperatures are required to protect the thermal degradation of the incorporated biologically active compound [30]. Therefore, Vanić [31] believes that the presence of phospholipids (in this case, phosphatidylcholine) in the matrix of a lipid nanoparticle enhances its ability to penetrate, and thus it can help to improve the activity of the incorporated active ingredient.

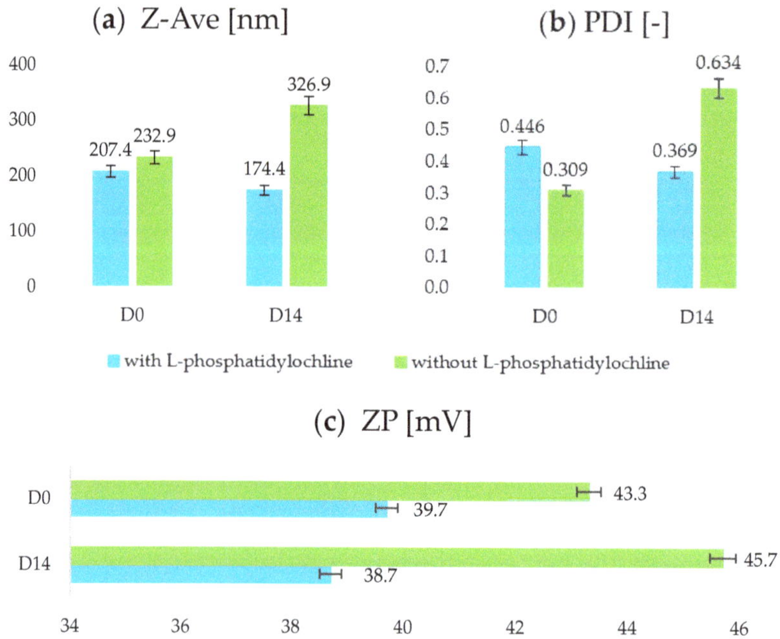

Figure 1. Comparison of physicochemical parameters: Z-Ave (**a**), PDI (**b**), and ZP (**c**) of dispersions of lipid nanoparticles containing and not containing phosphatidylcholine; the pHs of the solutions were within the range 5.33–5.47.

2.1.3. Selection of the Surfactant System

To select the appropriate surfactant system, five samples of lipid nanoparticle dispersions were made, and their physicochemical parameters were determined on the day of production of the dispersion (D0) and after 14 days of storage at room temperature (D14). The obtained results are presented in Table 2.

When analyzing the results obtained in D0, it was noted that in samples 1–4 (containing CTAB), the Z-Ave parameter, reflecting the particle size of the dispersion, obtained satisfactory values < 300 nm. In contrast, the value of the mean particle size determined for sample No. 5 significantly exceeded the desired size range. On this basis, the variant of the dispersion not containing CTAB was not repeated anymore, but the substance was left in the final dispersion formulation at 0.5 wt.%. When considering PDI, it was noted that for samples No. 2 (containing Tween® 80 and CTAB) and No. 3 (with Lutrol® F68, sodium cholate, and CTAB), PDI was significantly higher than in the other samples. Elevated PDI values were noted in both D0 and D14. The most promising, considering the analysis of physicochemical parameters, were sample No. 1, containing Tween® 80, sodium cholate, and CTAB, and sample No. 4, with Lutrol® F68 and CTAB present in the formulation. The results obtained at D0 were analyzed and then compared with those determined at D14. For both samples, the Z-Ave value remained similar (comparing measurements from D0 and D14), and any difference was within the statistical error. A more significant change was observed in the PDI values—sample 1 showed a 25% decrease in the PDI value from the baseline, while for sample 4, the polydispersity index value remained at a similar level. The zeta potential results for all samples, both at D0 and at D14, remained at the desired level of > | ±30 mV |. At values higher than | ±30 mV |, the system was considered physically stable [32]. According to Dabrowska et al. [33], high and positive zeta potential values were influenced by the presence of the surfactant CTAB and its cationic nature. The negatively charged skin surface and positively charged lipid nanoparticles can interact electrostatically with each other and affect the increased permeation of active ingredients through the different layers of the epidermis [34].

Table 2. The optimization of the surfactant system—the changes in the physicochemical parameters (Z-Ave, PDI, ZP) of the tested dispersions of lipid nanoparticles over time (data are expressed as mean ± standard deviation).

	Surfactant System	Results	D0	D14
1.	Tween® 80 sodium cholate CTAB	Z-Ave [nm] PDI [−] ZP [1] [mV]	165.1 ± 0.5 0.379 ± 0.001 42.3 ± 1.3	173.9 ± 0.9 0.286 ± 0.002 42.7 ± 0.4
2.	Tween® 80 – CTAB	Z-Ave [nm] PDI [−] ZP [1] [mV]	200.0 ± 0.7 0.403 ± 0.004 45.6 ± 0.2	185.7 ± 1.5 0.417 ± 0.013 43.9 ± 2.3
3.	Lutrol® F68 sodium cholate CTAB	Z-Ave [nm] PDI [−] ZP [1] [mV]	203.5 ± 2.4 0.456 ± 0.012 43.6 ± 1.1	475.8 ± 19.8 0.489 ± 0.042 39.6 ± 5.2
4.	Lutrol® F68 – CTAB	Z-Ave [nm] PDI [−] ZP [1] [mV]	192.1 ± 0.4 0.396 ± 0.005 39.7 ± 0.4	180.9 ± 0.5 0.374 ± 0.019 38.1 ± 0.7
5.	Lutrol® F68 sodium cholate –	Z-Ave [nm] PDI [−] ZP [1] [mV]	5618.0 ± 33.0 0.380 ± 0.022 –	– – –

[1] The pHs of the solutions are within the range 5.01–5.47.

Based on the analysis of the results additionally correlated with the available literature, the optimal surfactant system was selected: Tween® 80 (1.5 wt.%) and sodium cholate (0.1 wt.%). The necessity of using CTAB (0.5 wt.%) as a component of the lipid phase of the optimized dispersion was also confirmed.

Regarding the optimization discussed, the final composition and methodology for obtaining solid lipid nanoparticles incorporated with retinol and pentapeptide were established. As a result of the numerous trials and verification of physicochemical parameters, Imwitor® 900 K was included as a solid lipid in the composition of the SLN dispersion (based on the lipid screening performed). Subsequently, it was decided that L-phosphatidylcholine should be added due to the need to ensure the sufficient stability of the designed dispersion. As a last step, the surfactant system was verified—the one composed of CTAB, Tween® 80 and sodium cholate was selected. The final step was the addition of active substances to the dispersion—pentapeptide and retinol—in amounts that complied with current regulations for the cosmetics industry. The qualitative–quantitative composition of the SLN dispersion thus established the starting point for its use as a component of selected cosmetic products.

2.2. Assessment of Encapsulation Efficiency and Loading Capacity of Retinol and Oligopeptide

Encapsulation efficiency (EE) and loading capacity (LC) are key parameters for the physicochemical characterization of lipid nanoparticles, usually determined by spectrophotometric or chromatographic methods. EE is defined as the amount of active compound that has been incorporated into lipid nanoparticles, relative to the real amount of active substance used during the production of lipid nanocarriers. LC, on the other hand, expresses the ratio of the amount of the loaded active ingredient to the total weight of the nanoparticle's lipid components, through which it may indirectly characterize the matching of the lipid system used with the active substance [35]. The value of EE and LC is directly proportional to the lipophilicity of the incorporated raw material—for lipophilic substances, EE is usually in the 90–98% range, while hydrophilic compounds tend to reach relatively lower values due to their weak affinity for the lipid matrix [36]. Since LC determines the percentage of the active substance that can be bound to the lipid phase, this parameter for water-soluble substances will also assume lower values compared to LC values determined for lipid-soluble ingredients.

The encapsulation efficiency of retinol (91.9 ± 0.3%), as well as the loading capacity (52.5 ± 0.1%) determined for that compound, were definitely within the expected ranges of values typical for compounds of lipophilic nature. The high affinity of retinol to the lipid structures of the obtained carriers greatly facilitated the incorporation of this compound into their matrix. It is also worth noting that the high LC value of more than 50% directly confirmed the correct selection of the quantitative ratio of retinol relative to the components of the lipid matrix of the nanoparticles. These results can be considered satisfactory, especially when comparing them with the works of other authors on the production of retinol-containing lipid nanocarriers. For example, Boskabadi and co-workers [37] reported much lower EE values—at a maximum of 66%—by encapsulating retinol into SLNs using the ultrasound technique. For the second active ingredient—pentapeptide-18—the following results were indicated: EE (75.7 ± 0.4%) and LC (21.6 ± 0.1%) significantly exceeded the values usually obtained for water-soluble compounds, which could be related to the method used to produce solid lipid nanoparticles based on W/O/W multiple emulsion. It should be mentioned that in this method, the active substances are incorporated into the internal aqueous phase, which highly reduces their undesirable penetration into the outer aqueous phase of the surfactant, which in turn could adversely affect the level of encapsulation efficiency. The EE and LC values (determined for the tested active substances) confirmed the usefulness of the high shear homogenization based on multiple emulsions for the efficient incorporation of cosmetic raw materials of both a hydrophilic and hydrophobic nature in the case of the described study of retinol and pentapeptide-18.

2.3. Evaluation of Stability

The evaluation of the stability of lipid nanoparticle dispersions was necessary because the dispersion would eventually become one of the components of the cosmetic formulation. The choice of elevated temperature conditions (45 °C) was related to the aging tests of the cosmetic product, which were carried out in the following stages of this study [33]. The results obtained for the analyzed dispersions are shown in Table 3.

Table 3. The changes in the values of the mean particle size (Z-Ave), the polydispersity index (PDI), and the zeta potential (ZP) of the lipid nanoparticle dispersions (B—blank; R—retinol; P+R—oligopeptide and retinol) stored under different temperature conditions (4, 25, 45 °C) for 28 days (data are expressed as mean ± standard deviation).

		Temperature	D0	D7	D14	D21	D28
1. (B) blank	Z-Ave [nm]	4 °C	165.1 ± 1.1	173.5 ± 0.9	183.5 ± 0.4	206.3 ± 0.5	172.3 ± 1.1
		25 °C		165.9 ± 0.4	204.5 ± 1.2	177.8 ± 0.5	253.1 ± 9.8
		45 °C		147.8 ± 0.8	600.8 ± 12.3	667.4 ± 9.8	731.6 ± 7.4
	PDI [−]	4 °C	0.379 ± 0.007	0.276 ± 0.008	0.319 ± 0.023	0.419 ± 0.02	0.344 ± 0.007
		25 °C		0.255 ± 0.006	0.386 ± 0.032	0.334 ± 0.008	0.439 ± 0.013
		45 °C		0.498 ± 0.039	0.422 ± 0.027	0.384 ± 0.023	0.427 ± 0.022
	ZP [1] [mV]	4 °C	42.7 ± 0.9	43.0 ± 1.1	40.3 ± 0.4	39.5 ± 0.9	38.6 ± 0.4
		25 °C		41.1 ± 0.9	42.3 ± 0.8	42.7 ± 0.8	33.1 ± 0.3
		45 °C		41.0 ± 0.6	45.0 ± 0.3	43.5 ± 0.5	44.6 ± 1.3

Table 3. *Cont.*

		Temperature	D0	D7	D14	D21	D28
2. (R) retinol	Z-Ave [nm]	4 °C	164.0 ± 0.5	163.9 ± 0.4	268.4 ± 0.5	224.8 ± 5.4	429.4 ± 12.3
		25 °C		164.5 ± 1.1	202.1 ± 1.2	168.6 ± 0.3	157.9 ± 1.6
		45 °C		111.3 ± 1.4	213.4 ± 4.4	128.0 ± 0.23	216.7 ± 6.4
	PDI [−]	4 °C	0.287 ± 0.008	0.311 ± 0.018	0.566 ± 0.022	0.341 ± 0.011	0.485 ± 0.036
		25 °C		0.292 ± 0.008	0.358 ± 0.011	0.364 ± 0.015	0.327 ± 0.020
		45 °C		0.365 ± 0.005	0.515 ± 0.014	0.436 ± 0.012	0.626 ± 0.017
	ZP [1] [mV]	4 °C	43.6 ± 1.4	37.6 ± 0.7	44.5 ± 0.1	38.1 ± 0.8	39.9 ± 0.4
		25 °C		42.1 ± 1.2	42.6 ± 0.3	39.6 ± 1.3	53.5 ± 5.2
		45 °C		41.1 ± 0.3	43.1 ± 0.4	35.6 ± 1.1	45.5 ± 0.4
3. (P+R) oligopeptide retinol	Z-Ave [nm]	4 °C	156.5 ± 0.8	242.3 ± 7.2	228.3 ± 1.4	259.3 ± 8.7	248.3 ± 7.4
		25 °C		148.2 ± 0.8	151.8 ± 0.7	157.4 ± 1.8	134.7 ± 0.3
		45 °C		117.7 ± 0.9	107.2 ± 0.4	158.5 ± 2.9	231.1 ± 1.4
	PDI [−]	4 °C	0.284 ± 0.002	0.438 ± 0.009	0.489 ± 0.019	0.491 ± 0.022	0.445 ± 0.022
		25 °C		0.269 ± 0.006	0.261 ± 0.002	0.320 ± 0.034	0.269 ± 0.017
		45 °C		0.366 ± 0.003	0.272 ± 0.004	0.463 ± 0.007	0.464 ± 0.006
	ZP [1] [mV]	4 °C	45.6 ± 0.2	40.9 ± 0.4	43.3 ± 0.5	44.5 ± 0.6	43.8 ± 0.3
		25 °C		44.7 ± 0.7	45.8 ± 0.3	44.0 ± 1.9	42.7 ± 1.2
		45 °C		38.1 ± 2.1	45.1 ± 0.2	37.2 ± 0.9	44.5 ± 1.6

[1] The pHs of the solutions within the range 4.37–5.47.

It was concluded that the most favorable results of physicochemical parameters (in terms of Z-Ave and PDI) were obtained for sample 3, in which oligopeptide and retinol were present. In addition, it was noted that regardless of the presence of active substances, the samples retained the best parameters at 25 °C. Analysis of the data for individual samples stored for 28 days at different temperatures yielded the following observations:

I The changes in the value of the mean particle size of the dispersion depended on the presence of the active substances. For non-incorporated lipid nanoparticles (B), over the assumed 28 days, the Z-Ave value increased depending on the storage temperature by 4% at 4 °C, by 35% at 25 °C and by 77% at 45 °C ($p < 0.05$) (Figure 2a). The retinol-loaded SLNs (R) recorded Z-Ave values 62% (4 °C) and 24% (45 °C) higher ($p < 0.05$) compared to the results obtained at D0 and D28. On the other hand, at 25 °C, there was a 4% decrease in the value from 164.0 ± 0.5 to 157.9 ± 1.6 nm (Figure 2b). For sample 3 (P+R) enriched with two active substances, the following values of Z-Ave at day 28 were obtained: 248.3 ± 7.4 nm (+37%), 134.7 ± 0.3 nm (−16%) and 231.1 ± 1.4 nm (+32%) for storage temperatures of 4, 25 and 45 °C, respectively ($p < 0.05$) (Figure 2c). Significantly, despite the noticeable changes in Z-Ave values for dispersion 3, regardless of storage conditions, the value of 300 nm indicative of stable dispersion in this case was not exceeded [38].

II In the case of the polydispersity index (PDI), an increase in the value of this parameter was noted over time and under different storage conditions of the samples. The most noticeable changes in PDI values were noted for samples 1 (B) and 2 (R) stored at 45 °C for 28 days, changing by +35% and +54%, respectively ($p < 0.05$) (Figure 2a,b). Sample 3 (P+R) showed 36% and 39% increases in PDI at 4 °C and 45 °C, respectively ($p < 0.05$). The exception was sample 3, stored at 25 °C, for which the polydispersity index value decreased slightly (by 6%) from 0.284 ± 0.002 [−] (D0) to 0.269 ± 0.017 [−] (D28, $p < 0.05$) (Figure 2c). The PDI value for this sample did not exceed the value of 0.3 [-], considered in this study to be a value representing a stable dispersion [38].

III There were no statistically significant changes in the zeta potential values of the samples tested. There was also no correlation between the presence or absence of bioactive substances and variable storage conditions and the ZP results. In all three cases (B, R, P+R), the values were above the desired value of +30 mV, both at D0 and D28. After 28 days of stability testing, the parameters of sample 1 (B) stored at the three assumed temperatures were in the range 38.6 ± 0.4–44.6 ± 1.3 mV. The zeta potential values for sample 2 (R) were 39.9 ± 0.4 mV, 53.5 ± 5.2 mV, and 45.5 ± 0.4 mV (D28) for temperatures of 4, 25, and 45 °C, respectively. In contrast, the ZP results for sample 3 (P+R) oscillated between 42.7 ± 1.2 and 44.5 ± 1.6 mV (D28). It was shown that the incorporation of active substances (oligopeptide and retinol) did not affect the ZP values determined for the samples of lipid nanoparticle dispersions. The cationic surfactant (CTAB), present in the composition of the dispersion, was primarily responsible for the positive zeta potential value [39].

IV A comparison of the tested samples (B, R, P+R) containing different systems of encapsulated active substances indicated that the presence of these compounds influenced smaller variations in the values of parameters related to dispersion stability. This was, for example, evident by comparing the Z-Ave results obtained for samples stored at 25 °C, where sample 1(B) showed a 35% increase in mean particle size, while sample 2 (R) achieved an increase of 4%, and sample 3 (P+R) showed a 16% decrease in Z-Ave.

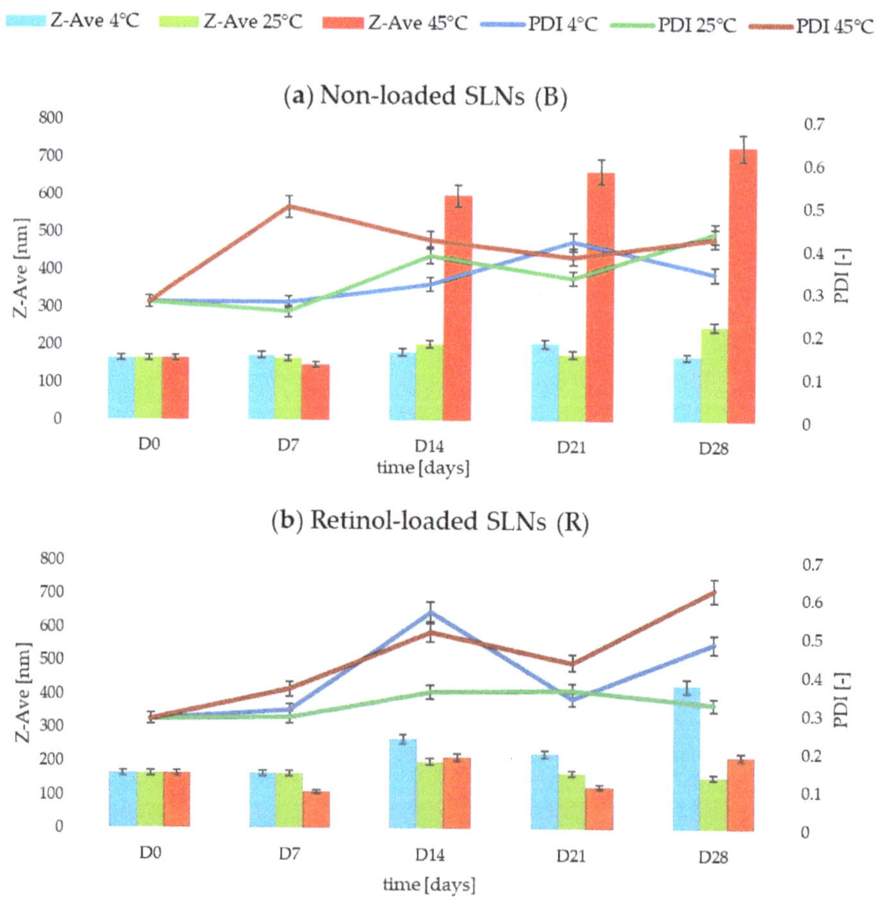

Figure 2. *Cont.*

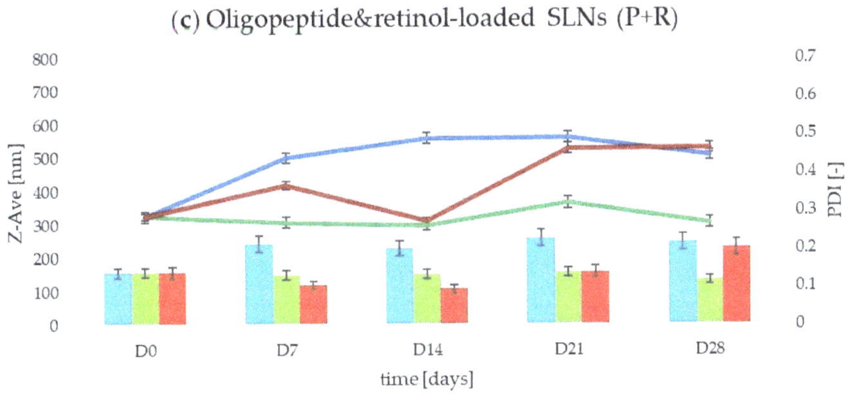

Figure 2. A comparison of the physicochemical parameters—Z-Ave, PDI, and ZP—of the lipid nanoparticle dispersions—B—blank (**a**); R—retinol (**b**); P+R—oligopeptide and retinol (**c**)—stored under different temperature conditions (4, 25, 45 °C) for 28 days.

In the course of stability testing, the following ranges of values of physicochemical parameters were obtained for the lipid nanoparticle dispersion containing oligopeptide and retinol: Z-Ave: 134.7 ± 0.3–248.3 ± 7.4 nm; PDI: 0.269 ± 0.017–0.464 ± 0.006 [−]; ZP: 42.7 ± 1.2–44.5 ± 1.6 mV. The results were obtained after 28 days of storage under three different temperature conditions (4, 25, 45 °C). The storage conditions recommended for the designed dispersion were found to be room temperature, and the slight variations in temperature conditions that occur should not affect the stability of the dispersion. The results were determined to be favorable for the use of SLN as a raw material in cosmetic formulations, where highly stable lipid carriers are necessary [18,40].

2.4. Characterization of Lipid Matrices

The solid state of the lipid nanoparticles and the procedure for the preparation of SLN-type lipid nanoparticles were confirmed by DSC and XRD techniques. For developed SLNs, melting peak values decreased compared to bulk Imwitor® 900 K (65.4 °C), but still, all three dispersions (B; R; P+R) showed onset and melting points higher than 40 °C, which is a prerequisite when it comes to designing lipid nanoparticles for dermal application [41]. Table 4 shows the DSC parameters, including melting point, onset and enthalpy, for all the tested samples.

Table 4. DSC parameters of bulk lipid (Imwitor® 900 K), non-incorporated (B), retinol-loaded (R), and oligopeptide and retinol-loaded SLNs (P+R).

Sample	Melting Peak [°C]	Onset [°C]	Enthalpy [J/g]
Imwitor® 900 K	65.4	54.7	164.5
(B) blank	62.3	51.7	26.8
(R) retinol	61.6	50.7	32.1
(P+R) oligopeptide + retinol	61.7	50.9	33.6

Lipid nanoparticles are classified as crystalline materials; hence, their melting behavior and polymorphic form can be characterized using the DSC method. Imwitor® 900 K, used for the production of lipid nanoparticles, is always represented on the thermogram in the form of a single peak with a maximum at about 65 °C. Hence, in bulk form, it mainly occurs in the stable β form. The process of formation of the lipid nanoparticles is associated with the transition of the bulk lipid to the colloidal state through the β′ form but back to the α form, which is usually reflected as a shift in the peak maximum toward lower values. By analyzing the DSC parameters, this manifested, as mentioned above, as a decrease in the melting point and onset temperature of dispersion of lipid nanoparticles by about 3–4 °C for SLNs, both non-loaded and loaded with bioactive components. The similar melting behavior for the dispersions tested (B; R; P+R) confirmed the effective incorporation of retinol and oligopeptide into the lipid nanoparticles not affecting the polymorphic form and stability of the dispersion system. Since the decrease in the values of the DSC parameters shown in Table 4, relative to bulk Imwitor®, was not significant, the presumed desirable type of SLN-type lipid nanoparticles was confirmed. In contrast, a much more pronounced decrease in the values of these parameters is usually observed with NLCs (nanostructured lipid carriers). For example, in the work of Averina et al. where liquid lipid (*Pinus sibirica* seed oil) was added to the lipid matrix, the decrease in melting point and onset temperature relative to solid lipid were significantly sizable, which may have resulted in the formation of lipid nanostructures with lower stabilities [42].

The DSC data were confirmed by XRD analysis. Figure 3 shows the diffractograms for bulk Imwitor® 900 K and unloaded and loaded SLN dispersions. The crystallization profile of the solid lipid revealed two scattered peaks at the angles 2Θ = 19.5° and 23.2°, which indicated the presence of the metastable polymorphic form β′. The effect of the SLN production process itself on the polymorphic state of the lipid was analyzed for both non-incorporated and incorporated formulations. From the patterns, it was noticed that for all three SLN dispersions, the β form was no longer determined. Instead, the observed characteristic single peak on the diffractograms of the dispersions of the lipid nanoparticles was studied. It was attributed to the α form in which the lipid-forming lipid nanoparticle was found due to the transition to a higher energy level after the production of LNs. Analogous results were obtained by Silva et al. when working on SLNs also based on Imwitor® 900 K as a solid lipid [43]. Despite the use of different methods for lipid nanoparticle production, the DSC parameters and crystallization profiles did not differ significantly from those determined for the nanocarriers developed within the framework of this paper. Thus, it can be concluded that the melting behavior of lipid nanoparticles depends mainly on the solid lipid used in their production and the polymorphic form, which affects the values of the determined DSC parameters and XRD results.

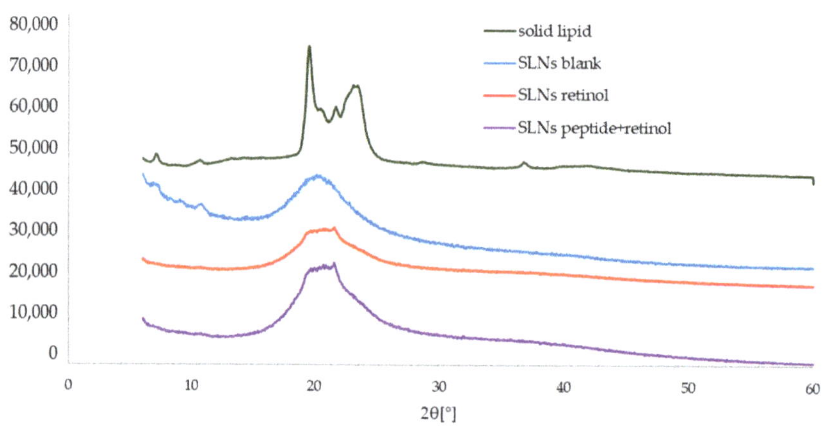

Figure 3. A comparison of the X-ray diffraction patterns of the solid lipid and the studied dispersions of SLNs (the diffractograms are shifted by a constant value of 14,000 arbitrary units in relation to the previous diffractogram).

2.5. Assessment of the Effect of Cosmetic Products on Selected Skin Parameters

Skin's hydration is one of the main parameters determining its proper functioning and healthy appearance. Numerous biochemical processes depend on the condition of the epidermis and on its barrier properties [44], and symptoms such as roughness and scaliness are the result of insufficient hydration [45]. The ability of the epidermis to retain water is determined by the condition of the protective barrier. The degree of hydration is closely related to the water content within the dermis. As a result of the difference in osmotic pressure, water passes through successive layers of the epidermis and evaporates into the environment [46]. This process is called transepidermal water loss (TEWL) and is a natural process that occurs within the structures of the skin [47]. The EYEN cream, containing SLNs incorporated with retinol and oligopeptide, showed stronger moisturizing properties compared to the EYEB cream, which is only a base without lipid nanoparticles. The results are shown in Figure 4. After 8 weeks of using EYEN, the degree of skin hydration in the eye area increased by 5% ($p < 0.05$) compared to baseline (W0), while measurements made for the cosmetic base (EYEB) showed changes in hydration at a statistically insignificant level. These results corresponded with the values of the TEWL parameter—there was a 7% decrease for the EYEN product and a 3% decrease for EYEB ($p < 0.05$). The noticeable difference in the obtained values of the degree of skin hydration may have been related to the presence of solid lipid nanoparticles, which, exhibiting occlusive properties, may have influenced the reduction in water evaporation from the epidermis [5].

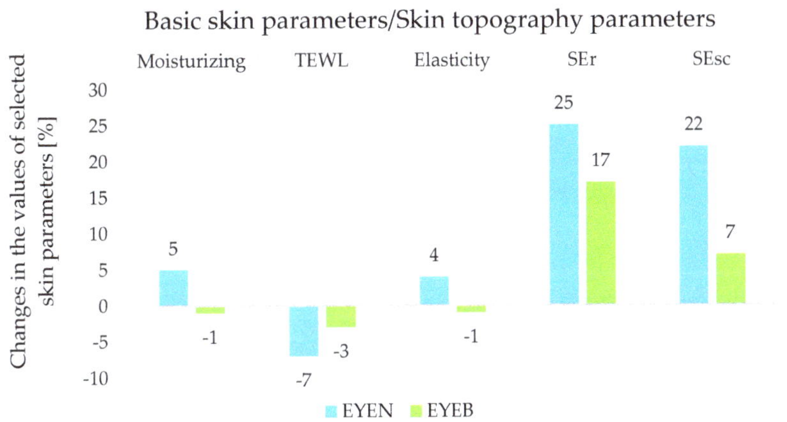

Figure 4. Changes in the values of basic skin parameters and selected skin topography parameters determined during an in vivo study for night eye creams after 8 weeks of application.

The analysis of SEr, a parameter characterizing skin roughness, and SEsc, a parameter of scaliness, is also important for the diagnosis of epidermal conditions and the degree of keratolytic processes (Figure 4). The SEr value recorded after 8 weeks of EYEN application indicated an increase in the degree of roughness by 25%, while for EYEB, the increase was 17% ($p < 0.05$) compared to the results obtained at W0. According to the manufacturer's instructions, the lower the SEr value, the rougher the skin [48]. In addition, the SEsc parameter for the EYEN product increased its value by 22% (T0→T8), while for EYEB, it increased by only 7% ($p < 0.05$). The lower the SEsc value, the less desquamation of the stratum corneum occurred, indicating a higher level of skin hydration [48]. The analyzed results testified to an accelerated process of the desquamation of the stratum corneum and its increased degree (for EYEN), while these phenomena could also affect the roughness of the epidermal surface. Certainly, the desquamation was related to retinol contained in SLNs, which actively affects the skin by accelerating cell renewal and increasing proliferation while, at the same time, enhancing the differentiation of keratinocytes and synthesis of collagens I, III, and VII [49,50]. Importantly, however, it had little effect on hydration parameters and TEWL, the results of which are described above.

Another important parameter was also analyzed, namely skin elasticity, reflecting the anti-aging effectiveness of the cosmetic product. Measurements of the flabbiness and deformability of the skin in the eye area showed that after an 8-week application of the EYEN cream, the value of the parameter increased by 4% ($p < 0.05$), while in the case of the EYEB product, it decreased by 1% with respect to the results from the day of the study, which was considered a statistically insignificant change (Figure 4). The anti-aging properties of the EYEN product containing SLN with active substances (retinol and peptide) were unequivocally confirmed. Satisfactory elasticity values may also have been influenced by proper epidermal hydration. Dry and dehydrated skin tends to be less elastic and more rigid compared to skin characterized by an adequate degree of hydration [46].

Analysis of skin macrorelief images using the profilometric method made it possible to assess the number, depth, and length of wrinkles present on the characterized skin surface in the eye area. The results of this study are summarized in Figure 5. After 8 weeks of using EYEN and EYEB creams, the total wrinkle area was reduced by 20% and 11%, respectively ($p < 0.05$). Moreover, there was also a reduction in the mean length and maximum depth of wrinkles when the EYEN preparation was applied. Mean length was shown to be reduced by 17% and maximum wrinkle depth by 4% ($p < 0.05$). On the other hand, in the case of the cosmetic base (EYEB), there was a 9% decrease in the value corresponding to the length of wrinkles ($p < 0.05$) and a 1% increase in their maximum depth, with the magnitude being within the statistical error; hence, our conclusion was that there was no improvement in this parameter for EYEB.

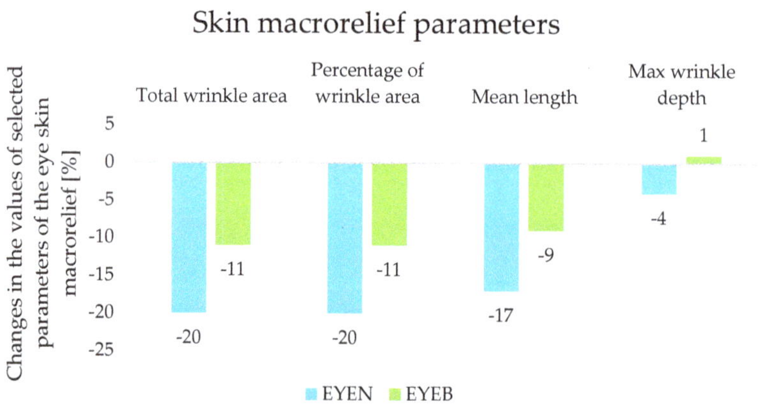

Figure 5. Changes in the values of skin macrorelief parameters determined during an in vivo study for night eye creams after 8 weeks of application.

The results obtained after an 8-week application of the tested cosmetic formulations undoubtedly indicated an improvement in the appearance of the skin and a reduction in the appearance of lines and wrinkles on its surface. Hence, the anti-wrinkle properties of the EYEN cosmetic product containing solid lipid nanoparticles encapsulated with retinol and oligopeptide were confirmed. The effectiveness of retinol-loaded lipid nanoparticles was also confirmed by Ferreira et al. They noted lower irritation potential and a lower degree of skin peeling of SLNs enriched with retinol compared to the use of retinoid in its classic form [6]. Salvioni et al., on the other hand, mentioned the occlusive properties of SLNs, which had an influence on the moisture level of the skin [29].

It should be mentioned here that the skin around the eyes is very thin (about 0.35 mm on the eyelids) [45], which makes its susceptible to loss of elasticity and hydration, and thus the formation of wrinkles relatively common [51]. Based on the in vivo study, it was concluded that the properties of the EYEN cream can fully meet the needs of the delicate skin around the eyes.

3. Materials and Methods

3.1. Materials

Four different solid lipids were selected for lipid screening—Compritol® 888 ATO (glyceryl dibehenate), Precirol® ATO 5 (glyceryl distearate) from Gattefossé (Lyon, France), Imwitor® 900 K (glyceryl monostearate) and Witepsol® H15 (a blend of tri-, di- and monoglycerides of fatty acids) from IOI Oleo GmbH (Hamburg, Germany). The dispersions of the lipid nanoparticles were produced by using glycerol (99.5%; Chempur, Piekary Śląskie, Poland), CTAB (hexadecyltrimethylammonium bromide, ≥98%; Chemat, Gdańsk, Poland), L-phosphatidylcholine (from soybean, Type II-S, 14–29% choline basis) and Lutrol® F68 (>90%) from Sigma-Aldrich (Steinheim, Germany), as well as sodium cholate (99%; Acros Organics, Antwerp, Belgium) and Tween® 80 (polysorbate 80; Pol-Aura, Zawroty, Poland). The incorporated active compounds—Retinol GS50 (retinol, 50%) and Leuphasyl® peptide (pentapeptide-18/Tyr-D-Ala-Gly-Phe-Leu, 99.5%)—were purchased from DSM (Kaiseraugst, Switzerland) and Lipotec (Barcelona, Spain), respectively.

3.2. Optimization of the Composition of Lipid Nanoparticles—Selected Factors

The optimization of the composition of the lipid nanoparticles started with the selection of a retinol-compatible solid lipid, as the composition of the lipid matrix is crucial for the stability of the SLN [23]. First, appropriate amounts of pre-selected solid lipids and retinol were weighed into glass vials. The components were then heated to a temperature above the melting point of the solid lipid. In the final step, the resulting mixtures were observed at 15 min, 30 min, 1 h, 24 h, and 72 h after the system was solidified. The experiment aimed to obtain a compatible and homogeneous mixture from the combination of retinol and solid lipid. Four solid lipids were used in this study: Compritol® 888 ATO (glyceryl dibehenate), Precirol® ATO 5 (glyceryl distearate), Imwitor® 900 K (glyceryl monostearate) and Witepsol® H15 (a blend of tri-, di- and monoglycerides of fatty acids). Solid lipids found to be compatible with the tested active ingredient after 72h were subjected to further testing as components (1.0 wt.%) of lipid nanoparticle dispersions.

The next optimization step was to verify the necessity of including phosphatidylcholine in the dispersion. This compound belongs to the group of glycerophospholipids (the most common polar lipid compounds found in nature), mainly in the outer parts of the cell membrane [52]. Phosphatidylcholine is also credited with the ability to ensure the stability of the lipid nanoparticle dispersions containing it [53]. In the course of this study, lipid nanoparticles differing regarding the presence (0.25 wt.%) or absence of the raw material were compared.

A combination of four different surfactants was also used during optimization. Included in this study were surfactants belonging to the non-ionic group—Tween® 80 (polysorbate 80) and Lutrol® F68 (polyoxyethylene–polyoxypropylene block copolymer), anionic–sodium cholate, and cationic-hexadecyltrimethylammonium bromide (CTAB). Optimization at the surfactant system selection stage consisted solely of varying the composition of the external aqueous phase for producing the target lipid nanoparticles. Various surfactant systems were tested based on a combination of 1.5 wt.% non-ionic surfactant with or without 0.1 wt.% sodium cholate while assuming a constant CTAB content of 0.5 wt.%. One trial was also performed, during which the use of CTAB in the lipid phase was abandoned.

3.3. Lipid Nanoparticles—Method of Production

Solid lipid nanoparticles were produced by applying a high shear homogenization (HSH) method based on multiple emulsion (W/O/W). The first step was to introduce the components of the external aqueous phase (I) into a glass beaker (0.1 wt.% sodium cholate and 1.5 wt.% Tween® 80 (polysorbate 80) were dispersed in distilled water) and heat them to 35 °C while stirring at 250–300 rpm for 15 min. Then, the system was cooled to room temperature. At the same time, the components of the lipid phase (II)—1.0 wt.% Imwitor® 900 K (glyceryl monostearate), 0.5 wt.% CTAB (hexadecyltrimethylammonium bromide) and 0.25 wt.% L-phosphatidylcholine—were introduced into a glass beaker and heated to the melting point of the solid lipid. Next, 2.0 wt.% of retinol was added to the properly heated lipid phase (II) and cooled to 60 °C. The next step was to gradually pour the heated glycerol into a beaker with the lipid phase (II) and subject the resulting mixture to continuous stirring for 10 min. At the same time, an aqueous solution of 0.1 wt.% Leuphasyl® peptide (pentapeptide-18/Tyr-D-Ala-Gly-Phe-Leu), which was the internal aqueous phase (III), was prepared in a glass vial. The solution was heated to 60 °C and then added to the lipid phase (II). The key step was to subject the system to high shear homogenization (Ultra-Turrax® DI 25 Basic, IKA-Werke GmbH, Staufen im Breisgau, Germany) at 8000 rpm for 15 min. The resulting W/O emulsion was combined in the final step with the external aqueous phase (I) under continuous stirring at 500 rpm for 15 min.

3.4. Lipid Nanoparticles—Physicochemical Parameters

The three basic physicochemical parameters of lipid nanoparticle dispersions were determined using a Zetasizer Nano ZS (Malvern Instruments, Malvern, UK). The mean particle size (Z-Ave), polydispersity index (PDI), and zeta potential (ZP) were taken into account. Measurements were performed on aqueous solutions of SLN dispersion with appropriate concentrations and pH values. The measurement procedure was repeated three times for each sample, and the arithmetic mean and standard deviation were calculated from the obtained results.

3.5. Encapsulation Efficiency and Loading Capacity

The encapsulation efficiency (EE) and loading capacity (LC) of retinol and oligopeptide in solid lipid nanoparticles were determined indirectly by the centrifugation technique, assessing the free retinol and pentapeptide-18 (non-incorporated) contents by UV-Vis spectroscopy (Varian Cary 50 Bio UV-Vis Spectrophotometer, Varian Inc./Agilent Technologies, Santa Clara, CA, USA). The spectroscopic method was previously validated and involved the quantification of non-encapsulated active substances present in the external aqueous phase of the surfactant of LNs dispersion. The samples were prepared by measuring 1 mL of the dispersion into an Eppendorf Tubes® test tube, placing it in the angular rotor of an MPW-350R laboratory centrifuge (MPW MED. INSTRUMENTS, Warsaw, Poland), and centrifuging at 4000 rpm for 30 min. In the next step, 20 μL of the supernatant (the aqueous solution of the external phase separated by centrifugation) was taken into the flask, which was subsequently increased up to 10 mL with a 50:50 mixture of phosphate buffer (pH = 5.8) and ethanol. The sample was then shaken vigorously for 5 min. The final step was to convert the concentration results into encapsulation efficiency and loading capacity (expressed as a percentage), in accordance with the following equations [54]:

$$EE[\%] = \frac{total\ amount\ of\ active\ substance - amount\ of\ non\text{-}incorporated\ active\ substance}{total\ amount\ of\ active\ substance} \quad (1)$$

$$LC\ [\%] = \frac{total\ amount\ of\ active\ substance - amount\ of\ non\text{-}incorporated\ active\ substance}{total\ amount\ of\ lipid} \quad (2)$$

3.6. Stability Test

The confirmation of the stability of the lipid nanoparticle dispersions was carried out by verifying changes in the values of three parameters (Z-Ave, PDI, and ZP; Zetasizer Nano ZS instrument (Malvern Instruments, Malvern, UK) over time and under varying temperature conditions. The stability test was performed for the target optimized dispersions of SLNs: sample 1 (B, blank)—non-incorporated lipid nanoparticles; sample 2 (R)—SLNs containing 2.0 wt.% of retinol; sample 3 (P+R)—SLNs loaded with 0.1 wt.% of pentapeptide-18 and 2.0 wt.% of retinol. The experiment was conducted on the day that the lipid nanoparticles were received (D0) and after 7, 14, 21, and 28 days of storage at 4, 25, and 45 °C. The statistical analysis was carried out using Statistica 10.0 software. The level of statistical significance was set at $p < 0.05$.

3.7. X-ray Diffraction (XRD) and Differential Scanning Calorimetry (DSC) Analysis

The characterization of the lipid matrices of SLNs was performed by using X-ray diffraction (XRD; D8 Advance powder diffractometer combined with a Johansson monochromator/Bruker, Billerica, MA, USA) and differential scanning calorimetry (DSC; DSC 8500 differential scanning calorimeter/PerkinElmer, Waltham, MA, USA). Both samples of solid lipid and dispersions of lipid nanoparticles were analyzed. The procedure of sample preparation and the detailed conditions of DSC and XRD analysis for lipid nanoparticle dispersions were described in earlier papers [55]. In brief, the principle of DSC measurement was to gradually heat a sample of solid lipid or SLN dispersions from 25 °C to 90 °C in a flow of nitrogen (20 mL/min) at a scanning rate of 5 °C per minute, maintaining the sample at 90 °C for 1 min, and then cooling it to 25 °C with similar parameters. On the other hand, during XRD analysis, samples were dried at room temperature before measurement. For each sample, the procedure was carried out within the high-angle range ($2\Theta = 6.0$–$60.0°$).

3.8. Cosmetic Products/Eye Creams—Preparation

This study was performed on two cosmetic preparations—two night eye creams produced by Dottore Polska Sp. z o.o., Poznan, Poland. The formulation of the cosmetic products was in accordance with Regulation (EC) No 1223/2009/EC of the European Parliament and the Council of 30 November 2009 on cosmetic products with subsequent updates. One of the cosmetic formulations contained a tested SLN dispersion, which was added in the last technological stage with continuous mixing to ensure the uniform consistency of the cosmetic mass. The exact quantitative composition and manufacturing technology of the eye creams remain the sole knowledge of Dottore Polska Sp. z o.o. Two cosmetic products of analogous composition (O/W emulsions with 24.5 wt.% of lipid phase), differing only in the presence of lipid nanoparticle dispersion, were used in a comparative study, which is discussed within the scope of this paper. The INCI compositions of night eye creams investigated are shown below.

3.8.1. EYEN (Anti-Aging Night Eye Cream Containing 1.0 wt.% of Lipid Nanoparticles Loaded with Retinol and Pentapeptide-18)

Ingredients (INCI): Aqua, Butylene Glycol, Coco-Caprylate/Caprate, Pentylene Glycol, Propanediol, Cetyl Alcohol, Glyceryl Stearate, Simmondsia Chinensis Seed Oil, Triisostearin, Glyceryl Stearate Citrate, Butyrospermum Parkii Butter, Theobroma Cacao Seed Butter, Glycerin, Cetyl Ricinoleate, Persea Gratissima Oil, retinol, Sodium Hyaluronate, Acetyl Tetrapeptide-5, Palmitoyl Tripeptide-5, Palmitoyl Dipeptide-5 Diaminobutyroyl Hydroxythreonine, Tetradecyl Aminobutyroylvalylaminobutyric Urea Trifluoroacetate, Pentapeptide-18, Phosphatidylcholine, Cetrimonium Bromide, sodium cholate, Potassium Cocoyl Hydrolyzed Rice Protein, Sodium Cocoyl Amino Acids, Caprylic/Capric Triglyceride, Tocopheryl Acetate, Tocopherol, Helianthus Annuus Seed Oil, Raspberry Ketone, Stearyl Alcohol, Polysorbate 80, Xanthan Gum, Magnesium Chloride, Tetrasodium Glutamate Diacetate, Sodium Hydroxide, Trisodium NTA, and Parfum.

3.8.2. EYEB (Anti-Aging Night Eye Cream without Lipid Nanoparticles)

Ingredients (INCI): Aqua, Butylene Glycol, Coco-Caprylate/Caprate, Pentylene Glycol, Propanediol, Cetyl Alcohol, Glyceryl Stearate, Simmondsia Chinensis Seed Oil, Triisostearin, Glyceryl Stearate Citrate, Butyrospermum Parkii Butter, Theobroma Cacao Seed Butter, Glycerin, Cetyl Ricinoleate, Persea Gratissima Oil, Sodium Hyaluronate, Acetyl Tetrapeptide-5, Palmitoyl Tripeptide-5, Palmitoyl Dipeptide-5 Diaminobutyroyl Hydroxythreonine, Tetradecyl Aminobutyroylvalylaminobutyric Urea Trifluoroacetate, Potassium Cocoyl Hydrolyzed Rice Protein, Sodium Cocoyl Amino Acids, Caprylic/Capric Triglyceride, Tocopheryl Acetate, Tocopherol, Helianthus Annuus Seed Oil, Raspberry Ketone, Stearyl Alcohol, Xanthan Gum, Magnesium Chloride, Tetrasodium Glutamate Diacetate, Sodium Hydroxide, Trisodium NTA, and Parfum.

3.9. Evaluation of the Effectiveness of Cosmetic Products—In Vivo Study

Application tests were carried out on a group of 20 female volunteers aged 30–65 years old with signs of aging in the eye area, such as wrinkles and skin sagging. Before this study, the volunteers were informed about the implications of this study (in the form of information material) and signed an informed consent form to participate in the tests. Each participant received two cosmetic products: EYEN (an eye cream containing lipid nanoparticles) and EYEB (a formulation devoid of active substances encapsulated in lipid carriers). The cosmetic formulations tested were to be applied once a day at night. During this time, the application of cosmetics of similar or analogous compositions and actions was prohibited. Sun exposure, including tanning beds, was also to be avoided (due to the presence of retinol in the cream formulation). Any adverse effects were to be reported to the principal investigator.

The duration of the in vivo study was 8 weeks and approved by the Bioethical Commission of Poznan University of Medical Sciences, Poland, on 12 October 2023 (768/23). Skin parameters were measured before the application of the products on the day that this study started (W0) and after 8 weeks (W8) of use. To assess skin condition, the following equipment from Courage + Khazaka electronic GmbH (Cologne, Germany) was used: Tewameter® TM 300 (transepidermal water loss level, TEWL); Corneometer® CM 825 (epidermal hydration level); Cutometer® MPA 580 (skin elasticity); Visioscan® VC 98 (skin topography parameters); Visioline® VL 650 (skin macrorelief parameters). The Wilcoxon test for pairs of observation results was used for statistical analysis. The level of significance was $p < 0.05$.

4. Conclusions

The use of solid lipid nanoparticles encapsulated with retinol and pentapeptide (Tyr-D-Ala-Gly-Phe-Leu) in the cosmetic industry is perfectly in line with the trend of skin minimalism. The main advantage of using lipid nanocarriers is the provision of assisted transepidermal transport, which enables the delivery of active substances with prolonged action directly to the deeper layers of the epidermis. The additive nature of the active compounds incorporated into SLNs provides a broader spectrum of action and offers more functions than retinol and peptide used in therapy alone. Through numerous trials to optimize the SLN formulation, it was proven that Imwitor® 900 K was most compatible as a solid lipid. The presence of L-phosphatidylcholine as a polar lipid compound also proved crucial for dispersion. In contrast, the surfactant group represented by the combination of Tween® 80, sodium cholate, and CTAB proved to be the most effective. The physicochemical parameters of the designed lipid nanoparticles enriched with retinol and oligopeptide, such as a particle size (Z-Ave) of less than 200 nm, a polydispersity index (PDI) of up to 0.3, and a zeta potential (ZP) of >30 mV, met the requirements of their functionality for cosmetic application. The efficiency of transepidermal transport (using SLNs) was confirmed by in vivo studies, indicating the potential of these lipid nanoparticles as a cosmetic raw material in terms of reducing wrinkles and supporting the regeneration of the delicate skin around the eyes. It should be mentioned that there are few cosmetic products on the cosmetic market that use lipid nanoparticle technology. Available solid lipid nanoparticles usually contain only one active ingredient, such as retinol itself. Peptides incorporated in the lipid matrix are virtually absent in cosmetics. The presented research proved the effectiveness of SLNs containing simultaneously lipophilic retinol and hydrophilic oligopeptide, which is an innovative solution not previously used in the cosmetics industry.

Author Contributions: Conceptualization, M.P., M.M. and I.N.; methodology, M.P. and M.M.; validation, M.M.; formal analysis, M.P. and M.M.; investigation, M.P. and M.M.; resources, I.N.; writing—original draft preparation, M.P. and M.M.; writing—review and editing, M.P., M.M. and I.N.; supervision, W.J. and I.N.; funding acquisition, I.N. All authors have read and agreed to the published version of the manuscript.

Funding: This research was partially funded by the Ministry of Science and Higher Education via the fourth edition of the Minister's program entitled 'Implementation PhD', grant no. DWD/4/33/2020.

Institutional Review Board Statement: This study was conducted according to the guidelines of the Declaration of Helsinki and approved by the Bioethical Commission of Poznan University of Medical Sciences (768/23, 12 October 2023).

Informed Consent Statement: Informed consent was obtained from all subjects involved in this study.

Data Availability Statement: The raw data supporting the conclusions of this article will be made available by the authors upon request. The general procedure for the emulsions' formation is only available to non-professionals.

Acknowledgments: The authors would like to thank IOI Oleo GmbH (Hamburg, Germany) for providing free samples of the solid lipids used in the course of this study.

Conflicts of Interest: Authors M.P. and W.J. were employed by Dottore Polska Sp. z o.o. The remaining authors declare that the research was conducted in the absence of any commercial or financial relationships that could be construed as potential conflicts of interest.

References

1. Kim, B.; Cho, H.-E.; Moon, S.H.; Ahn, H.-J.; Bae, S.; Cho, H.-D.; An, S. Transdermal delivery systems in cosmetics. *Biomed. Dermatol.* **2020**, *4*, 10. [CrossRef]
2. Anil, L.; Kannan, K. Microemulsion as drug delivery system for peptides and proteins. *J. Pharm. Sci. Res.* **2020**, *10*, 16–25.
3. Czerwonka, W.; Puchalska, D.; Lipińska, M.; Habrat, A. Mechanizmy i metody przenikaniasubstancji czynnych przez barierę lipidową skóry. *Kosmetol. Estet.* **2018**, *7*, 667–670.
4. Jaworska, M.; Sikora, E.; Ogonowski, J. Czynniki Wpływające Na Penetrację Składników Aktywnych Przez Skórę Factors Influencing the Percutaneous Penetration of Active Ingerdients. *Wiadomości Chem.* **2011**, *65*, 3–4.
5. Ghasemiyeh, P.; Mohammadi-Samani, S. Potential of nanoparticles as permeation enhancers and targeted delivery options for skin: Advantages and disadvantages. *Drug Des. Dev. Ther.* **2020**, *14*, 3271–3289. [CrossRef] [PubMed]
6. Ferreira, R.; Napoli, J.; Enver, T.; Bernardino, L.; Ferreira, L. Advances and challenges in retinoid delivery systems in regenerative and therapeutic medicine. *Nat. Commun.* **2020**, *11*, 4265. [CrossRef]
7. Zasada, M.; Budzisz, E. Retinoids: Active molecules influencing skin structure formation in cosmetic and dermatological treatments. *Adv. Dermatol. Allergol.* **2019**, *36*, 392–397. [CrossRef]
8. Charoenputtakhun, P.; Opanasopit, P.; Rojanarata, T.; Ngawhirunpat, T. All-trans retinoic acid-loaded lipid nanoparticles as a transdermal drug delivery carrier. *Pharm. Dev. Technol.* **2014**, *19*, 164–172. [CrossRef]
9. Zasada, M.; Adamczyk, A.; Witamina, A. Budowa i mechanizm działania. Vitamin A. Structure and mechanism of action. Kosmetologia estetyczna. *Kosmetol. Estet.* **2018**, *7*, 517–521.
10. Ferreira, M.S.; Magalhães, M.C.; Sousa-Lobo, J.M.; Almeida, I.F. Trending Anti-Aging Peptides. *Cosmetics* **2020**, *7*, 91. [CrossRef]
11. Resende, D.; Ferreira, M.S.; Sousa-Lobo, J.M.; Sousa, E.; Almeida, I.F. Usage of synthetic peptides in cosmetics for sensitive skin. *Pharmaceuticals* **2021**, *14*, 702. [CrossRef] [PubMed]
12. *SCCS/1576/16*; Scientific Committee on Consumer Safety SCCS. 2016. Available online: https://ec.europa.eu/health/scientific_committees/consumer_safety/docs/sccs_o_199.pdf (accessed on 16 September 2024).
13. Aguilar-Toalá, J.E.; Hernández-Mendoza, A.; González-Córdova, A.F.; Vallejo-Cordoba, B.; Liceaga, A.M. Potential role of natural bioactive peptides for development of cosmeceutical skin products. *Peptides* **2019**, *122*, 13. [CrossRef] [PubMed]
14. Limbert, G.; Masen, M.A.; Pond, D.; Graham, H.K.; Sherratt, M.J.; Jobanputra, R.; McBride, A. Biotribology of the ageing skin—Why we should care. *Biotribology* **2018**, *17*, 75–90. [CrossRef]
15. Almeida, A.J.; Souto, E. Solid lipid nanoparticles as a drug delivery system for peptides and proteins. *Adv. Drug Deliv. Rev.* **2007**, *59*, 478–490. [CrossRef]
16. Errante, F.; Ledwoń, P.; Latajka, R.; Rovero, P.; Papini, A.M. Cosmeceutical Peptides in the Framework of Sustainable Wellness Economy. *Front. Chem.* **2020**, *8*, 572923. [CrossRef] [PubMed]
17. Information material from Lipotec®. Leuphasyl. 2020; pp. 1–15.
18. Almeida, A.J.; Runge, S.; Müller, R.H. Peptide-loaded solid lipid nanoparticles (SLN): Influence of production parameters. *Int. J. Pharm.* **1997**, *149*, 255–265. [CrossRef]
19. Wissing, S.A.; Müller, R.H. Cosmetic applications for solid lipid nanoparticles (SLN). *Int. J. Pharm.* **2003**, *254*, 65–68. [CrossRef]
20. Lasoń, E.; Ogonowski, J. Stałe Nanocząsteczki Lipidowe-charakterystyka, zastosowanie i otrzymywanie. *Chemik* **2011**, *65*, 960–967.
21. Wosicka, J. Stałe nanocząstki lipidowe i nanostrukturalne nośniki lipidowe w nowoczesnych kosmetykach. *Pol. J. Cosmetol.* **2009**, *12*, 23–37.
22. Yadav, N.; Khatak, S.; Vir, U.; Sara, S. Solid lipid nanoparticles—A review. *Int. J. Appl. Pharm.* **2013**, *5*, 8–18.

23. Banerjee, J.; Pillai, S. Solid lipid matrix mediated nanoarchitectonics for improved oral bioavailability of drugs. *Expert Opin. Drug Metab. Toxicol.* **2019**, *15*, 499–515. [CrossRef]
24. Abdelbary, G.; Fahmy, R.H. Diazepam-Loaded solid lipid nanoparticles: Design and characterization. *AAPS PharmSciTech* **2009**, *10*, 211–219. [CrossRef] [PubMed]
25. Müller, R.H.; Runge, S.A.; Ravelli, V.; Thünemann, A.F.; Mehnert, W.; Souto, E.B. Cyclosporine-loaded solid lipid nanoparticles (SLN®): Drug-lipid physicochemical interactions and characterization of drug incorporation. *Eur. J. Pharm. Biopharm.* **2008**, *68*, 535–544. [CrossRef] [PubMed]
26. Jenning, V.; Schafer-Korting, M.; Gohla, S. Vitamin A-loaded solid lipid nanoparticles for topical use: Drug release properties. *J. Control. Release* **2022**, *66*, 115–126. [CrossRef]
27. Fangueiro, J.F.; Andreani, T.; Egea, M.A.; Garcia, M.L.; Souto, S.B.; Souto, E.B. Experimental factorial design applied to mucoadhesive lipid nanoparticles via multiple emulsion process. *Colloids Surf. B Biointerfaces* **2012**, *100*, 84–89. [CrossRef]
28. Molet-Rodriguez, A.; Martin-Belloso, O.; Salvia-Trujillo, L. Formation and Stabilization of W1/O/W2 Emulsions with Gelled Lipid Phases. *Molecules* **2021**, *26*, 312. [CrossRef] [PubMed]
29. Salvioni, L.; Morelli, L.; Ochoa, E.; Labra, M.; Fiandra, L.; Palugan, L.; Prosperi, D.; Colombo, M. The emerging role of nanotechnology in skincare. *Adv. Colloid Interface Sci.* **2021**, *293*, 102437. [CrossRef]
30. Musielak, E.; Feliczak-Guzik, A.; Nowak, I. Synthesis and Potential Applications of Lipid Nanoparticles in Medicine. *Materials* **2022**, *15*, 682. [CrossRef] [PubMed]
31. Vanic, Z. Phospholipid vesicles for enhanced drug delivery in dermatology. *J. Drug Discov. Dev. Deliv.* **2015**, *2*, 8.
32. Saupe, A.; Rades, T. *Nanocarrier Technologies*; Springer: Berlin/Heidelberg, Germany, 2006.
33. Dąbrowska, M.; Souto, E.B.; Nowak, I. Lipid nanoparticles loaded with iridoid glycosides: Development and optimization using experimental factorial design. *Molecules* **2021**, *26*, 3161. [CrossRef]
34. Gonzalez-Rodriguez, M.L.; Rabasco, A.M. Charged liposomes as carriers to enhance the permeation through the skin. *Drug Deliy* **2011**, *8*, 857–871. [CrossRef] [PubMed]
35. Souto, E.B.; Müller, R.H. *Lipid Nanoparticles: Effect on Bioavailability and Pharmacokinetic Changes*; Springer: Berlin/Heidelberg, Germany, 2010. [CrossRef]
36. Guimarães, K.L.; Ré, M.I. *Lipid Nanoparticles as Carriers for Cosmetic Ingredients: The First (SLN) and the Second Generation (NLC)*; Springer: Berlin/Heidelberg, Germany, 2011. [CrossRef]
37. Boskabadi, M.; Saeedi, M.; Akbari, J.; Morteza-Semnani, K.; Hashemi, S.M.H.; Babaei, A. Topical Gel of Vitamin A Solid Lipid Nanoparticles: A Hopeful Promise as a Dermal Delivery System. *Adv. Pharm. Bull.* **2021**, *11*, 663–674. [CrossRef] [PubMed]
38. Dąbrowska, M.A. Optymalizacja właściwości fizykochemicznych oraz aplikacyjnych formulacji kosmetycznych zawierających wybrane glikozydy irydoidowe. Ph.D. Thesis, AMU Poznań, Poznan, Poland, 2019; p. 167.
39. Hallan, S.S.; Sguizzato, M.; Esposito, E.; Cortesi, R. Challenges in the physical characterization of lipid nanoparticles. *Pharmaceutics* **2021**, *13*, 549. [CrossRef] [PubMed]
40. Dąbrowska, M.; Nowak, I. Lipid nanoparticles loaded with selected iridoid glycosides as effective components of hydrogel formulations. *Materials* **2021**, *14*, 4090. [CrossRef]
41. Saupe, A.; Wissing, S.A.; Lenk, A.; Schmidt, C.; Müller, R.H. Solid lipid nanoparticles (SLN) and nanostructured lipid carriers (NLC)—Structural investigations on two different carrier systems. *Bio-Med. Mater. Eng.* **2005**, *15*, 393–402.
42. Averina, E.S.; Seewald, G.; Müller, R.H.; Radnaeva, L.D.; Popov, D.V. Nanostructured lipid carriers (NLC) based on Siberian pine (*Pinus sibirica*) seed oil. *Pharmazie* **2010**, *65*, 25–31.
43. Silva, A.; González-Mira, E.; García, M.; Egea, M.; Fonseca, J.; Silva, R.; Santos, D.; Souto, E.; Ferreira, D. Preparation, characterization and biocompatibility studies on risperidone-loaded solid lipid nanoparticles (SLN): High pressure homogenization versus ultrasound. *Colloids Surf. B Biointerfaces* **2011**, *86*, 158–165. [CrossRef]
44. Kilpatrick-Liverman, L.; Mattai, J.; Tinsley, R.; Wu, J. Mechanisms of Skin Hydration. In *Cosmetic Science and Technology*; CRC Press: Boca Raton, FL, USA, 2014; Chapter 7; pp. 81–92.
45. Adamski, Z.; Kaszuba, A. *Dermatologia dla kosmetologów*; Edra Urban & Partner: Wrocław, Poland, 2019; ISBN 9788376091471.
46. Noszczyk, M. *Kosmetologia pielęgnacyjna i lekarska*; PZWL Wydawnictwo Lekarskie: Warsaw, Poland, 2012; ISBN 9788320061062.
47. Machado, M.; Salgado, T.M.; Hadgraft, J.; Lane, M.E. The relationship between transepidermal water loss and skin permeability. *Int. J. Pharm.* **2010**, *384*, 73–77. [CrossRef]
48. Manual Instruction from Couriage+Khazaka®. Visioscan VC 98. 2017.
49. Bojarowicz, H.; Płowiec, A. Wpływ witaminy A na kondycję skóry. *Probl. Hig. Epidemiol.* **2010**, *91*, 352–356.
50. Czarnota, A. Retinoidy. Mechanizm działania, właściwości oraz zakres stosowania w dermatologii i kosmetologii. *Kosmetol. Estet.* **2018**, *7*, 371–376.
51. Suligowska, I. *Pielęgnowanie oczu i ich oprawy. Poradnik dla ucznia*; Instytut Technologii Eksploatacji-Państwowy Instytut Badawczy: Radom, Poland, 2006.
52. Carreira, A.C.; Santos, T.C.; Lone, M.A.; Zupančič, E.; Lloyd-Evans, E.; de Almeida, R.F.M.; Hornemann, T.; Silva, L.C. Mammalian sphingoid bases: Biophysical, physiological and pathological properties. *Prog. Lipid Res.* **2018**, *75*, 100988. [CrossRef] [PubMed]
53. Zhang, B.; Zhou, X.; Miao, Y.; Wang, X.; Yang, Y.; Zhang, X.; Gan, Y. Effect of phosphatidylcholine on the stability and lipolysis of nanoemulsion drug delivery systems. *Int. J. Pharm.* **2020**, *583*, 119354. [CrossRef] [PubMed]

54. Gao, S.; Tian, B.; Han, J.; Zhang, J.; Shi, Y.; Lv, Q.; Li, K. Enhanced transdermal delivery of lornoxicam by nanostructured lipid carrier gels modified with polyarginine peptide for the treatment of carrageenan-induced rat paw edema. *Int. J. Nanomed.* **2019**, *2*, 6135–6150. [CrossRef] [PubMed]
55. Marzec, M.; Dąbek, P.; Witkowski, A.; Monedeiro, F.; Pomastowski, P.; Buszewski, B.; Nowak, I. Lipid Constituents of Diatoms (Halamphora) as Components for Production of Lipid Nanoparticles. *Pharmaceutics* **2022**, *14*, 1171. [CrossRef] [PubMed]

International Journal of
Molecular Sciences

MDPI

Article

Topical Delivery of Dual Loaded Nano-Transfersomes Mediated Chemo-Photodynamic Therapy against Melanoma via Inducing Cell Cycle Arrest and Apoptosis

Yiping Guo [1,2,†], Wenxiao Zhong [2,†], Cheng Peng [1,2,*] and Li Guo [1,2,*]

1 State Key Laboratory of Southwestern Chinese Medicine Resources, Chengdu University of Traditional Chinese Medicine, Chengdu 611137, China; yguo8@ucmerced.edu
2 School of Pharmacy, Chengdu University of Traditional Chinese Medicine, Chengdu 611137, China; zhongwenxiao@stu.cdutcm.edu.cn
* Correspondence: pengcheng@cdutcm.edu.cn (C.P.); guoli@cdutcm.edu.cn (L.G.)
† These authors contributed equally to this work.

Abstract: Melanoma is a malignant skin cancer associated with high mortality rates and drug resistance, posing a significant threat to human health. The combination of chemotherapy and photodynamic therapy (PDT) represents a promising strategy to enhance antitumor efficacy through synergistic anti-cancer effects. Topical delivery of chemotherapeutic drugs and photosensitizers (PS) offers a non-invasive and safe way to treat melanoma. However, the effectiveness of these treatments is often hindered by challenges such as limited skin permeability and instability of the PS. In this study, transfersomes (TFS) were designed to facilitate transdermal delivery of the chemotherapeutic drug 5-Fluorouracil (5-FU) and the PS Imperatorin (IMP) for combined chemo-photodynamic therapy for melanoma. The cytotoxic and phototoxic effects of TFS-mediated PDT (TFS-UVA) were investigated in A375 cells and nude mice. The study also demonstrated that TFS-UVA generated intracellular ROS, induced G2/ M phase cell cycle arrest, and promoted cell apoptosis. In conclusion, this study indicated that 5-FU/ IMP-TFS serves as an effective transdermal therapeutic strategy for chemo-PDT in treating melanoma.

Keywords: melanoma; chemo-photodynamic therapy; cell cycle; apoptosis

1. Introduction

Melanoma originates from abnormal melanocytes in the basal layer of the epidermis and has experienced a significant increase in recent years, with an estimated 99,700 new cases of melanoma expected in the United State in the year of 2024 [1–3]. Although surgical excision remains the primary treatment for early-stage melanoma, it can result in severe side effects such as large skin defects [4]. Melanoma is known for its resistance to conventional treatments including chemotherapy, radiotherapy, as well as innovative treatments such as immunochemotherapy and bio-chemotherapy [5–7]. Therefore, it is necessary to develop more effective therapeutic strategies to fill the gap for better melanoma treatments.

5-Fluorouracil (5-FU) is a chemotherapeutic agent known for its broad-spectrum anti-cancer activities against colorectal, breast, and skin cancers. The therapeutic mechanism involves cell apoptosis by blocking thymidylate synthase, thereby disrupting DNA synthesis [8]. However, systemic delivery of 5-FU via oral and intravenous routes can lead to gastrointestinal and cardiac toxicities [9]. To solve this issue, transdermal delivery of 5-FU has been introduced to restrict systemic toxicities and enhance local drug concentration. However, with poor skin permeability and low local absorption, the applications of 5-FU were limited by its hydrophilic nature [10]. The root of *Angelica dahurica* (Chinese name: Baizhi), has been used in Traditional Chinese Medicine to treat skin diseases for over 1700 years [11]. Imperatorin (IMP), the primary active component found in *Angelica dahurica*, exhibits strong pharmacological effects such as anti-cancer and anti-inflammatory properties [12]. Researchers have demonstrated its concentration-dependent inhibition manner and synergistic effects with cisplatin against melanoma growth [13,14]. In the murine model, IMP administered intraperitoneally under UVA illumination showed significant inhibition of melanoma cell viability [15]. This suggests that IMP could serve as a novel Photodynamic Therapy (PDT) photosensitizer for treating melanoma. Chemo-PDT represents a promising approach to overcoming the limitations of single-treatment modalities and enhancing total treatment efficiency [16]. PDT is a safe and effective non-invasive treatment used in dermatological conditions. Topical PDT involves the application of PS to the skin and specific wavelengths of light for the generation of reactive oxygen species to induce cell apoptosis through oxidative damage [17]. PDT is widely used for skin diseases such as psoriasis [18], vitiligo [19], squamous cell carcinoma of the skin, and melanoma [20,21]. However, challenges remain as the effectiveness of PDT is often hindered by skin penetration and PS instability [22].

Transfersomes (TFS) are a type of liposome characterized by their highly elastic vesicles [23]. Composed of phospholipids and edge-activating agents (EA), TFS destabilize the lipid bilayer to increase vesicle deformability [24,25]. Compared to liposomes, TFS have a smaller particle size and more elastic liposomes, enabling them to deform and squeeze through the interstitial spaces of stratum corneum cells without compromising their structures [26]. This property enhances its skin permeability and local drug concentration during topical administration [27,28]. Additionally, TFS share a similar phospholipid bilayer structure with liposomes, which improves drug solubility and stability. The vesicle structure enhances the effective encapsulation of both hydrophilic and hydrophobic drugs [29]. TFS has shown its benefits in delivering drugs to localized lesions for the treatment of melanoma [30].

In this study, we focused on developing an efficient chemo-PDT using TFS as a nanodrug carrier for treating melanoma. The optimized TFS formulation was determined using the Box-Behnken-Design (BBD) and characterized by particle size, potential, Fourier transform infrared spectroscopy, and transmission electron microscopy (TEM). Drug release kinetics were evaluated using the dialysis method and skin permeation properties were assessed using the Franz diffusion cell method in vitro, respectively. The therapeutic efficacy of chemo-PDT was evaluated through both in vitro and in vivo experiments. We additionally explored the effects of photocytotoxicity on the cell cycle and cell apoptosis in A375 cells. These results signify that this transdermal drug delivery system can be a probable chemo-PDT candidate for the treatment of melanoma.

2. Results

2.1. The BBD Studies

The BBD optimization strategy was developed with three main factors: drug/lipid ratio (*X*1), the SPC/TW-80 ratio (*X*2), and the hydration temperature (*X*3), where the co-encapsulation efficiency (*Y*) was set as the experimental response value. Three different levels of the factors are defined in Table 1, and the 3D response surface plot is shown in Figure 1. Optimization data can be found in Supplementary Materials: Tables S1 and S2. The results were analyzed by ANOVA using Design-Expert 8 software (R^2 = 0.9795), and *Y* was calculated using the following equation:

$$\begin{aligned} Y = 0.85 - 2.96 \times 10^{-3} \times X1 + 4.412 \times 10^{-3} \times X2 + 0.022 \times X3 - 1.702 \times 10^{-3} \times X1X2 \\ -3.143 \times 10^{-3} \times X1X3 + 5.623 \times 10^{-3} \times X2X3 - 0.021 \times X1^2 - 0.029 \times X2^2 \\ -0.032 \times X3^2 \end{aligned} \quad (1)$$

Table 1. The design of the factor levels.

Factors	Definition	Levels		
		−1	0	1
Drug/Lipid Ratio	X1	1:12.5	1:15.0	1:17.5
SPC/TW-80 Ratio	X2	6:4	7:3	8:2
Hydration Temp.	X3	35 °C	40 °C	45 °C

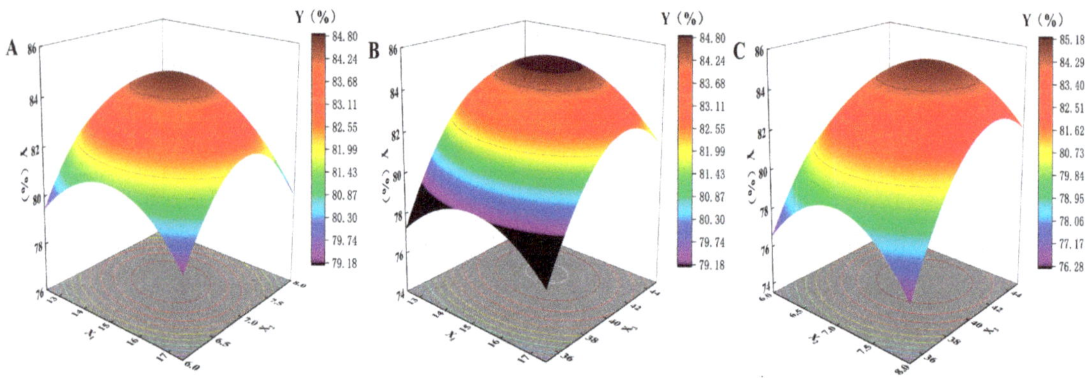

Figure 1. 3D response surface plot of encapsulation efficiency. (**A**) Drug/lipid ratio and SPC/TW-80 ratio; (**B**) drug/lipid ratio and hydration temperature; (**C**) SPC/TW-80 ratio and hydration temperature.

2.2. TFS Preparation, Characterization and Optimization

Studies have shown that lipid vesicles with smaller particle sizes enhance the penetration of encapsulated drugs into the skin [31]. The optimized formulation of 5-FU/IMP-TFS had a smaller particle size of 50 ± 1.92 nm (Figure 2C) and a uniform size distribution with a PDI value of 0.20 ± 0.02 (less than 0.3). Vesicle size reduction of TFS may be due to the addition of surfactants and using a probe sonicator. Zeta potential is a predictor of formulation stability. High zeta potential values reduce aggregation by electrostatic interactions and enhance formulation stability. The optimized zeta potential of 5-FU/IMP-TFS was −13 ± 1.5 mV (Figure 2B), which is consistent with the results reported by Afzal Hussain [32]. The HPLC analysis revealed that the loadings of 5-FU and IMP were 2.49 ± 0.02% and 2.81 ± 0.07%, respectively. The encapsulation efficiency of 5-FU and IMP in the formulations were 79.71 ± 0.65% and 89.89 ± 2.16%, respectively. These results indicate that TFS has a small particle size and good dispersibility, making it capable of forming a stable co-delivery system for hydrophilic (5-FU) and hydrophobic drugs (IMP). The methodology for HPLC analysis is shown in Figures S1 and S2, and Tables S3–S8.

TEM revealed the microstructure of 5-FU/IMP-TFS (Figure 2A). With 80 kV accelerating voltage, 80 k magnification, and a work distance of 11.8 mm, the TEM image presented the membranes of 5-FU/IMP-TFS vesicles, with a clearly visible phospholipid bilayer and a smooth, structurally intact surface. Due to its bilayer membrane structure, TFS can co-deliver lipophilic and hydrophilic drugs [33]. The FTIR spectra (Figure 2D) confirmed the successful preparation of TFS. The characteristic absorption peaks of IMP were observed at 3106 cm^{-1} (=CH), 1718 cm^{-1} (C=O), and 1145 cm^{-1} (C-O), while those of 5-FU were observed at 3045 cm^{-1} (=CH), 1650 cm^{-1} (C=O), and 1251 cm^{-1} (C-F). The physical mixture of 5-FU and IMP exhibited a superposition of characteristic peaks. This suggested successful encapsulation of 5-FU and IMP in the TFS.

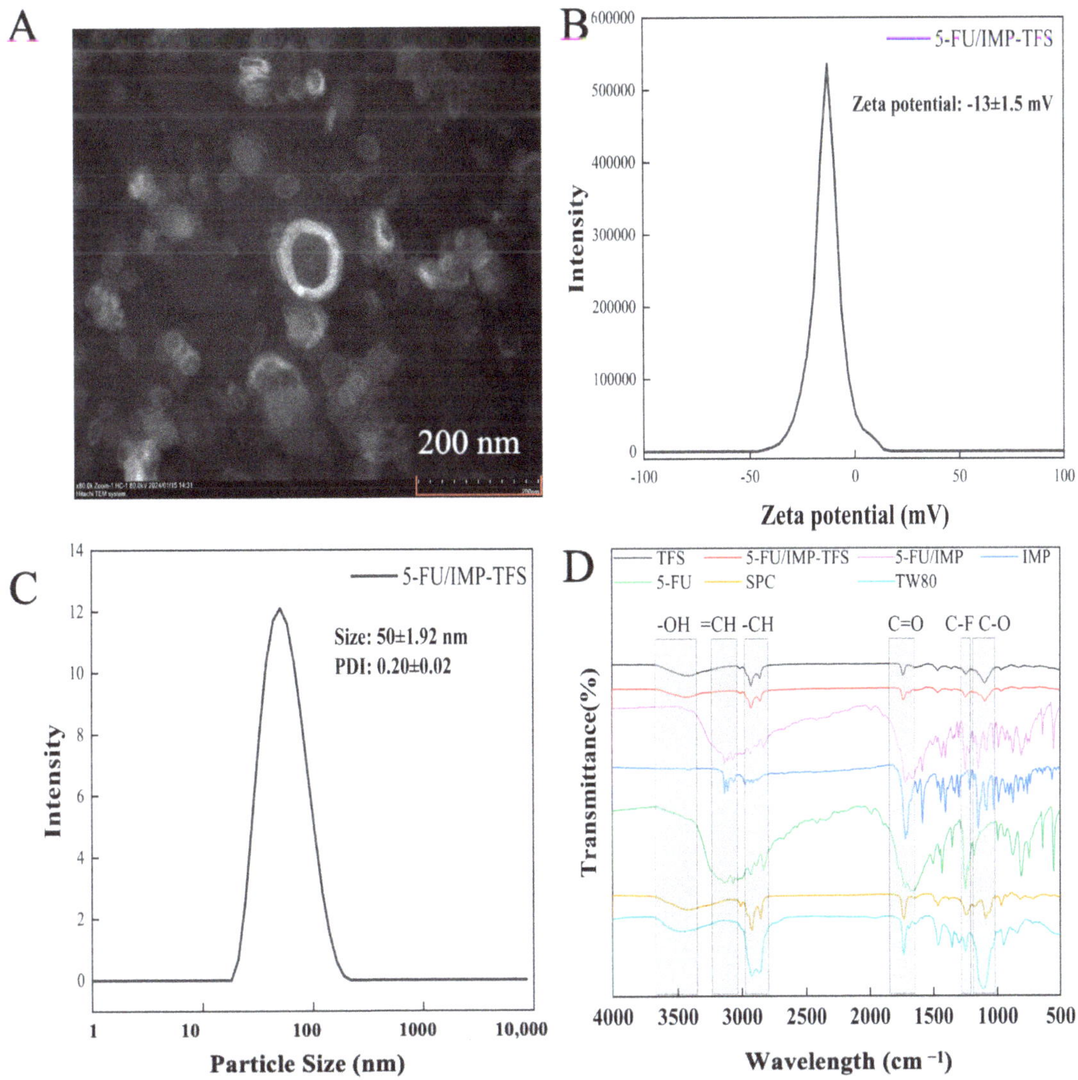

Figure 2. Characterization of TFS. (**A**) TEM image of 5-FU/IMP-TFS; (**B**) the zeta potential of TFS; (**C**) the particle size distribution of TFS; (**D**) the FT-IR spectroscopy of each group.

2.3. In Vitro Drug Release and Ex Vivo Skin Permeation Studies

This study evaluated the in vitro drug release profiles of both drugs in free form or TFS-loaded form (single-loaded or co-loaded) in a water/ethanol release medium (*v*/*v* 70:30) (Figure 3A). The free drug released around 80% of the medication within an hour, while the drug-loaded TFS gradually released 90% of 5-FU and 80% of IMP over a 24 h period. The results showed that drugs using TFS as a carrier have a certain sustained release effect compared to the free drug. There was no significant difference in the release rate between single-drug loading and co-drug loading.

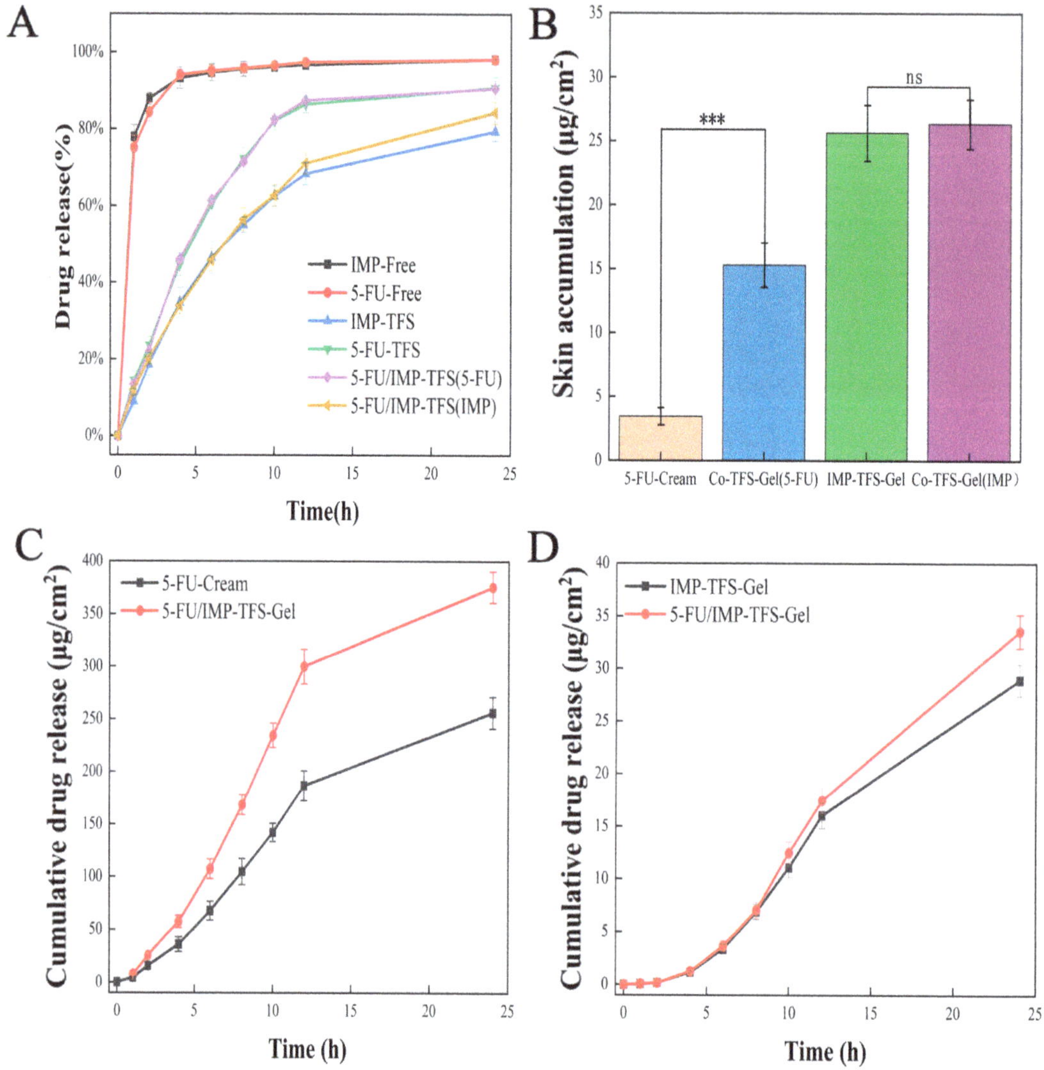

Figure 3. Drug release studies, skin permeation, and deposition studies. (**A**) The drug release profiles. (**B**) Drug deposited into the skin after treatment with different formulations. (**C**) Rat skin permeation profiles of 5-FU loaded formulations. (**D**) Rat skin permeation profiles of IMP loaded formulations. (*** $p < 0.001$, ns indicates statistically no significance).

TFS formulations are fluid and cannot be applied topically. Therefore, TFS was mixed with a Carbopol gel matrix for topical administration [34]. Carbopol gel is a hydrophilic gel matrix that is easily applied and absorbed into the skin and is often used as a vehicle for topical dermal administration [35]. Incorporating the TFS liquid formulation into a semi-solid gel for topical delivery can improve drug stability and patient compliance. Permeation studies were conducted using a Franz diffusion cell (Figure 3) to evaluate the quantitative percutaneous permeation of FU-Cream and TFS-gel formulations across excised rat abdominal skin. The fluxes for 5-FU were 11.71 ± 1.1 μg/cm^2/h and 17.24 ± 2.1 μg/cm^2/h for the FU-Cream and 5-FU/IMP-TFS-Gel, respectively. The cumulative penetration amount per unit area of FU-Cream and 5-FU/IMP-TFS-Gel in 24 h was 256.19 ± 15.00 μg/cm^2 and 375.63 ± 14.84 μg/cm^2, respectively (Figure 3C). For IMP, the fluxes were 1.30 ± 0.08 μg/cm^2/h and 1.49 ± 0.09 μg/cm^2/h for the IMP-TFS-Gel and 5-FU/IMP-TFS-Gel, respectively. The IMP-TFS-Gel and 5-FU/IMP-TFS-Gel had a cumulative penetration amount of 28.91 ± 1.50 μg/cm^2 and 33.58 ± 1.60 μg/cm^2 per unit area, respectively, in 24 h (Figure 3D). Following the permeation studies, the gel formulations underwent drug deposition study in the same experimental conditions. For drug deposition, FU-Cream, Co-TFS-Gel (FU), IMP-TFS-Gel, and Co-TFS-Gel (IMP) demonstrated a total skin accumulation of 3.45 ± 0.67 μg, 15.30 ± 1.73 μg, 25.65 ± 2.19 μg, and 26.36 ± 1.793 μg, respectively (Figure 3B). Specifically, the steady-state rate of 5-FU in TFS was 1.47 times greater than the cream group, and the retention amount was 4.43 times that of the cream. However, there was no significant difference for IMP. The results of drug release and the skin permeation tests indicated that TFS gel was more effective for on-site skin delivery compared to conventional creams. The stratum corneum barrier effect was reduced and skin penetration was enhanced due to the small particle size and hyper-deformability of TFS [36]. Additionally, drug deposition in the skin was improved by the slow release of TFS and the reticular structure of the gel matrix.

2.4. Cytotoxicity and Phototoxicity Studies In Vitro

This study evaluated the effects of the cytotoxicity and phototoxicity of the drugs on HaCaT and A375 cells using the CCK-8 assay. 5-FU is known as a hydrophilic drug that faces challenges crossing the lipophilic cell membranes, resulting in low intracellular concentrations [37]. IMP is a lipophilic compound with low water solubility, poor bioavailability, and rapid degradation in cell culture media [27]. In this study, TFS was performed with anti-cancer and photosensitizing agents in a phospholipid bilayer to decrease restriction and facilitate cellular uptake. In Figure 4, results show that drug-loaded TFS formulations are more cytotoxic than the free drug. It is noteworthy that K-TFS formulations without drug loading did not reduce cell viability under any conditions, which supports the contention that safe nanocarriers can enhance cytotoxicity. Studies have proven that solar ultraviolet radiation (UVR) is a leading factor in many skin diseases [38]. However, a study conducted by Otilia Gag showed that UVA (10 J/ cm^2) did not have a cytotoxic effect on HaCaT and A375 cells, whereas UVB (0.5 J/cm^2) significantly reduced cell viability [39]. In addition, our studies confirmed that there was no significant toxicity in HaCaT and A375 cells under UVA (1.5 J/cm^2) irradiation. Furthermore, the combination of UVA with furanocoumarins and their derivatives has been reported to inhibit melanoma cells [40,41].

Although 5-FU restricts DNA synthesis and inhibits cell growth, its lack of selectivity for cancer cells can harm normal cells and cause side effects such as gastrointestinal reactions and dermatitis [42]. We developed a 48-h dark toxicity study of 5-FU and IMP on normal HaCaT cells (Figure 4A). The cell survival rate of both IMP and IMP-TFS was over 95%. The half maximal inhibitory concentration (IC_{50}) values of 5-FU and 5-FU/IMP on HaCaT cells were 10.92 and 35.36 μM (calculated from the 5-FU concentration), respectively. The study results indicate that IMP did not exhibit cytotoxic effects on HaCaT cells. Furthermore, the co-delivery of 5-FU and IMP resulted in a significant reduction in the cytotoxicity of 5-FU on HaCaT cells [43]. An assessment of the antitumor activity of 5-FU and IMP was conducted on melanoma A375 cells using a 48-h dark toxicity test (Figure 4B). Our study discovered that both 5-FU and IMP reduced the viability of A375 cells in a concentration-dependent manner. The IC_{50} values for 5-FU and 5-FU/IMP were 36.89 and 21.64 μM (calculated from the 5-FU concentration), respectively. Although IMP and IMP-TFS had limited antitumor effects, the combination of 5-FU and IMP significantly enhanced the antitumor effects. In summary, the combination of IMP and 5-FU reduces toxicity to normal cells and enhances the inhibitory effect on tumor cells.

Chemo-photodynamic therapy can overcome the limitations of monotherapy and improve antitumor activity [44,45]. In studies of cellular phototoxicity, the cellular activity of PDT decreased gradually as the concentration of IMP increased (Figure 4C). IMP-TFS+L damaged both normal and tumor cells, but the viability of HaCaT cells at the highest concentration of IMP-TFS+L was still above 85%, indicating biosafety. The IC_{50} for 5-FU-TFS, 5-FU/IMP-TFS, and 5-FU/IMP-TFS+L were 78.8, 60.2, and 35.24 μM (calculated from the 5-FU concentration), respectively (Figure 4D). Among these, 5-FU/IMP-TFS+L exhibited the strongest antitumor effect on A375 cells. The results indicate that PDT has biosafety and significant antitumor activity, and the combination of chemo-photodynamic therapy can further enhance the antitumor efficacy.

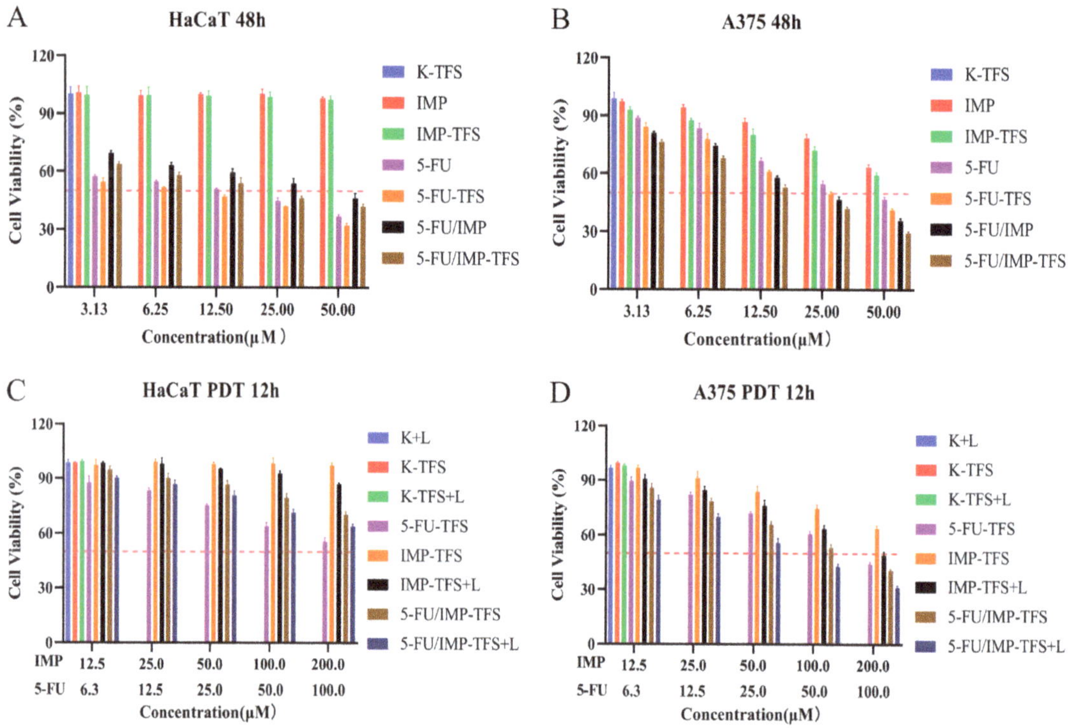

Figure 4. Cytotoxicity and phototoxicity. (**A**) Dark toxicity study of cell viability of HaCaT cell line for 48 h. (**B**) Dark toxicity study of cell viability of A375 cell line for 48 h. (**C**) Phototoxicity study of cell viability for HaCaT cell line for 12 h (1.5 J/cm^2, UVA). (**D**) Phototoxicity study of cell viability of A375 cell line for 12 h (1.5 J/cm^2, UVA). Red dash lines reflect the half maximal inhibitory concentrations of each group.

2.5. *Intracellular Reactive Oxygen Species Generation*

In chemo-photodynamic therapy, the photosensitizers IMP can react with molecular oxygen and generate cytotoxic reactive oxygen species (ROS) which induces oxidative damage to accelerate cell death with the excitation of UVA. To verify and quantify the ROS excited from the photochemistry of 5-FU/IMP-TFS (aka. TFS) and UVA, intracellular ROS generation was measured by staining with DCFH-DA assay and analyzed by flow cytometry. As shown in Figure 5, no obvious change can be observed in ROS generation for the groups in the dark. The quantification of intracellular ROS is presented in Figure 5A. When A375 cells were exposed to UVA, the difference in ROS generation between groups was obvious, as shown in Figure 5B. The TFS-UVA group could produce much more ROS than the TFS-only group. This result indicates that increasing intracellular ROS was induced by the photodynamic activation between IMP and UVA, resulting in the combination therapeutic efficacy in treating A375 cells.

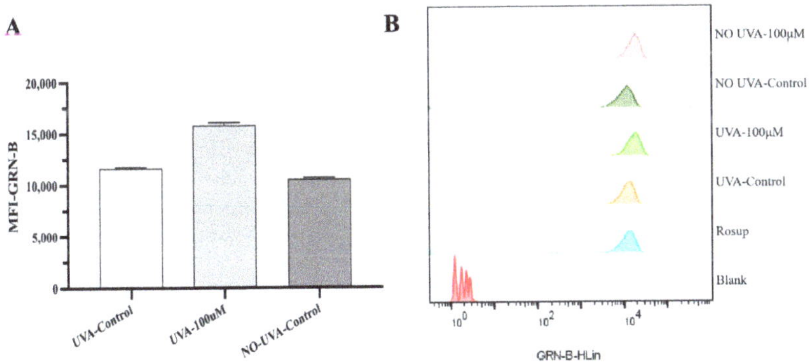

Figure 5. Effect of TFS/ UVA photodynamic treatments on ROS generation in A375 cells. (**A**) The quantification of intracellular ROS generated in different groups. (**B**) Cells were stained with DCFH-DA and the fluorescence intensity was measured using the flow cytometry.

2.6. Cell Apoptosis Analysis

The percentage of apoptotic A375 cells was assessed using Annexin V-fluorescein isothiocyanate (FITC) and propidium iodide (PI) double staining for flow cytometry analysis. High Annexin V and low PI staining (Q3) indicated early apoptosis of cells. in contrast, strong staining signals of both Annexin V label and PI (Q2) showed the cells were in the stage of necrosis or late apoptosis. As shown in Figure 6, 14.38% of UVA-irradiated cells were found to be apoptotic (early apoptosis plus late apoptosis) when incubated with TFS. However, the UVA-only and the drug-only groups showed no significant cell apoptosis. This assay indicated that chemo-photodynamic therapy could induce apoptosis and necrosis in A375 cells.

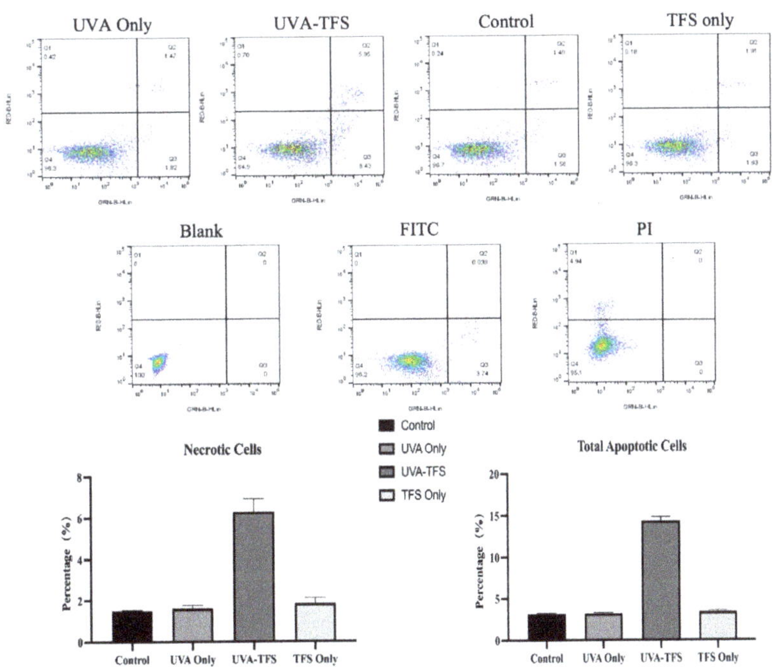

Figure 6. TFS/ UVA photodynamic treatments induced apoptosis in A375 cells. Cells were stained with Annexin V-FITC and PI before flow cytometric analysis.

2.7. Cell Cycle Analysis

PI was used to measure the cell DNA content. As shown in Figure 7, the treatment of UVA did not alter the cell cycle compared to the control group. Cells treated with TFS and UVA-TFS showed a significant increase in G2/M phases. These results indicated that chemo-photodynamic therapy could impact cell cycle progression and cause cell cycle arrest in the G2/M phase, leading to the inhibition of the cells.

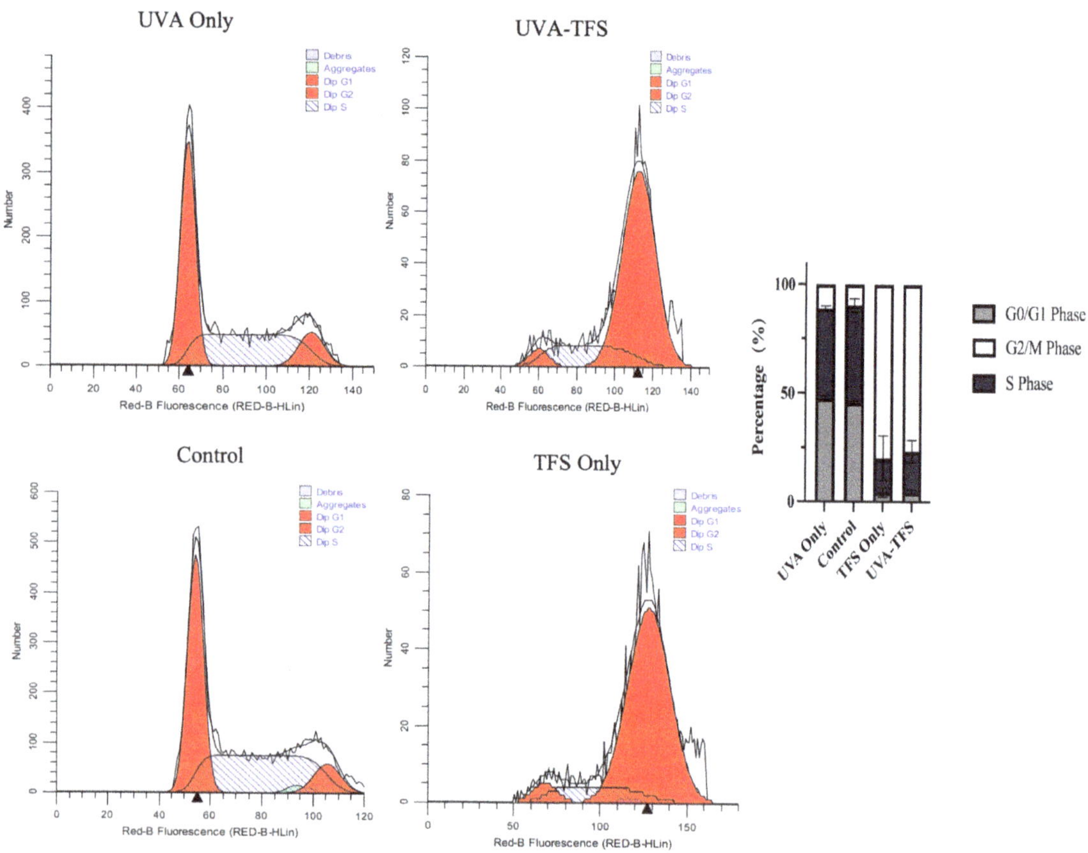

Figure 7. TFS/ UVA photodynamic therapy-induced cell cycle arrest of A375 cells. Cells were stained with PI and the cell cycle distribution was measured by flow cytometric analysis. (Triangles indicate the most PI-stained DNA content measured by flow cytometry).

2.8. Antitumor Therapeutic Efficacy In Vivo

A375 tumor-bearing nude mice models were used to validate the antitumor effect of topical PDT co-administered with 5-FU and IMP. Once the tumor volume reached 150 mm^3, the A375 tumor-bearing nude mice were randomly divided into 7 groups and treated with different protocols. All groups displayed an inhibitory effect on tumor growth except the model and PDT groups (Figure 8A). Photographs of tumor tissues treated with the different regimens are presented in Figure 8B. These results are consistent with the changes in tumor volume. The group treated with 5-FU/IMP-TFS-Gel and UVA Laser exhibited the highest tumor inhibition effect, with an IRT of 95.23% (Figure 8C). No significant weight loss was observed in any group during the treatment period. This indicated that the treatment regimens had a good biosafety profile (Figure 8D). To evaluate the antitumor effect in vivo, we conducted H&E staining and histological studies of tumor tissues. The tumor cells in the model and PDT groups were densely packed with large nuclei and active nuclear division. In contrast, the tumor cells in the treatment group exhibited necrotic cell disintegration, loss of cell membrane structure, and residual cellular debris. The level of necrosis in tumor cells varied across the different treatment groups, with the highest level of apoptosis observed in the 5-FU/IMP-TFS-Gel with UVA Laser treatment group.

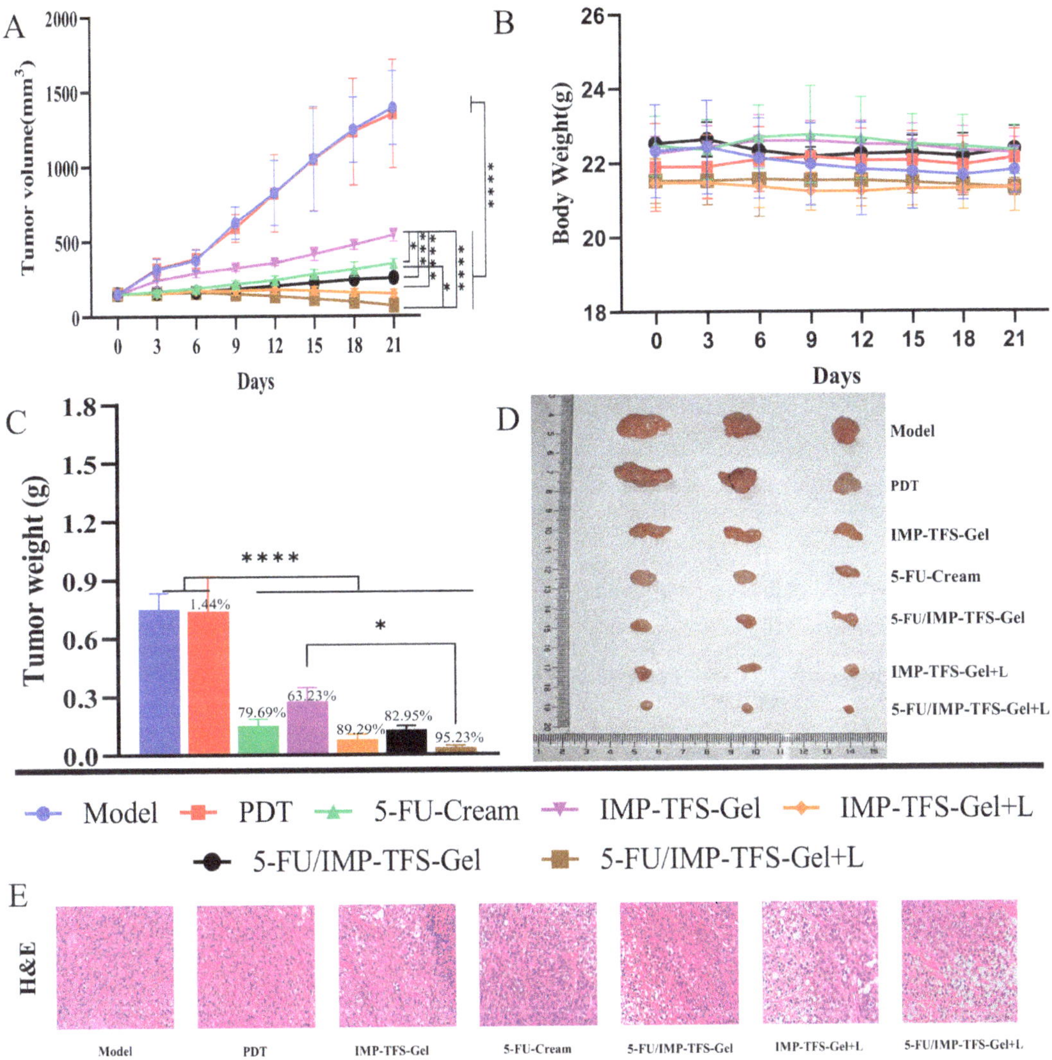

Figure 8. Antitumor therapeutic efficacy in vivo. (**A**) Tumor growth inhibition curves in mice with different treatment regimens ($n = 3$). (**B**) Photos of tumors. (**C**) The weight of tumors. (**D**). Body weight of mice during treatment. (**E**) H&E staining of tumor tissues captured at ×400 magnification. Data are presented as mean ± SD, $n = 3$; * $p < 0.05$, *** $p < 0.001$, **** $p < 0.0001$.

3. Discussion

The encapsulation of drugs within the nano drug delivery systems (NDDS) has been extensively studied over the past few decades and has demonstrated great potential in cancer treatments [46,47]. Systemic administration is typically achieved through intravenous injection, providing rapid onset of action and improved drug bioavailability [48]. However, due to biological barriers, only a fraction of the therapeutic dosage from intravenous administrations reaches the tumor sites and residues [49]. Therefore, systemic administration may require high doses or multiple administrations, which can lead to side effects [50]. Administering drugs locally for skin cancer offers advantages such as non-invasive administration during long-term treatment and increased local drug concentration with minimal toxicity [16]. Additionally, local delivery of multi-functional nanoparticles for photodynamic and photothermal therapy against tumor cells could exhibit enhanced antitumor efficiency [51]. The electrostatic interaction and photochemistry of inorganic materials and ligands would contribute to tumor inhibition [52].

In this study, nano-transfersomes co-loaded with 5-FU and IMP were designed, and their transdermal delivery was further explored. TFS-gel could improve local drug concentration and skin permeability. The strongest antitumor activity was observed in 5-FU/IMP-TFS-Gel with UVA Laser, which may be attributed to several factors. TFS gel offers a promising option for NDSS, with better penetration and patient adherence to skin cancer treatments. The ability to deliver PS agents and chemotherapeutic drugs altogether to the cancerous sites shows a great potential for skin diseases. Additionally, the topical delivery system increases the local concentration of the drug at the affected site, resulting in a rise in PS concentration and enhanced topical PDT. Results have indicated that chemo-PDT significantly inhibited the growth and proliferation of melanoma. However, current studies were limited as antitumor activities were only carried out in cell lines like HaCaT and A375, but not in skin cells or organs. Notably, the evaluation of antitumor agents has not only involved human cancer cells (A375 and A431), but also the human normal skin cell lines (TE353.sk). The study showed that proapoptotic agents like saponins can induce cell apoptosis in cancer cells in a dose-dependent manner. However, they did not induce the same effect in TE353.sk normal cells [53]. Similarly, TE353.sk cells and LoVo cells were employed in a study to investigate the photokilling selectivity and efficiency of TiO_2 nanoparticles in PDT. The results indicated that combined chemo-PDT could kill all LoVo cancer cells within 90 min, while it killed only 39% of TE353.sk normal cells [54]. The TE353.sk normal human dermal fibroblasts (NHDF) were utilized for testing the proapoptotic and prooxidative effects of gallium sensitizers compared to melanoma cancer cells and keratinocytes [55]. This study has proven that these second-generation sensitizers complemented with metal ions did not inhibit the viability of fibroblasts and keratinocytes, but significantly reduced the viability of cancer cells. This indicates that an ideal antitumor agent should provide advanced therapeutic effects in cancer cells with limited toxicity to normal cells. To validate the effectiveness of Chem-PDT agents such as 5-FU/IMP-TFS, further studies are crucial, particularly those focusing on selective targeting and evaluating cytotoxicity across normal skin cells and keratinocytes. Additionally, it would act selectively efficiently in cancer cells while being less active in normal cells. While chemo-PDT shows promise in treating melanoma, ongoing studies are needed to ensure it achieves the desired efficacy without compromising normal cells. Future research should refine these therapies to enhance selectivity and minimize toxicity, thereby providing safer and more effective strategies for cancer treatment.

4. Materials and Methods

4.1. Materials

Imperatorin (purity > 98%) was purchased from Push Bio-Technology Co., Ltd. (Chengdu, China), and 5-FU (purity > 98%) was obtained from Macklin Biochemical Co., Ltd. (Shanghai, China). Soybean lecithin was purchased from A. V. T. Pharmaceutical Co., Ltd. (Shanghai, China). Methanol (HPLC grade) was obtained from Thermo Fisher Scientific (Shanghai, China). HQ fetal bovine serum (FBS) was purchased from TransGen Biotech Co., Ltd. (Beijing, China). DMEM cell culture medium was purchased from Gibco (MA, USA). Cell Counting Kit-8 (CCK-8) was obtained from Labgic Technology Co., Ltd. (Beijing, China).

Ultra-pure water was produced using an ULUPURE integral water purification system. All other chemical reagents were of analytical grade.

4.2. Cell Culture and Animals

Human immortalized keratinocyte cell lines, HaCaT, and human malignant melanoma cells, A375, were obtained from ATCC, cultured in DMEM with 10% FBS and 1% penicillin/streptomycin maintained at 37 °C in a 5% CO_2 incubator. Male SD rats (200 g) and female BALB/c-Nu mice (4–5 weeks, 18–22 g) were obtained from SiPeiFu Biological Technology Co., Ltd. (Beijing, China). All animal experiments were conducted under the guidelines approved by the Institutional Animal Care and Use Committee (IACUC) of Chengdu University of Traditional Chinese Medicine.

4.3. Preparation of TFS

TFS encapsulating IMP and 5-FU were prepared using the thin film hydration method. Soybean lecithin (SPC), Tween-80 (TW-80), and IMP were dissolved in methanol/chloroform (v/v = 1:2), and the organic solvent was removed using a rotary evaporator. The thin lipid film was then dissolved with UP water containing 5-FU by shaking in the water bath for 30 min and the suspension was ultrasonicated (300 W) for 5 min to retain homogeneous size. The TFS liquid was further centrifuged at 10,000 rpm for 10 min and then the unincorporated IMP and 5-FU were removed by filtration through a 0.22 μm cellulose nitrate membrane. Finally, TFS co-loaded with 5-FU and IMP was stored at 4 °C and labeled as 5-FU/IMP-TFS.

4.4. Optimization of TFS by BBD

Based on the scope to most efficiently relate the response variables and experimental results, BBD (Box-Behnken Design, Design-Expert 8.0.6) was used to optimize the preparation process of 5-FU/IMP-TFS and determine the suitable formulation. The independent variables include the drug/lipid ratio (X1), the SPC/TW-80 ratio (X2), and the hydration temperature (X3), with co-encapsulation efficiency (Y) as the experimental response variable.

4.5. Characterization of TFS

The particle size, polydispersity index (PDI), and zeta potential of the TFS were measured using a Malvern Zettaliter (Nano ZS, Malvern Panalytical Ltd., Malvern, UK) dynamic light scattering detector at room temperature. Before measurement, samples were diluted 1:10 (v/v) with UP water. The sample morphology was evaluated by TEM (HT7800, Hitachi, Tokyo, Japan). To verify the successful preparation of the TFS, its mid-infrared spectrum was measured using FT-IR (Clry610, Agilent, Santa Clara, CA, USA). The samples were prepared using the potassium bromide press method and the scanning wavelengths ranged from 4000 to 400 cm^{-1}.

4.6. Encapsulation Efficiency (EE%) and Drug Loading (DL%)

The EE% and DL% of 5-FU/IMP-TFS were analyzed with the HPLC (Agilent 1260, Santa Clara, CA, USA). The samples were centrifuged for 20 min at 5000 rpm and 4 °C. Then, samples were subjected to column chromatography using 95:5 (v/v) water/methanol to yield 5-FU, and 30:70 (v/v) water/methanol to yield IMP. The analysis was completed using a C18 analytical column (4.6 mm × 150 mm, 5 μm), with a column temperature set at 30 °C. The injection volume was 10 μL, and the flow rate was 1.0 mL/min. The detection wavelengths were 265 nm and 300 nm, respectively. The EE% and DL% were calculated using the following equations:

$$EE\% = \frac{The\ weight\ of\ drug\ encapsulated}{The\ weight\ of\ the\ drug\ added} \times 100\%$$

$$DL\% = \frac{The\ weight\ of\ drug\ mearsured}{The\ weight\ of\ all\ the\ materials} \times 100\%$$

4.7. Drug Release, Skin Permeation, and Deposition Studies

The release profile of TFS was assessed by performing the cumulative in vitro studies in a water/ethanol medium (v/v = 70:30). TFS suspension was added to dialysis bags, and the samples were immersed in 40 mL of the release medium. The temperature was maintained at 37 ± 0.5 °C with a 100 rpm rotation rate. At predetermined time intervals (1, 2, 4, 6, 8, 10, 12, and 24 h), 1 mL of dissolution medium was withdrawn and replaced with an equal volume of fresh medium. The drug concentration was quantified using HPLC.

The ex vivo skin permeation was determined using a Franz diffusion cell. The cleaned skin was clamped between the donor and receptor compartments of the cell (0.64 cm^2). The skin was mounted carefully with the stratum corneum facing the donor compartment, and the dermal side was exposed to the receptor medium. The skin in the donor compartment was treated with TFS gel and 5-FU cream and the receptor compartment (5 mL) was filled with a water/ethanol receptor medium (v/v = 70:30). The diffusion cell assembly was maintained at 32 ± 1 °C, with the contents of the receptor compartment stirred at 100 rpm using a magnetic stirrer. Samples were withdrawn at regular intervals (1, 2, 4, 6, 8, 10, 12, and 24 h) through the sampling port and replaced with an equal volume of fresh diffusion medium to maintain sink conditions throughout the experiment. Drug content was analyzed using HPLC. The experiment was conducted in triplicate.

To assess drug deposition in the skin, the mounted skin on the diffusion cell was carefully removed and washed with receptor medium. The skin was cut into small pieces and placed in a grinding tube. Methanol was added for tissue grinding, followed by ultrasonic treatment for 10 min. The mixture was then centrifuged at 10,000 rpm for 5 min, and the supernatant was filtered through a 0.22 μm cellulose nitrate membrane. The drug content of the samples was analyzed using the HPLC.

4.8. Preparation of TFS Gel

The TFS gel was prepared by incorporating the optimized TFS into the dispersion of Carbopol 940. A precise weight of 1% Carbopol 940 was slowly added to TFS suspension and stirred continuously (1000 rpm) at room temperature overnight until evenly dispersed. Triethanolamine was added dropwise with continuous stirring to achieve TFS gel.

4.9. Cell Culture

Cells in the logarithmic growth phase were seeded into 96-well plates at a density of 5×10^3 cells per well in 10% FBS medium. After 24 h of incubation, the medium was replaced with 1% FBS medium containing different formulations for 48 h. Plates designated as controls were kept in the dark, while the others were irradiated with a UVA irradiator (365 nm) at a dose of 1.5 J/ cm^2. After 12 h of incubation in 1% FBS medium, 10 μL of CCK-8 reagent was added, and the cells were further incubated for 1 h at 37 °C in the dark. Optical density was then measured at 450 nm using a microplate reader (Gen5, BioTek, Winooski, VT, USA). The experiments were performed in triplicate.

4.10. Animal Experiment Design

Female BALB/c-Nu mice (4–5 weeks, 18–22 g) were divided into 7 groups for different treatments. Cultured A375 cells were injected subcutaneously into each mouse at a concentration of 5×10^6 cells. Antitumor tests were conducted with an equivalent tumor volume (around 150 mm^3). Tumor volume was measured every three days and calculated as the following equation:

$$Tumor\ volume = length \times width^2/2 \quad (2)$$

The 7 groups were (n = 6): model, PDT, 5-FU-Cream, IMP-TFS-Gel, 5-FU/IMP-TFS-Gel, IMP-TFS-Gel Laser, and 5-FU/IMP-TFS-Gel Laser treated groups. The concentration of 5-FU or IMP in all preparations was 0.5% w/w. A spatula was used to apply the cream or gel formulation (200 mg) once daily for 21 consecutive days to the area (1 cm × 2 cm) where the tumor cells were injected. For PDT experiments, the residual drug was wiped off the administration site with a 75% alcohol cotton 4 h after application and administration. The site was then illuminated with UVA (6 J/cm^2) daily. The inhibition rate of tumor growth (IRT) was calculated using the following formula: IRT = 100% × (mean tumor weight of the control group—mean tumor weight of the experimental group) / mean tumor weight of the control group. The mice were sacrificed 21 days after treatment, and their tumors were harvested and fixed in 10% neutral buffered formalin for further study. Histological analysis was performed using H&E staining, and tumors were sectioned into thin slices and processed routinely into paraffin.

4.11. Detection of the Generation of ROS

A375 cells were placed in 6-well plates at the density of 5 × 10^4 cells per well for 24 h. After incubation with 5-FU/IMP-TFS (100 μM) for 4 h, cells were washed three times using a PBS buffer and added fresh medium. One plate was exposed to UVA (power dose = 1.5 J/cm^2) for 10 min and another was kept in the dark incubation. After 12 h of incubation, the cells were washed and added with ROS probe 2′,7′-dichloroflurescin diacetate (DCFH-DA, Sigma-Aldrich, St. Louis, MO, USA). The cells were incubated for 30 min at 37 °C and the fluorescence intensity was measured using the flow cytometry with Green-B (GRN-B, 488 nm) fluorophores detection channel (Guava EasyCyteTM, Luminex, Austin, TX, USA).

4.12. Cell Cycle and Apoptosis Analysis

5 × 10^4 A375 cells were seeded in 6-well plates for 24 h and treated with 5-FU/IMP-TFS (100 μM) for 4 h. The cells were washed three times using PBS buffer and fresh medium. One plate received UVA exposure (power dose = 1.5 J/cm^2) for 10 min and another plate was incubated in the dark. The cells were harvested, washed, and fixed with ethanol at −20 °C for 20 h, and the suspension was centrifuged and washed twice with PBS (4 °C). Cells were stained with Guava cell cycle reagent according to the manufacturer's instructions. Stained cells were analyzed by flow cytometry with Red-B fluorophores detection channel (Guava EasyCyteTM, Luminex, Austin, TX, USA).

Cell apoptosis was measured using the propidium iodide (PI) and Annexin V-FITC apoptosis assay kit. Cells were treated, harvested, and washed with PBS before being stained with PI/ Annexin V kit according to the manufacturer's instructions. Stained cells were analyzed by flow cytometry with a Green-B, Red-B (488 nm) fluorophores detection channel (Guava EasyCyteTM, Luminex, Austin, TX, USA).

4.13. Statistical Analysis

All quantitative data are shown as mean ± SD, n ≥ 3. Statistical analyses were performed using GraphPad Prism 9.5.1 (MacOS) and Design-Expert 8.0.6 software and * $p < 0.05$, ** $p < 0.01$, *** $p < 0.001$, **** $p < 0.0001$.

5. Conclusions

In summary, we have developed a 5-FU/ IMP transdermal delivery system combined with chemo-photodynamic therapy for melanoma. The morphology of 5-FU/IMP-TFS was revealed with a particle size of 50 ± 1.92 nm and −13 ± 1.5 mV zeta potential. The in vitro drug release profiles and skin permeation studies were further evaluated. The steady-state rate of 5-FU in TFS was 1.47 times greater than conventional cream, and the retention was 4.43 times than that of the cream group. The study utilized 5-FU as the chemotherapeutic agent while IMP served as the photosensitizer, and 5-FU/IMP was co-encapsulated in the transdermal delivery vehicle TFS and embedded in Carbopol gel. The TFS gel has been shown to significantly improve skin permeability and increase local drug concentration compared to conventional creams. At the cellular level, the combination of 5-FU and IMP significantly reduced the toxicity of 5-FU in normal cells while enhancing the inhibition of tumor cells. Additionally, 5-FU/IMP-TFS, acting as both a chemotherapy and photodynamic agent, could improve anti-cancer efficiency when exposed to UVA laser sources. We also verified the results in nude mouse tumor models. The ROS generation, cell cycle, and cell apoptosis studies have proven the enhancement made by the combined chemo-photodynamic therapy. Overall, this study indicates that TFS, a transdermal delivery vehicle co-loaded with 5-FU and IMP, represents a novel therapeutic option for synergistic anti-cancer treatment.

Supplementary Materials: The following supporting information can be downloaded at: https://www.mdpi.com/article/10.3390/ijms25179611/s1.

Author Contributions: Conceptualization, Y.G. and L.G.; methodology, Y.G. and W.Z.; software, Y.G. and W.Z.; validation, Y.G., W.Z. and L.G.; formal analysis, Y.G. and W.Z.; investigation, Y.G. and W.Z.; resources, Y.G., C.P. and L.G.; data curation, Y.G. and W.Z.; writing—original draft preparation, Y.G. and W.Z.; writing—review and editing, Y.G.; visualization, Y.G. and W.Z.; supervision, C.P. and L.G.; project administration, C.P. and L.G.; funding acquisition, Y.G., C.P. and L.G. All authors have read and agreed to the published version of the manuscript.

Funding: This research was co-funded by the Sichuan Science and Technology Program (2024NSFSC1839) and the Xinglin Scholar Research Promotion Project of Chengdu University of TCM (BSH2021025).

Institutional Review Board Statement: The study was approved by the Animal Research Ethics Committee of Chengdu University of Traditional Chinese Medicine (protocol code: No. 2023-183 and the date of approval was 12 October 2023).

Informed Consent Statement: Not applicable.

Data Availability Statement: The data are available upon request.

Conflicts of Interest: The authors declare no conflicts of interest.

References

1. Long, G.V.; Swetter, S.M.; Menzies, A.M.; Gershenwald, J.E.; Scolyer, R.A. Cutaneous melanoma. *Lancet* **2023**, *402*, 485–502. [CrossRef]
2. Rahman, S.; Haque, T.N.; Sugandhi, V.V.; Saraswat, A.L.; Xin, X.; Cho, H. Topical Cream Carrying Drug-Loaded Nanogels for Melanoma Treatment. *Pharm. Res.* **2023**, *40*, 2291–2301. [CrossRef]
3. Siegel, R.L.; Giaquinto, A.N.; Jemal, A. Cancer statistics, 2024. *CA Cancer J. Clin.* **2024**, *74*, 12–49. [CrossRef]
4. Eggermont, A.M.; Spatz, A.; Robert, C. Cutaneous melanoma. *Lancet* **2014**, *383*, 816–827. [CrossRef] [PubMed]
5. Chen, K.G.; Leapman, R.D.; Zhang, G.; Lai, B.; Valencia, J.C.; Cardarelli, C.O.; Vieira, W.D.; Hearing, V.J.; Gottesman, M.M. Influence of melanosome dynamics on melanoma drug sensitivity. *J. Natl. Cancer Inst.* **2009**, *101*, 1259–1271. [CrossRef]
6. Fang, D.; Nguyen, T.K.; Leishear, K.; Finko, R.; Kulp, A.N.; Hotz, S.; Van Belle, P.A.; Xu, X.; Elder, D.E.; Herlyn, M. A tumorigenic subpopulation with stem cell properties in melanomas. *Cancer Res.* **2005**, *65*, 9328–9337. [CrossRef] [PubMed]
7. Gottesman, M.M. Mechanisms of cancer drug resistance. *Annu. Rev. Med.* **2002**, *53*, 615–627. [CrossRef] [PubMed]
8. Goette, D.K. Topical chemotherapy with 5-fluorouracil. A review. *J. Am. Acad. Dermatol.* **1981**, *4*, 633–649. [CrossRef]
9. Safwat, M.A.; Soliman, G.M.; Sayed, D.; Attia, M.A. Fluorouracil-Loaded Gold Nanoparticles for the Treatment of Skin Cancer: Development, in Vitro Characterization, and in Vivo Evaluation in a Mouse Skin Cancer Xenograft Model. *Mol. Pharm.* **2018**, *15*, 2194–2205. [CrossRef]
10. Crisóstomo, L.; Carvalho, G.S.G.; Leal, L.; de Araújo, T.G.; Nogueira, K.A.B.; da Silva, D.A.; de Oliveira Silva Ribeiro, F.; Petrilli, R.; Eloy, J.O. Sorbitan Monolaurate-Containing Liposomes Enhance Skin Cancer Cell Cytotoxicity and in Association with Microneedling Increase the Skin Penetration of 5-Fluorouracil. *AAPS PharmSciTech* **2022**, *23*, 212. [CrossRef]
11. Zhao, H.; Feng, Y.L.; Wang, M.; Wang, J.J.; Liu, T.; Yu, J. The Angelica dahurica: A Review of Traditional Uses, Phytochemistry and Pharmacology. *Front. Pharmacol.* **2022**, *13*, 896637. [CrossRef] [PubMed]
12. Nasser, M.I.; Zhu, S.; Hu, H.; Huang, H.; Guo, M.; Zhu, P. Effects of imperatorin in the cardiovascular system and cancer. *Biomed. Pharmacother.* **2019**, *120*, 109401. [CrossRef] [PubMed]
13. Kim, T.; Hyun, C.G. Imperatorin Positively Regulates Melanogenesis through Signaling Pathways Involving PKA/CREB, ERK, AKT, and GSK3β/β-Catenin. *Molecules* **2022**, *27*, 6512. [CrossRef] [PubMed]
14. Wróblewska-Łuczka, P.; Grabarska, A.; Florek-Łuszczki, M.; Plewa, Z.; Łuszczki, J.J. Synergy, Additivity, and Antagonism between Cisplatin and Selected Coumarins in Human Melanoma Cells. *Int. J. Mol. Sci.* **2021**, *22*, 537. [CrossRef]
15. Kimura, Y.; Sumiyoshi, M.; Sakanaka, M.; Taniguchi, M.; Baba, K. In vitro and in vivo antiproliferative effect of a combination of ultraviolet-A and alkoxy furocoumarins isolated from Umbelliferae medicinal plants, in melanoma cells. *Photochem. Photobiol.* **2013**, *89*, 1216–1225. [CrossRef] [PubMed]

16. Guo, Y.; Liu, H.; Xiao, H.; Yuan, M.; Liu, Y.; Sedlařík, V.; Chin, W.C.; Liu, J.; Guo, L.; Li, C. Self-assembled Camptothecin derivatives—Curcuminoids conjugate for combinatorial chemo-photodynamic therapy to enhance antitumor efficacy. *J. Photochem. Photobiol. B* **2021**, *215*, 112124. [CrossRef] [PubMed]
17. Dragicevic, N.; Predic-Atkinson, J.; Nikolic, B.; Pajovic, S.B.; Ivkovic, S.; Adzic, M. Nanocarriers in topical photodynamic therapy. *Expert Opin. Drug Deliv.* **2024**, *21*, 279–307. [CrossRef]
18. Doppalapudi, S.; Jain, A.; Chopra, D.K.; Khan, W. Psoralen loaded liposomal nanocarriers for improved skin penetration and efficacy of topical PUVA in psoriasis. *Eur. J. Pharm. Sci.* **2017**, *96*, 515–529. [CrossRef]
19. Keyal, U.; Bhatta, A.K.; Zhang, G.; Wang, X.L. Present and future perspectives of photodynamic therapy for cutaneous squamous cell carcinoma. *J. Am. Acad. Dermatol.* **2019**, *80*, 765–773. [CrossRef]
20. Jain, R.; Paul, M.; Padaga, S.G.; Dubey, S.K.; Biswas, S.; Singhvi, G. Dual-Drug-Loaded Topical Delivery of Photodynamically Active Lipid-Based Formulation for Combination Therapy of Cutaneous Melanoma. *Mol. Pharm.* **2023**, *20*, 3653–3671. [CrossRef]
21. Clemente, N.; Miletto, I.; Gianotti, E.; Sabbatini, M.; Invernizzi, M.; Marchese, L.; Dianzani, U.; Renò, F. Verteporfin-Loaded Mesoporous Silica Nanoparticles' Topical Applications Inhibit Mouse Melanoma Lymphangiogenesis and Micrometastasis In Vivo. *Int. J. Mol. Sci.* **2021**, *22*, 13443. [CrossRef] [PubMed]
22. Champeau, M.; Vignoud, S.; Mortier, L.; Mordon, S. Photodynamic therapy for skin cancer: How to enhance drug penetration? *J. Photochem. Photobiol. B* **2019**, *197*, 111544. [CrossRef]
23. Benson, H.A. Transfersomes for transdermal drug delivery. *Expert Opin. Drug Deliv.* **2006**, *3*, 727–737. [CrossRef] [PubMed]
24. Simrah Hafeez, A.; Usmani, S.A.; Izhar, M.P. Transfersome, an ultra-deformable lipid-based drug nanocarrier: An updated review with therapeutic applications. *Naunyn Schmiedebergs Arch. Pharmacol.* **2024**, *397*, 639–673. [CrossRef]
25. Phatale, V.; Vaiphei, K.K.; Jha, S.; Patil, D.; Agrawal, M.; Alexander, A. Overcoming skin barriers through advanced transdermal drug delivery approaches. *J. Control. Release* **2022**, *351*, 361–380. [CrossRef] [PubMed]
26. Jiang, T.; Wang, T.; Li, T.; Ma, Y.; Shen, S.; He, B.; Mo, R. Enhanced Transdermal Drug Delivery by Transfersome-Embedded Oligopeptide Hydrogel for Topical Chemotherapy of Melanoma. *ACS Nano* **2018**, *12*, 9693–9701. [CrossRef]
27. Lin, H.; Xie, Q.; Huang, X.; Ban, J.; Wang, B.; Wei, X.; Chen, Y.; Lu, Z. Increased skin permeation efficiency of imperatorin via charged ultradeformable lipid vesicles for transdermal delivery. *Int. J. Nanomed.* **2018**, *13*, 831–842. [CrossRef] [PubMed]
28. Souto, E.B.; Macedo, A.S.; Dias-Ferreira, J.; Cano, A.; Zielińska, A.; Matos, C.M. Elastic and Ultradeformable Liposomes for Transdermal Delivery of Active Pharmaceutical Ingredients (APIs). *Int. J. Mol. Sci.* **2021**, *22*, 9743. [CrossRef]
29. Cosco, D.; Paolino, D.; Maiuolo, J.; Marzio, L.D.; Carafa, M.; Ventura, C.A.; Fresta, M. Ultradeformable liposomes as multidrug carrier of resveratrol and 5-fluorouracil for their topical delivery. *Int. J. Pharm.* **2015**, *489*, 1–10. [CrossRef]
30. Demartis, S.; Rassu, G.; Murgia, S.; Casula, L.; Giunchedi, P.; Gavini, E. Improving Dermal Delivery of Rose Bengal by Deformable Lipid Nanovesicles for Topical Treatment of Melanoma. *Mol. Pharm.* **2021**, *18*, 4046–4057. [CrossRef]
31. Verma, D.D.; Verma, S.; Blume, G.; Fahr, A. Particle size of liposomes influences dermal delivery of substances into skin. *Int. J. Pharm.* **2003**, *258*, 141–151. [CrossRef] [PubMed]
32. Hussain, A.; Haque, M.W.; Singh, S.K.; Ahmed, F.J. Optimized permeation enhancer for topical delivery of 5-fluorouracil-loaded elastic liposome using Design Expert: Part II. *Drug Deliv.* **2016**, *23*, 1242–1253. [CrossRef] [PubMed]
33. Chen, T.; Chen, H.; Jiang, Y.; Yan, Q.; Zheng, S.; Wu, M. Co-Delivery of 5-Fluorouracil and Paclitaxel in Mitochondria-Targeted KLA-Modified Liposomes to Improve Triple-Negative Breast Cancer Treatment. *Pharmaceuticals* **2022**, *15*, 881. [CrossRef] [PubMed]
34. Wang, J.; Zhao, Y.; Zhai, B.; Cheng, J.; Sun, J.; Zhang, X.; Guo, D. Phloretin Transfersomes for Transdermal Delivery: Design, Optimization, and In Vivo Evaluation. *Molecules* **2023**, *28*, 6790. [CrossRef]
35. Yuan, M.; Niu, J.; Xiao, Q.; Ya, H.; Zhang, Y.; Fan, Y.; Li, L.; Li, X. Hyaluronan-modified transfersomes based hydrogel for enhanced transdermal delivery of indomethacin. *Drug Deliv.* **2022**, *29*, 1232–1242. [CrossRef]
36. Cosco, D.; Celia, C.; Cilurzo, F.; Trapasso, E.; Paolino, D. Colloidal carriers for the enhanced delivery through the skin. *Expert Opin. Drug Deliv.* **2008**, *5*, 737–755. [CrossRef]
37. Hussain, A.; Samad, A.; Ramzan, M.; Ahsan, M.N.; Ur Rehman, Z.; Ahmad, F.J. Elastic liposome-based gel for topical delivery of 5-fluorouracil: In vitro and in vivo investigation. *Drug Deliv.* **2016**, *23*, 1115–1129. [CrossRef]
38. Cadet, J.; Douki, T. Formation of UV-induced DNA damage contributing to skin cancer development. *Photochem. Photobiol. Sci.* **2018**, *17*, 1816–1841. [CrossRef]
39. Gag, O.; Dinu, Ş.; Manea, H.; Marcovici, I.; Pînzaru, I.; Popovici, R.; Crăiniceanu, Z.; Gyori, Z.; Iovănescu, G.; Chiriac, S. UVA/UVB Irradiation Exerts a Distinct Phototoxic Effect on Human Keratinocytes Compared to Human Malignant Melanoma Cells. *Life* **2023**, *13*, 1144. [CrossRef]
40. Marrelli, M.; Giordano, F.; Perri, M.R.; Amodeo, V.; Baldino, N.; Lupia, C.; Uzunov, D.; Musolino, V.; Conforti, F.; Panno, M.L. Phytochemical Profile and In Vitro Antioxidant and Photobiological Properties of Different Extracts from Prangos ferulacea Lindl. *Antioxidants* **2023**, *12*, 384. [CrossRef]
41. Marrelli, M.; Perri, M.R.; Amodeo, V.; Giordano, F.; Statti, G.A.; Panno, M.L.; Conforti, F. Assessment of Photo-Induced Cytotoxic Activity of Cachrys sicula and Cachrys libanotis Enriched-Coumarin Extracts against Human Melanoma Cells. *Plants* **2021**, *10*, 123. [CrossRef] [PubMed]

42. Rapa, S.F.; Magliocca, G.; Pepe, G.; Amodio, G.; Autore, G.; Campiglia, P.; Marzocco, S. Protective Effect of Pomegranate on Oxidative Stress and Inflammatory Response Induced by 5-Fluorouracil in Human Keratinocytes. *Antioxidants* **2021**, *10*, 203. [CrossRef] [PubMed]
43. Deng, M.; Xie, L.; Zhong, L.; Liao, Y.; Liu, L.; Li, X. Imperatorin: A review of its pharmacology, toxicity and pharmacokinetics. *Eur. J. Pharmacol.* **2020**, *879*, 173124. [CrossRef] [PubMed]
44. Huang, Y.; Lai, H.; Jiang, J.; Xu, X.; Zeng, Z.; Ren, L.; Liu, Q.; Chen, M.; Zhang, T.; Ding, X.; et al. pH-activatable oxidative stress amplifying dissolving microneedles for combined chemo-photodynamic therapy of melanoma. *Asian J. Pharm. Sci.* **2022**, *17*, 679–696. [CrossRef] [PubMed]
45. Huang, L.; Chen, X.; Bian, Q.; Zhang, F.; Wu, H.; Wang, H.; Gao, J. Photosensitizer-stabilized self-assembling nanoparticles potentiate chemo/photodynamic efficacy of patient-derived melanoma. *J. Control. Release* **2020**, *328*, 325–338. [CrossRef]
46. Xu, S.; Liu, C.; Zang, S.; Li, J.; Wang, Y.; Ren, K.; Li, M.; Zhang, Z.; He, Q. Multifunctional self-delivery micelles targeting the invasion-metastasis cascade for enhanced chemotherapy against melanoma and the lung metastasis. *Asian J. Pharm. Sci.* **2021**, *16*, 794–805. [CrossRef]
47. Gilani, S.J.; Jahangir, M.A.; Chandrakala; Rizwanullah, M.; Taleuzzaman, M.; Shahab, M.S.; Shakeel, K.; Aqil, M.; Imam, S.S. Nano-Based Therapy for Treatment of Skin Cancer. *Recent Pat. Anti-Infect. Drug Discov.* **2018**, *13*, 151–163. [CrossRef]
48. Moller, P.L.; Sindet-Pedersen, S.; Petersen, C.T.; Juhl, G.I.; Dillenschneider, A.; Skoglund, L.A. Onset of acetaminophen analgesia: Comparison of oral and intravenous routes after third molar surgery. *Br. J. Anaesth.* **2005**, *94*, 642–648. [CrossRef]
49. Li, H.; Wang, Y.; Tang, Q.; Yin, D.; Tang, C.; He, E.; Zou, L.; Peng, Q. The protein corona and its effects on nanoparticle-based drug delivery systems. *Acta Biomater.* **2021**, *129*, 57–72. [CrossRef]
50. Wen, W.; Wu, J.; Liu, L.; Tian, Y.; Buettner, R.; Hsieh, M.Y.; Horne, D.; Dellinger, T.H.; Han, E.S.; Jove, R.; et al. Synergistic antitumor effect of combined inhibition of EGFR and JAK/STAT3 pathways in human ovarian cancer. *Mol. Cancer* **2015**, *14*, 100. [CrossRef]
51. Akbar, A.; Khan, S.; Chatterjee, T.; Ghosh, M. Unleashing the power of porphyrin photosensitizers: Illuminating breakthroughs in photodynamic therapy. *J. Photochem. Photobiol. B* **2023**, *248*, 112796. [CrossRef] [PubMed]
52. Gaona-Esquivel, A.; Diana, S.H.-M.; Hernández-Rodríguez, Y.M.; Cigarroa-Mayorga, O.E. The role of Nd as a dopant in Mn3O4NPs on the heat induction of artificial breast tissue due to the irradiation of microwaves. *Mater. Chem. Phys.* **2022**, *292*, 126822. [CrossRef]
53. Simo, L.M.; Messi, L.M.; Mbing, J.N.; Muller, C.D.; Boyom, F.F.; Begoudé, A.B.; Pegnyemb, D.E.; Haddad, M.; Noté, O.P. A New Triterpenoid Saponin from Albizia zygia Induced Apoptosis by Reduction of Mitochondrial Potential Status in Malignant Melanoma Cells. *Planta Med.* **2023**, *89*, 86–98. [CrossRef] [PubMed]
54. Xu, J.; Sun, Y.; Huang, J.; Chen, C.; Liu, G.; Jiang, Y.; Zhao, Y.; Jiang, Z. Photokilling cancer cells using highly cell-specific antibody-TiO(2) bioconjugates and electroporation. *Bioelectrochemistry* **2007**, *71*, 217–222. [CrossRef]
55. Nackiewicz, J.; Kliber-Jasik, M.; Pogoda-Mieszczak, K.; Skonieczna, M. Gallium octacarboxyphthalocyanine hydroxide as a potential pro-apoptotic drug against cancer skin cells. *Bioorg. Chem.* **2024**, *152*, 107736. [CrossRef]

International Journal of
Molecular Sciences

MDPI

Article

Synthesis, Optimization and Molecular Self-Assembly Behavior of Alginate-g-Oleylamine Derivatives Based on Ugi Reaction for Hydrophobic Drug Delivery

Yanan Bu [1,2,3], Xiuqiong Chen [1,2,3], Ting Wu [1,2], Ruolin Zhang [1,2], Huiqiong Yan [1,2,3,*] and Qiang Lin [1,2,3]

1 Key Laboratory of Tropical Medicinal Resource Chemistry of Ministry of Education, College of Chemistry and Chemical Engineering, Hainan Normal University, Haikou 571158, China; buynedu@163.com (Y.B.); chenxiuqiongedu@163.com (X.C.); wt09250916@163.com (T.W.); 13021328869@163.com (R.Z.); linqianggroup@163.com (Q.L.)

2 Key Laboratory of Water Pollution Treatment & Resource Reuse of Hainan Province, College of Chemistry and Chemical Engineering, Hainan Normal University, Haikou 571158, China

3 Key Laboratory of Natural Polymer Functional Material of Haikou City, College of Chemistry and Chemical Engineering, Hainan Normal University, Haikou 571158, China

* Correspondence: 070114@hainnu.edu.cn; Tel.: +86-0898-65884995

Abstract: To achieve the optimal alginate-based oral formulation for delivery of hydrophobic drugs, on the basis of previous research, we further optimized the synthesis process parameters of alginate-g-oleylamine derivatives (Ugi-FOlT) and explored the effects of different degrees of substitution (DSs) on the molecular self-assembly properties of Ugi-FOlT, as well as the in vitro cytotoxicity and drug release behavior of Ugi-FOlT. The resultant Ugi-FOlT exhibited good amphiphilic properties with the critical micelle concentration (CMC) ranging from 0.043 mg/mL to 0.091 mg/mL, which decreased with the increase in the DS of Ugi-FOlT. Furthermore, Ugi-FOlT was able to self-assemble into spherical micellar aggregates in aqueous solution, whose sizes and zeta potentials with various DSs measured by dynamic light scattering (DLS) were in the range of 653 ± 25~710 ± 40 nm and −58.2 ± 1.92~−48.9 ± 2.86 mV, respectively. In addition, RAW 264.7 macrophages were used for MTT assay to evaluate the in vitro cytotoxicity of Ugi-FOlT in the range of 100~500 μg/mL, and the results indicated good cytocompatibility for Ugi-FOlT. Ugi-FOlT micellar aggregates with favorable stability also showed a certain sustained and pH-responsive release behavior for the hydrophobic drug ibuprofen (IBU). Meanwhile, it is feasible to control the drug release rate by regulating the DS of Ugi-FOlT. The influence of different DSs on the properties of Ugi-FOlT is helpful to fully understand the relationship between the micromolecular structure of Ugi-FOlT and its macroscopic properties.

Keywords: algal polysaccharides; polysaccharide derivatives; micellar aggregates; Ugi four-component reaction; hydrophobic pharmaceutical formulations

1. Introduction

In recent years, amphiphilic polymers have been verified to be an extremely important class of polymeric materials that can self-assemble into a variety of functional micro/nanostructures (such as micelles and vesicles) with potential applications in biomedical, biosensor and food applications [1,2]. A variety of natural, semisynthetic and synthetic hydrophilic polymers can be applied to synthesize amphiphilic polymers with different characteristics [3,4]. Among these polymers, natural polysaccharides, mainly derived from animals, plants and microorganisms, are considered to be one of the most prominent and promising candidates for the construction of amphiphilic polymers due to their important properties such as abundant sources, low cost, non-toxicity, good biocompatibility, chemical modifiability and in vivo degradation into harmless products via the enzymatic or non-enzymatic modes [5–8].

Sodium alginate (SA) is a linear polyanionic polysaccharide composed of two structural units, β-D-mannuronic acid (M) and its C-5 isomer α-L-guluronic acid (G), which are irregularly linked by β-(1, 4) glycosidic bonds. Its structure contains three different chain segment blocks, namely, homopolymer M chain segment block (-M-M-M-), homopolymer G chain segment block (-G-G-G-) and heteropoly chain segment block (-G-M-G-M-) [9–11]. SA is recognized as a safe natural polysaccharide with significant advantages such as biocompatibility, biodegradability, non-toxicity, non-allergy and easy modification [12–14]. Therefore, the development of SA-based amphiphilic polymers for the construction of self-assembled nanocarriers has broad application prospects in the field of drug delivery.

Recently, the Ugi four-component condensation reaction (Ugi-4CR) has attracted the attention of many researchers in polymer chemistry and involves the coupling of four chemical functional groups (aldehyde/ketone, amine, carboxylic acid and isocyanide compounds) [15,16]. In the current literature, most alginate-based polymers are prepared by using flammable liquid cyclohexyl isonitrile with high toxicity, heavy odor and high price, which is not conducive to environmental protection and health, and is not conducive to industrial production [17–19]. Through screening the isocyanide compounds currently on the market, it was found that p-phenylmethylsulfonyl isonitrile (p-TI) is a solid that is stable, almost odorless and can be stored at room temperature without decomposition, and its price is the most advantageous among isocyanates [20]. In previous work, our group also tried to substitute p-TI for the toxic cyclohexyl isonitrile to prepare amphiphilic alginate derivatives with good biocompatibility, which were used to construct controlled, poorly soluble drug release systems [21–23]. Therefore, using p-TI instead of cyclohexyl isonitrile to prepare amphiphilic polymers is of great research value for the development of polymeric biomaterials.

In the previous work, we preliminarily explored the synthesis of an amphiphilic sodium alginate-g-oleylamine derivative (Ugi-FOIT) with a DS of 20.4% via the Ugi-4CR reaction using SA as a hydrophilic main chain and bio-based oleylamine (unsaturated fatty amine) as a hydrophobic side chain and p-TI instead of cyclohexyl isonitrile [24]. On the basis of previous work, we focused on the influence of different DSs on the physicochemical properties of Ugi-FOIT to further analyze and understand the relationship between the micromolecular structure of Ugi-FOIT and its macroscopic properties, so as to achieve the optimal alginate-based hydrophobic drug formulation. In the present work, we further optimized the preparation process parameters of Ugi-FOIT to achieve its simple and efficient synthesis. Furthermore, the effects of different degrees of substitution (DSs) on the amphiphilic properties, molecular self-assembly properties and self-assembled nanomicelles' colloidal properties of Ugi-FOIT, and the in vitro cytotoxicity and drug release behavior of Ugi-FOIT were also mainly explored. The objective of this study was to lay a theoretical foundation for the development of optimal alginate-based oral formulations for biomedical applications based on the Ugi-4CR reaction.

2. Results and Discussion

2.1. Optimization of Process Parameters for the Synthesis of Ugi-FOIT

The Ugi-FOIT was synthesized by UGI-4CR reaction. According to the reaction mechanism of UGI-4CR and considering that p-TI is unstable and decomposes easily in acidic solution, all optimization experiments were based on the same molar amount of formaldehyde and oleylamine and a 20% excess of the p-TI dosage relative to the molar amount of oleylamine to ensure that the reaction could proceed completely [20]. On this basis, the effects of reaction parameters such as molar ratio of raw materials, reaction time, reaction temperature and pH value of solution on the DS of Ugi-FOIT were investigated. Through a series of experiments, the optimum reaction conditions for the preparation of Ugi-FOIT with high DS were determined, with the results shown in Table 1. The effects of reaction parameters such as molar ratio of raw materials, reaction temperature and pH value of solution on the DS of Ugi-FOIT under different reaction times are shown in Figure 1. It can be observed that with the increase in reaction time, the DS of Ugi-FOIT gradually increased and reached the peak value when the reaction was carried out for 24 h. This may be attributed to the fact that SA is a kind of polymer that has a large number of intramolecular and intermolecular hydrogen bonds and has a rigid structure, and the reaction takes a certain amount of time to destroy this structure. When the reaction time was extended to 36 h, the DS of Ugi-FOIT not only did not increase, but presented a slightly decreasing trend, which may mean that too long reaction time leads to a certain degree of degradation of the generated product in the solution, resulting in a slight decrease in its DS. From the point of view of energy saving and efficiency improvement, 24 h was the best reaction time. In addition, the effect of molar ratio of raw materials on the DS of Ugi-FO1T is shown in Figure 1A. When the molar ratio of raw material (N_{SA}:N_{OLA}:N_{FA}:$N_{p\text{-}TI}$) gradually increased from 1:0.4:0.4:0.5 to 1:1.2:1.2:1.4, the DS of Ugi-FOIT gradually rose from 9.5% to 20.4% after 24 h reaction. When the molar ratio of the raw material was sequentially increased to 1:1.4:1.4:1.6, the DS of UI-FOLT was 20.5%, and no significant improvement was observed. This phenomenon indicated that the increase in oleylamine and formaldehyde relative to the molar amount of SA was conducive to the enhancement in DS for Ugi-FOIT, but when the dosage of oleylamine and formaldehyde reached a certain amount, there was steric hindrance between the hydrophobic groups on the molecular chain of Ugi-FOIT, which probably affected the further grafting of oleylamine in the reaction. Considering the problem of removing unreacted raw materials in the later stage, we determined that the molar ratio of raw materials for preparing Ugi-FOIT with the highest DS was 1:1.2:1.2:1.4.

Table 1. DS and yield of Ugi-FOIT under different reaction parameters.

N_{SA}:N_{OLA}:N_{FA}:$N_{p\text{-}TI}$	Time (h)	Temp (°C)	pH	DS (%)	Yield (%)
1:0.4:0.4:0.5	24	25	3.6	9.5	61.8
1:0.8:0.8:1.0	24	25	3.6	14.6	62.7
1:1.2:1.2:1.4	24	25	3.6	20.4	74.0
1:1.4:1.4:1.6	24	25	3.6	20.5	78.5
1:1.2:1.2:1.4	24	37	3.6	19.1	76.4
1:1.2:1.2:1.4	24	25	4.8	16.4	68.2
1:1.2:1.2:1.4	24	25	6.0	13.1	51.5

Additionally, the reaction temperature affected the DS of Ugi-FOIT; the results are presented in Figure 1B. Within 12 h of the initial reaction, the DS of Ugi-FOIT obtained at 37 °C was 14.8%, which was higher than that at 25 °C (13.2%). When the reaction time was extended to 24 h, the DS of Ugi-FOIT at 25 °C was 20.4%, higher than that at 37 °C (19.1%), which may be attributed to the fact that the high reaction temperature was beneficial to improving the reaction rate, resulting in a higher DS of Ugi-FOIT at the initial stage. However, high reaction temperature would cause unnecessary side reactions, which was not conducive to increasing the DS of the products. Therefore, the optimum temperature for the synthesis of Ugi-FOIT reaction was 25 °C. Finally, the pH value of the solution also had a great influence on the DS of Ugi-FOIT. The effect of solution pH on the DS of Ugi-FOIT is displayed in Figure 1C. When the pH value of the solution was lower than 3.6, the alginate in the solution was protonated to form alginic acid that was insoluble in water, which is detrimental to the reaction [25]. However, when the pH value of the solution was increased, the DS of Ugi-FOIT displayed a decreasing trend; especially when the pH value of the solution was adjusted to 6.0, the DS of Ugi-FOIT decreased significantly. This was owing to the fact that the increase in the pH of the solution is not conducive to the protonation of the reaction intermediate imide. In addition, when the pH value of the solution was high, the carboxyl groups on the SA skeleton mainly existed as carboxyl anions, and the protonated carboxyl groups were less, resulting in their failure to fully participate in the Ugi-4CR reaction. Therefore, the optimal pH for the reaction was 3.6.

In short, the process conditions for the preparation of Ugi-FOIT with high DS were obtained through the experiment, that is, the molar ratio of the raw material (N_{SA}:N_{OLA}:N_{FA}:$N_{p\text{-}TI}$) was 1:1:2:1.2:1.4, the pH value of the solution was 3.6, the reaction temperature was 25 °C and the reaction time was 24 h. In order to explore the effects of different DSs on the physicochemical properties and drug release properties of Ugi-FOIT, we selected Ugi-FOIT with DS of 9.5%, 14.6% and 20.4% for follow-up experimental studies, and the corresponding derivatives were, respectively, named as Ugi-FOIT9.5, Ugi-FOIT14.6 and Ugi-FOIT20.4.

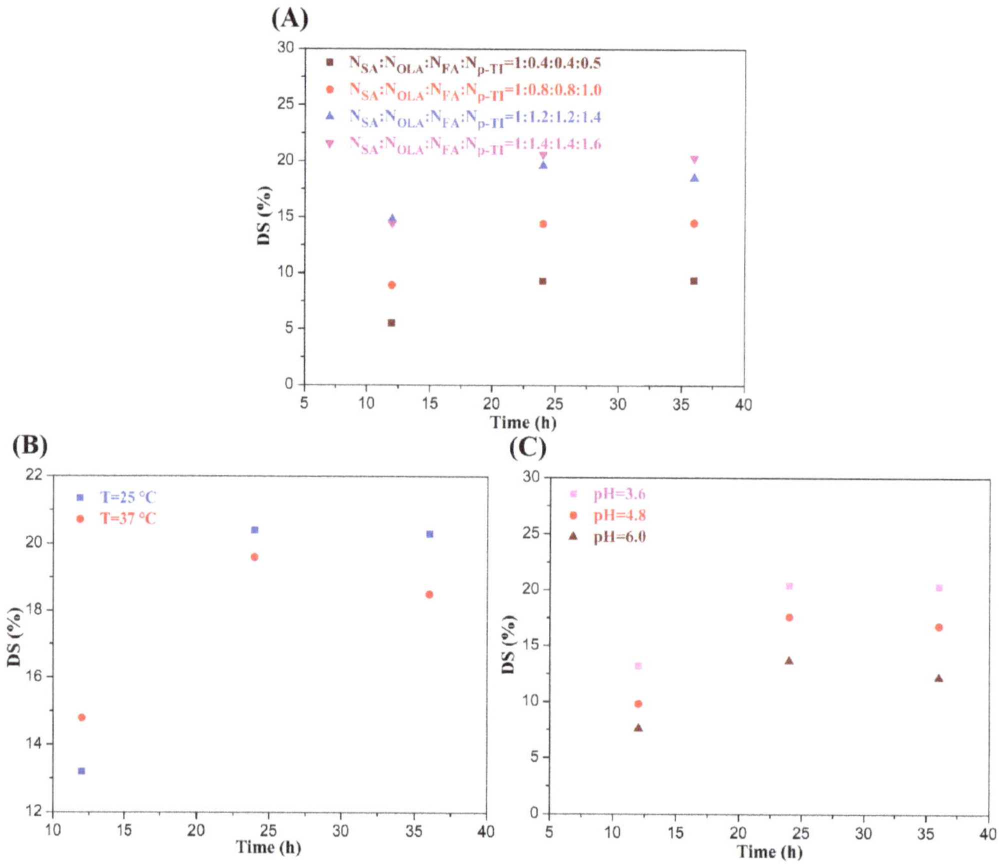

Figure 1. DS as a function of reaction time after coupling of oleylamine to SA varying (**A**) molar ratios of oleylamine to SA at 25 °C with pH value of 3.6, (**B**) temperatures using the feed molar ratio of 1:1.2:1.2:1.4 (N_{SA}:N_{OLA}:N_{FA}:$N_{p\text{-}TI}$) with pH value of 3.6 and (**C**) pH values using the feed molar ratio of 1:1.2:1.2:1.4 (N_{SA}:N_{OLA}:N_{FA}:$N_{p\text{-}TI}$) at 25 °C.

2.2. Structural Characteristics of Ugi-FOIT

As a hydrophobic segment, the oleylamine was grafted on the hydrophilic SA skeleton by UGI-4CR reaction. The contents of C, H and N elements in Ugi-FOIT were determined by the elemental analyzer. The contents of various elements in Ugi-FOIT9.5, Ugi-FOIT14.6 and Ugi-FOIT20.4 are shown in Table 2. As seen, the content of N element in raw material SA was close to zero, while the content of N element in Ugi-FOIT grafted with oleylamine significantly increased, so it can be speculated that the oleylamine reacted with SA in Ugi-FOIT via the Ugi-4CR reaction.

Table 2. Elemental contents and DS of Ugi-FOlT.

Sample	C (%)	H (%)	N (%)	DS (%)
SA	30.60	5.96	0.005	-
Ugi-FOlT9.5	33.78	7.26	0.87	9.5
Ugi-FOlT14.6	35.45	7.14	1.20	14.6
Ugi-FOlT20.4	38.34	6.97	1.57	20.4

Figure 2 shows the typical FT-IR of SA, OLA, SA + OLA mixture, and Ugi-FOlT. The raw SA exhibited a strong and wide absorption peak at 3429.28 cm^{-1}, which was attributed to the stretching vibration of hydroxyl groups. The two peaks located at 1620.94 and 1419.62 cm^{-1} belonged to the asymmetric and symmetrical stretching vibrations of carboxyl groups, respectively. Meanwhile, the peaks appearing at 2929.97 and 1034.54 cm^{-1} were assigned to the stretching vibrations of C-H and C-O-C in the uronic acid skeleton, respectively [26]. In addition to the characteristic absorption peak similar to that of SA, the large and wide peak for the synthesized Ugi-FOlT ascribed to the hydroxyl stretching vibration at 3421.25 cm^{-1} was sharped, indicating that intramolecular hydrogen bond on the glucuronic acid chain had been destroyed. At the same time, more obvious and extremely sharp absorption peaks appeared at 2924.95 and 2853.72 cm^{-1} owing to the antisymmetric and symmetric stretching vibration of methylenes in the oleylamine molecular structure ($CH_3(CH_2)_7CH{=}CH(CH_2)_8NH$-) [27]. The above results showed that oleylamine, formaldehyde and p-TI were successfully grafted onto SA's molecular skeleton, thus generating Ugi-FOlT by Ugi-4CR reaction.

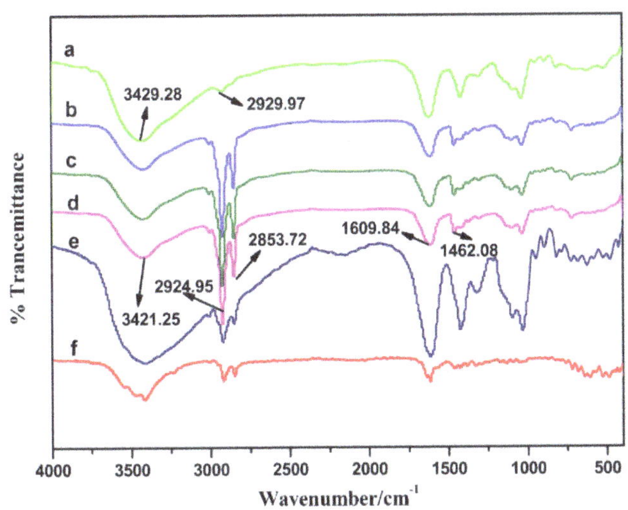

Figure 2. FT-IR spectra of (**a**) SA, (**b**) Ugi-FOlT9.5, (**c**) Ugi-FOlT14.6, (**d**) Ugi-FOlT20.4, (**e**) OLA and (**f**) SA + OLA mixture.

The chemical structure of Ugi-FOlT can be further confirmed by ^1H NMR (400 MHz) analysis. ^1H NMR spectra of oleylamine ($CDCl_3$), p-TI (D_2O), SA (D_2O) and Ugi-FOlT (D_2O) are shown in Figure 3. The strong proton signal peak at δ 4.7 ppm belongs to the proton peak of solvent D_2O. The proton signal peak at δ 3.6~5.1 ppm is attributed to the proton peak on the SA skeleton [28,29]. By comparison, it was found that the signal peaks of protons on the molecular skeleton of SA were retained in the spectra of Ugi-FOlT, indicating that the main skeleton structure of SA was retained in its structure during the modification. In addition, Ugi-FOlT revealed some additional proton signal peaks in the δ 7.0~8.0 ppm and δ 0.6~3.0 ppm ranges. Among them, the proton signal peak at δ 7.41 and 7.27 ppm was due to the proton peak on the benzene ring of p-TI, and the proton signal peak at δ 5.22 ppm was ascribed to the proton peak on the oleylamine's double bond (-CH=CH-). The proton signal peak at δ 2.27 ppm was attributed to the proton peak on methyl (-CH_3) of p-TI, and the proton signal peak at δ 1.92 ppm was assigned to the proton peak on the double-bonded methylene (=C-CH_2-) of oleylamine. The proton signal peaks at δ 1.3~1.2 ppm and 0.81 ppm were owing to the proton peaks of oleylamine's methylene (-CH_2-) and methyl (-CH_3), respectively [30]. The results further confirmed that formaldehyde, oleylamine and p-TI successfully reacted with SA via the Ugi-4CR reaction.

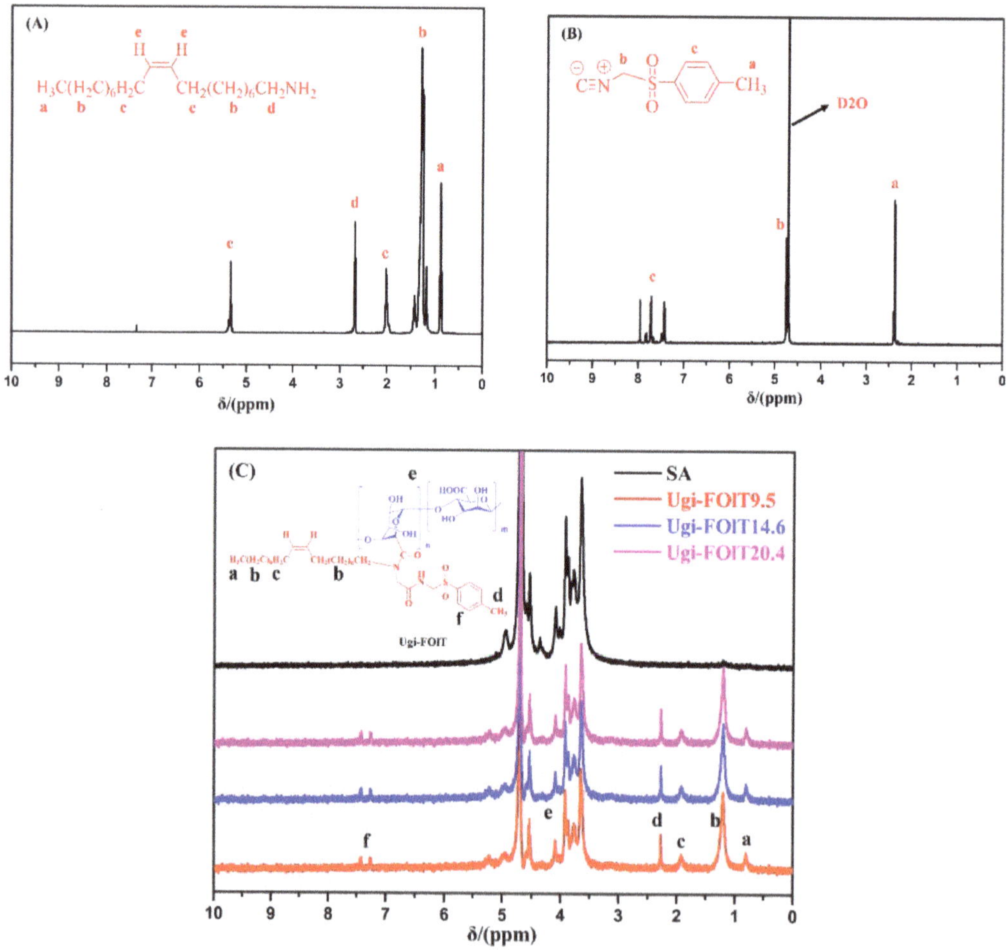

Figure 3. ^{1}H NMR spectra of (**A**) OLA ($CDCl_3$), (**B**) TOSMIC (D_2O) and (**C**) SA, Ugi-FOIT9.5, Ugi-FOIT14.6 and Ugi-FOIT20.4 (D_2O).

TGA is a common thermal analysis method that can be used to determine characteristics such as decomposition reaction, absorbed water content and thermal stability. The thermal stability performance of SA and Ugi-FOIT was analyzed by TGA and DTG measurements. The TGA and DTG curves of SA and UGI-FOLT are shown in Figure 4. It can be seen from the TGA curves that SA and Ugi-FOIT underwent two major stages of weightlessness. Among them, the first weight loss between 50 °C and 150 °C was due to the hydrogen bond between the polar groups (hydroxyl and carboxylic groups) and water, which caused the evaporation of the physical adsorption and wrapped water in the polymer. The second weight loss stage occurred in the range of 200~280 °C, which mainly resulted from the thermal cracking of the polymer molecular chain, which gradually cracked into CO_2, CO and H_2O [31–33]. When the heating temperature exceeded 500 °C, the polymer lost weight completely and a residue of about 20 wt. % was obtained, which may have been caused by carbides formed after the polymer was calcined at high temperatures.

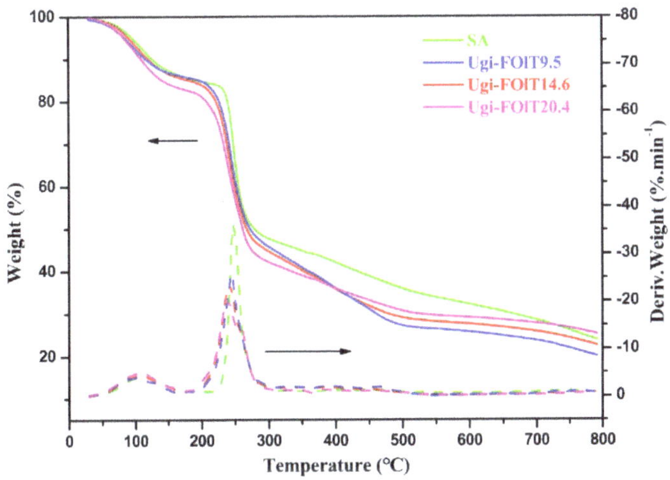

Figure 4. Thermogravimetric analysis and derivative thermograms of SA, Ugi-FOIT9.5, Ugi-FOIT14.6 and Ugi-FOIT20.4.

Moreover, there are two peaks in the DTG curves of SA and Ugi-FOIT, which also indicate that the polymer underwent two major weight loss stages, consistent with the results shown in the TGA curves. In addition, the DTG curves also visually display the initial cracking temperature of the polymer. It can be seen that the initial pyrolysis temperatures of Ugi-FOIT9.5, Ugi-FOIT14.6 and Ugi-FOIT20.4 in the second stage were 240, 238 and 232 °C, respectively, which were lower than that of SA (248 °C), indicating that the introduction of hydrophobic oleylamine group destroyed the intramolecular and intermolecular hydrogen bonds of SA, enhanced the flexibility of the molecular chain and reduced the rigid structure of the polymer, thus decreasing the thermal stability of Ugi-FOIT [34].

Figure 5 shows the XRD patterns of SA and Ugi-FOIT. The XRD pattern of SA exhibited weak and wide diffraction peaks at 2θ = 14.6° and 21.6°, which are attributed to the diffraction peaks of the hydrated crystal structure of SA [35]. In the XRD pattern of Ugi-FOIT, the diffraction peaks corresponding to SA disappeared, and only a strong diffraction peak in the amorphous region located at 2θ = 20° remained, implying that the Ugi-FOIT was the amorphous structure [10]. This further indicated that SA weakened the intermolecular and intramolecular hydrogen bonds during the grafting of the hydrophobic side chain of oleylamine, reduced the rigidity of the molecular chain and enhanced its flexibility, thus reducing the regularity of the polymer molecular chain and increasing the amorphous region in its structure. However, the DS of Ugi-FOIT had no significant effect on its XRD patterns.

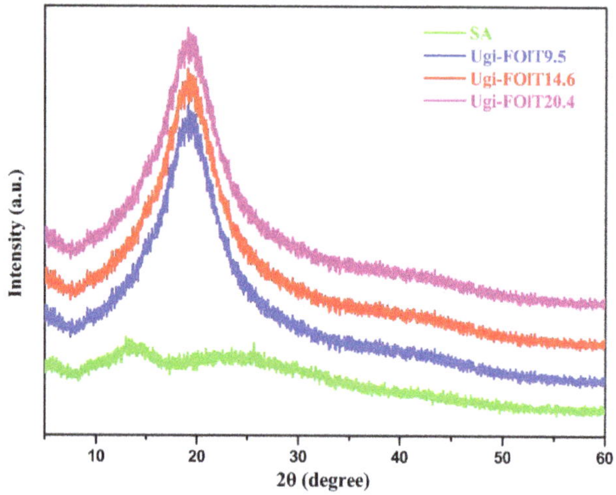

Figure 5. X-ray diffractograms of SA, Ugi-FOIT9.5, Ugi-FOIT14.6 and Ugi-FOIT20.4.

2.3. Self-Assembly Behavior of Ugi-FOIT

2.3.1. Amphiphilic Ugi-FOIT's CMC Value

Amphiphilic Ugi-FOIT polymers are expected to self-assemble into core–shell micellar aggregates in aqueous solution. In the process of self-assembly, the CMC value is an important parameter for characterizing the self-assembly behavior of amphiphilic polymers, which can be used to characterize the formation and thermodynamic stability of micellar aggregates [36]. The pyrene fluorescence probe method is the most commonly used method to measure the CMC value of amphiphilic polymers. With the increase in polymer concentration, pyrene, as a hydrophobic fluorescence probe, enters the hydrophobic core of the micellar structure through non-covalent interaction forces, and the hydrophilic environment in which it is in solution is transformed into a hydrophobic environment, thus causing sensitive changes in the fluorescence intensity ratio I_1/I_3 of pyrene [37]. In other words, the value of I_1/I_3 remains basically unchanged when the polymer concentration is lower than the CMC value, but when the polymer concentration reaches the CMC value, the value of I_1/I_3 decreases sharply, and the concentration corresponding to the inflection point of the resulting curve is the CMC value of the polymer [38]. The change curves of the ratio I_1/I_3 of pyrene fluorescence intensity with the concentration of Ugi-FOIT were shown in Figure 6A, which was similar to our previous results [24]. The CMC values of Ugi-FOIT9.5, Ugi-FOIT14.6 and Ugi-FOIT20.4 obtained by nonlinear fitting were 0.08, 0.06 and 0.043 mg/mL, respectively. They were much lower than the CMC value (2.30 mg/mL) of sodium dodecyl sulfate (SDS), a commonly used low molecular surfactant. For Ugi-FOIT with different DSs, the CMC value decreased with the increase in the DS, as shown in Figure 6C. This may be attributed to the fact that the increase in the grafting rate of hydrophobic groups in Ugi-FOIT was conducive to enhancing the hydrophobic interactions within and between molecules, and Ugi-FOIT was more likely to self-assemble to form micellar aggregates under strong hydrophobic forces, thus exhibiting lower CMC values [39].

In addition, the surface tension method (SFT) is another method for measuring the CMC value of amphiphilic polymers. When amphiphilic polymers begin to self-assemble to form micellar aggregates in aqueous solution, the surface tension at the water/air interface decreases sharply with the increase of their concentration [40]. Figure 6B displays the variations in surface tension of Ugi-FOIT with their concentration. The inflection point of the curves corresponded to the concentration of the polymer's CMC. The CMC values of Ugi-FOlT9.5, Ugi-FOlT14.6 and Ugi-FOlT20.4 were 0.091, 0.072 and 0.051 mg/mL, respectively, and the corresponding surface tensions were, respectively, 61.0, 60.5 and 58.5 mN/m. These results are similar to the CMC values determined by the pyrene fluorescence probe method.

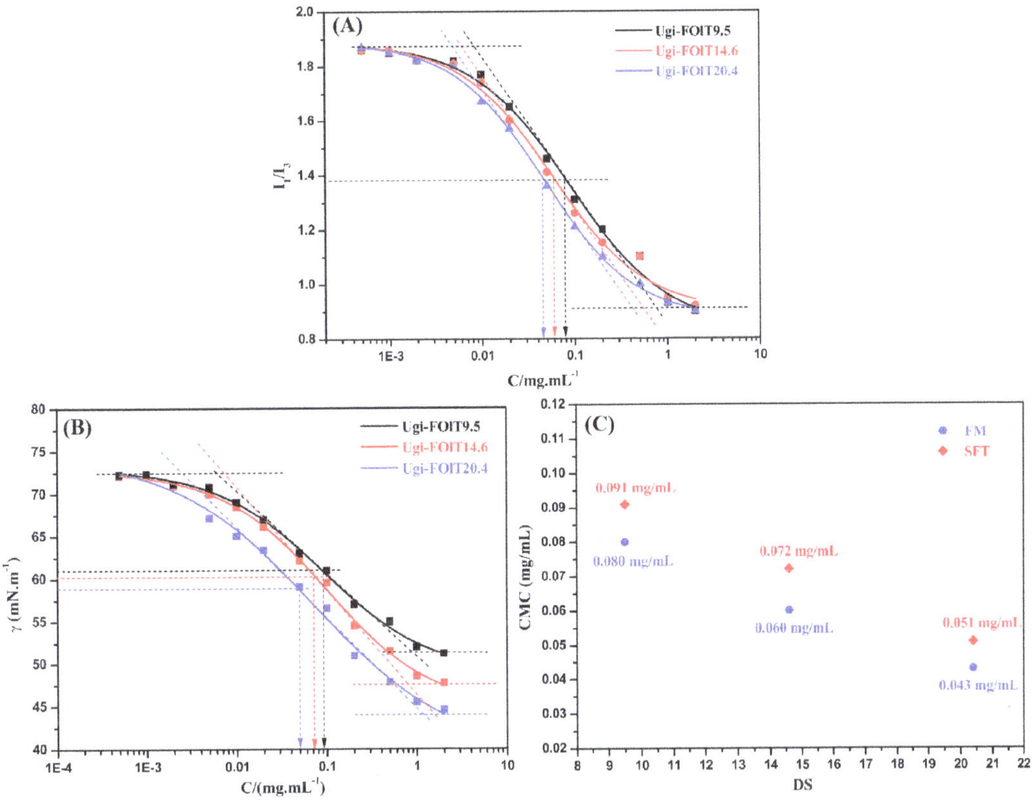

Figure 6. (**A**) Plots of pyrene fluorescence intensity ratio I_1/I_3 and (**B**) surface tension as a function of the concentration of Ugi-FOlT at 25 °C; (**C**) CMC values of Ugi-FOlT with different DSs determined by FM and SFT.

2.3.2. Colloidal Properties of Ugi-FOlT Self-Assembled Micelles

The structure of Ugi-FOlT was composed of a hydrophilic SA main chain and hydrophobic oleylamine chain segment. The hydrophobic long carbon chain aggregated inward under the intramolecular and intermolecular hydrophobic forces, and the hydrophilic SA main chain moved outward towards the aqueous solution under the action of hydrogen bonding, etc., and self-assembled to form core–shell micelle aggregates. The particle size, PDI and zeta potential of Ugi-FOlT micelle aggregates were determined by Malvern laser particle size meter, with the results shown in Table 3 and Figure 7.

Table 3. Sizes, PDI and Zeta potentials of Ugi-FOlT micellar aggregates ($n = 3$).

Sample	Size (nm)	PDI	Zeta Potential (mV)
Ugi-FOlT9.5	710 ± 40	0.38 ± 0.03	−58.2 ± 1.9
Ugi-FOlT14.6	670 ± 35	0.35 ± 0.05	−53.5 ± 2.1
Ugi-FOlT20.4	653 ± 25	0.31 ± 0.04	−48.9 ± 2.8

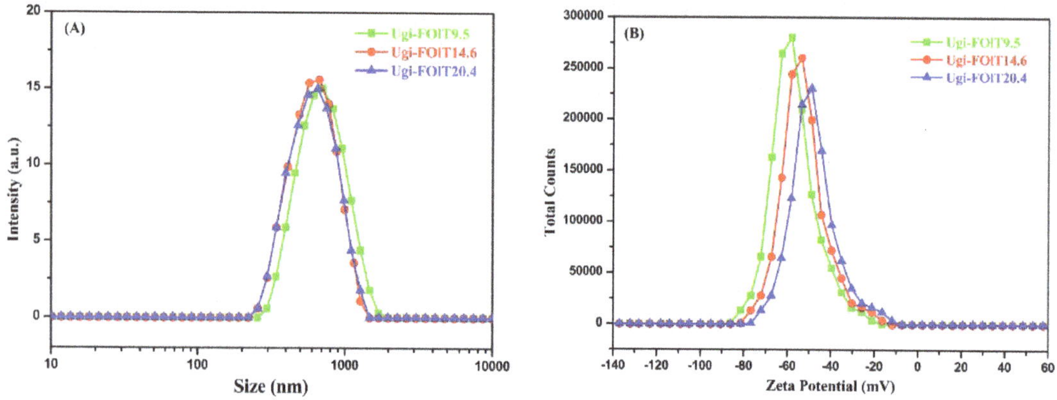

Figure 7. Size (**A**) and zeta potential (**B**) distribution of Ugi-FOIT9.5, Ugi-FOIT14.6 and Ugi-FOIT20.4 micellar aggregates.

As shown in Figure 7A, the particle size distribution of Ugi-FOIT micellar aggregates presented relatively wide particle size distribution, ranging from 250 nm to 1300 nm, which was narrower than that of octyl-grafted amphiphilic alginate amide derivative (OAAD) microcapsules (in the range of 100~20,000 nm) [26] and close to that of hexyl alginate ester derivative (HAED) self-aggregates (in the range of 130~1100 nm) [41]. The average particle sizes of Ugi-FOIT9.5, Ugi-FOIT14.6 and Ugi-FOIT20.4 micellar aggregates were 710 $\pm$ 40, 670 $\pm$ 35 and 653 $\pm$ 25 nm (PDI= 0.3~0.4), respectively, which were slightly higher than that of hydrophobically modified alginate derivative (HMAD) self-aggregated micelles [21] and oleoyl alginate ester (OAE) nanoparticles [42]. The main reason for the relatively wide particle size distribution and large average particle size of the above Ugi-FOIT micellar aggregates was that Ugi-FOIT was an ill-defined polymer with the broad molecular weight distribution that had a great influence on the particle size and distribution of its assembled micelles. Meanwhile, some possible micelle aggregation would also increase the hydrodynamic particle size of Ugi-FOIT micellar aggregates after filtration. Furthermore, the particle size of Ugi-FOIT micellar aggregates measured by Malvern laser particle size analyzer is the hydrodynamic particle size in aqueous media, which is defined as the actual diameter of the Ugi-FOIT micellar aggregates plus the size of the water shell formed by interactions between hydrophilic segments and the continuous water phase [43]. The Ugi-FOIT micellar aggregates in solution were surrounded by water molecules and were in an expanded state, making their particle size enlarged [44]. Moreover, Ugi-FOIT possesses negatively charged carboxyl groups, and the electrostatic repulsion of the negatively charged carboxyl groups make the Ugi-FOIT micellar aggregates expand, resulting in an enhancement in their particle size.

The average particle size of Ugi-FOIT20.4 was lower than that of Ugi-FOIT9.5 and Ugi-FOIT14.6, which may be attributed to the stronger hydrophobic and π–π interaction forces of Ugi-FOIT with high oleylamine substitution in aqueous solution, which led to the formation of more compact micelle aggregates [45,46].

A large number of reports have shown that micellar aggregates with higher zeta potential are less likely to agglomerate due to larger electrostatic repulsion between particles, and when the absolute value of the zeta potential of micellar aggregates in solution is higher than 30 mV, they are considered to have good stability [47]. In addition, micellar aggregates with negatively charged surfaces can avoid binding with proteins and cells in the body, which is conducive to extending their circulation time in the body [48]. On the contrary, the cell internalization of such particles would be impeded since the cell membrane is also negatively charged. The zeta potential distribution on the surface of Ugi-FOIT micellar aggregates is shown in Figure 7B. The zeta potential values of Ugi-FOIT9.5, Ugi-FOIT14.6 and Ugi-FOIT20.4 micellar aggregates were −58.2 ± 1.9, −53.5 ± 2.1 and −48.9 ± 2.8 mV, respectively, revealing high negative potential values. This was due to the presence of a large number of $-COO^-$ groups on the SA molecular skeleton. The results implies that the Ugi-FOIT micellar aggregates possessed good stability to some extent.

Figure 8 presented the TEM image of Ugi-FOIT micelle aggregates. It was observed that Ugi-FOIT exhibited spheroidal structures with different sizes in aqueous solution, indicating that the hydrophobic modified polymers were amphiphilic and able to self-assemble into micellar structures in aqueous solution. The particle size of Ugi-FOIT9.5, Ugi-FOIT14.6 and Ugi-FOIT20.4 micelle aggregates measured by TEM were 314, 355 and 362 nm, respectively, which were lower than that measured by DLS. This phenomenon was attributed to the fact that the particle size observed by TEM was the actual size of micelle aggregates in the completely dry state, and dry Ugi-FOIT micelle aggregates would shrink due to water evaporation. However, the particle size of Ugi-FOIT micellar aggregates measured by DLS is the hydrodynamic diameter of Ugi-FOIT micelle aggregates in solution, which were surrounded by water molecules, making them in the expanded state; the electrostatic repulsion of the negatively charged carboxyl groups made the Ugi-FOIT micellar aggregates expand, resulting in the enlarged particle size [44].

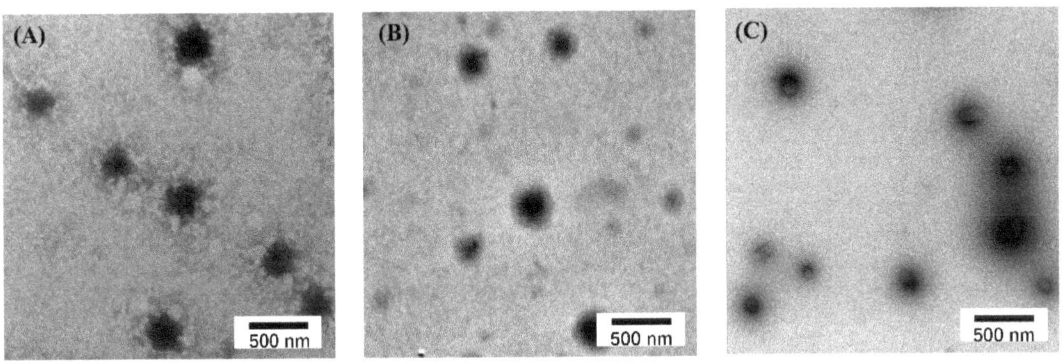

Figure 8. TEM images of (**A**) Ugi-FOIT9.5, (**B**) Ugi-FOIT14.6 and (**C**) Ugi-FOIT20.4 micelle aggregates.

The stability of Ugi-FOIT micelle aggregates in dispersive media can be analyzed by measuring the change in particle size with time. As shown in Figure 9, after incubating in PBS at pH 7.4 for 40 h, the particle sizes of both Ugi-FOIT9.5 and Ugi-FOIT14.6 micellar aggregates increased to varying degrees, while the particle sizes of Ugi-FOIT20.4 micellar aggregates increased slightly. The results show that the stability of the Ugi-FOIT20.4 micellar aggregates was better than that of Ugi-FOIT9.5 and Ugi-FOIT14.6 micellar aggregates, which may be attributed to the existence of more hydrophobic interactions between the hydrophobic chains of Ugi-FOIT with high DS, which was helpful in the formation of a more compact hydrophobic core, thereby enhancing the stability of its micellar aggregates [23]. Additionally, the optical graphs of blank Ugi-FOIT14.6 and IBU loaded Ugi-FOIT14.6 micellar aggregates incubated in PBS at pH 7.4 for 1 h and 40 h are displayed in Figure 10. It can be found that the drug-loaded Ugi-FOIT14.6 micelle aggregate solution exhibited a darker color than the blank Ugi-FOIT14.6 micelle aggregate solution, and after 40 h incubation, the appearance of the Ugi-FOIT14.6 micelle aggregate solution had no change, revealing its excellent stability.

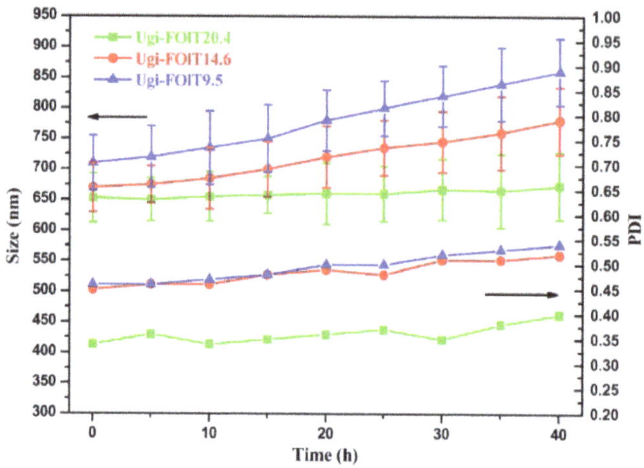

Figure 9. Size distribution of Ugi-FOIT micellar aggregates incubated in PBS at pH 7.4 for 40 h.

Figure 10. Optical graphs of blank Ugi-FOIT14.6 and IBU loaded Ugi-FOIT14.6 micellar aggregates incubated in PBS at pH 7.4 for 1 h and 40 h.

2.4. In Vitro Release Behavior of IBU for Ugi-FOIT Micelles

IBU is an early-developed non-steroidal antipyretic, analgesic and anti-inflammatory drug of phenylpropionic acid, which is the only recommended antipyretic drug for children by the WHO and the US FDA [31]. As a kind of hydrophobic model drug used in the preparation of oral formulations, IBU is easily soluble in organic solvents such as ethanol, ethylene glycol, propylene glycol and ether, but its solubility in water is very low. Poor water solubility may be one of the reasons affecting the absorption of IBU. Amphiphilic Ugi-FOIT could self-assemble into micellar aggregates with hydrophobic core and hydrophilic shell in aqueous solution, and IBU was able to be embedded in its hydrophobic microregion through physical adsorption and interaction between hydrophobic chains to achieve solubilization effect. The encapsulation efficiencies (EE) of Ugi-FOIT9.5, Ugi-FOIT14.6 and Ugi-FOIT20.4 micellar aggregates were 78.24%, 80.65% and 81.30%, respectively. It is known that the EE values of micellar aggregates are directly affected by the concentration of Ugi-FOIT, the mass ratio of IBU/Ugi-FOIT and the DS of Ugi-FOIT [23,44]. Among them, the DS of FOIT involved the grafting of hydrophobic side groups that reflected the affinity with hydrophobic anti-inflammatory drugs, which has the greatest influence on the loading of hydrophobic drugs [41]. With the increase in FOIT's DS, the EE values of micellar aggregates increased, which indicated that the hydrophobic inner cavity constructed by more hydrophobic side groups was more conducive to the encapsulation of drug.

The in vitro drug release behavior of Ugi-FOIT micellar aggregates of IBU under simulated physiological conditions was analyzed by dialysis [49]. The cumulative release curves of IBU for Ugi-FOIT micellar aggregates in PBS at 37 °C with pH 5.0 and 7.4 are shown in Figure 11. The results show that the drug-loaded micelle aggregates could release the drug slowly and continuously in PBS, but the pH value of the releasing medium and the DS of Ugi-FOIT had a certain effect on the cumulative release rate of IBU.

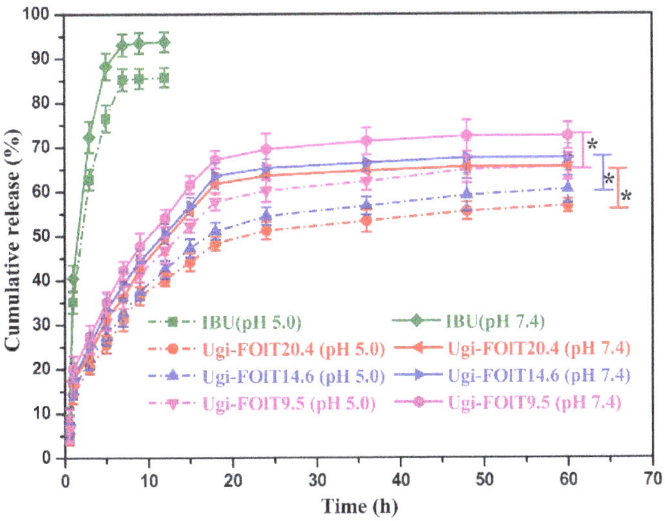

Figure 11. In vitro cumulative release profiles of free IBU and IBU loaded Ugi-FOIT micellar aggregates in PBS (pH 7.4, 5.0) containing Tween 80 (0.5%, *w/v*) at 37 °C. * $p < 0.05$: significantly different between pH 7.4 and 5.0 groups at 60 h.

In the PBS release medium with pH 7.4, the cumulative release rate of free IBU reached up to 93% within 7 h, which was attributed to the fact that free IBU was released only by the concentration gradient of the drug, while the concentration of free IBU in the dialysis bag was higher, so the release rate was faster. Within 3 h after the initial release of Ugi-FOIT micellar aggregates, about 23.4%~27.5% of the release amount was achieved by the free diffusion of IBU physically adsorbed on or near the surface of the micellar aggregates, and the release rate of IBU in this part was faster. The rest of the IBU distributed in the hydrophobic core of micelle aggregates needed to break down the hydrophobic force through concentration difference and release slowly and continuously along with the degradation and dissolution of the carrier material, so the cumulative release rate within 7 h was only 36.2%~42.3% [50]. In the sustained release after 3 h, the release rate of IBU in the Ugi-FOIT micellar aggregates was lower than that in the Ugi-FOIT14.6 and Ugi-FOIT9.5 micellar aggregates, which may be due to the self-assembly of Ugi-FOIT20.4 in aqueous solution to form more compact micellar aggregates. Its hydrophobic cavities were more hydrophobic towards IBU molecules, resulting in a slower rate of IBU release [51].

In addition, the cumulative release rate of IBU for Ugi-FOIT micellar aggregates in PBS with pH 7.4 was significantly higher than that in PBS with pH 5.0 at 60 h. This may be attributed to the higher degree of deprotonation of carboxyl functional groups on the Ugi-FOIT molecular chains in the alkaline release medium, which enhanced the hydrophilicity of micellar aggregates and made it easier to absorb water and expand. In addition, due to the formation of more negative carboxylate ions on the molecular chains of Ugi-FOIT, the increase in inter-chain repulsion would cause the polymer chain to form a tensile conformation, resulting in the relaxation of the polymer chains and the destruction of hydrophobic microregions, resulting in an accelerated release rate. However, in the more acidic release medium, more compact micellar aggregates were formed due to the protonation of carboxyl groups on the molecular chains of Ugi-FOIT, which slowed down the release rate of IBU. These results indicate that Ugi-FOIT micellar aggregates possessed a certain pH-responsive release behavior, which could be used as an effective oral formulation. Moreover, it was feasible to control the drug release rate by regulating the DS of hydrophobic segments in amphiphilic Ugi-FOIT.

2.5. In Vitro Cytotoxicity of Ugi-FOIT

Due to the special core–shell structures, the Ugi-FOIT micellar aggregates could trap hydrophobic drugs in their hydrophobic cores, which has a wide range of application prospects in drug delivery for oral formulations. However, the polymer micellar aggregates must have good biocompatibility to realize their applications in the biomedical field. The murine macrophage RAW264.7 cells are considered as a representative cell line for various types of macrophages routinely used in many studies. Since RAW264.7 cells have been reported to retain many of the characteristics of macrophages in vivo that have an important role in innate immune defense against microbial infections, which involves secretion of pro-inflammatory mediators and phagocytic activities, we used RAW264.7 cells to evaluate the cytotoxicity of Ugi-FOIT for oral formulations [52]. In this study, the MTT method was used to examine the biological toxicity of Ugi-FOIT with various DSs on RAW 264.7 macrophages, with the results displayed in Figure 12. As seen, after the Ugi-FOIT had been incubated with RAW 264.7 cells for 48 h in the concentration range of 0~500 μg/mL, the cell survival rate decreased slightly with the increase in polymer concentration but remained above 90%. This result indicates that Ugi-FOIT with various DSs had low cytotoxicity to RAW 264.7 cells, exhibiting good biocompatibility [53].

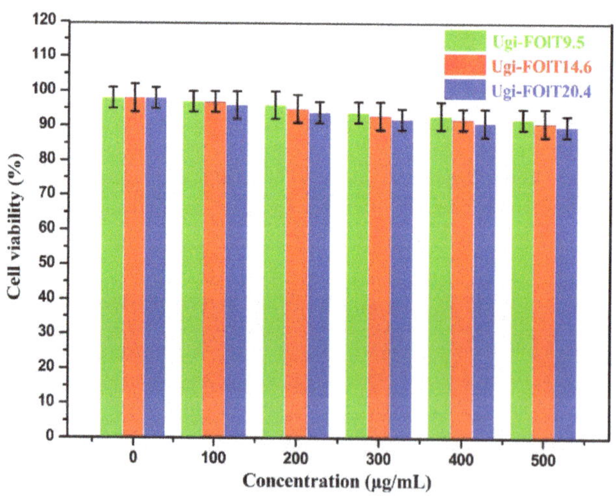

Figure 12. In vitro cytotoxicity of Ugi-FOIT at various concentrations against RAW 264.7 cells after incubation for 48 h.

3. Materials and Methods

3.1. Materials

Sodium alginate (SA, Mw = 135,600 Da, Mn = 102,000 Da) was prepared by NaOH neutralization of alginate based on our previous method and the molecular weight was measured by gel permeation chromatography (GPC, Waters e2695, Santa Barbara, CA,

USA) [11]. p-phenylmethylsulfonyl isonitrile (p-TI), oleylamine (90%), formaldehyde (37 wt. % in H_2O) and ibuprofen (IBU, 98%) were purchased from Shanghai Aladdin Chemical Reagent Co., Ltd. (Aladdin, Shanghai, China). Other reagents were obtained from Guangzhou Chemical Reagent Co., Ltd. (Guangzhou, China). They were of analytical grade and used without further purification. RAW 264.7 cells and biochemical reagents used in the experiments and assays for cytotoxicity were obtained from Gibco (Thermo Fisher Scientific, Carlsbad, CA, USA).

3.2. Synthesis and Optimization of Process Conditions of Ugi-FOlT

The synthesis of sodium alginate-g-oleylamine polymers (Ugi-FOlT) was performed as we previously reported [10]. The Ugi-4CR reaction of formaldehyde, oleylamine, SA and p-TI was carried out using a one-pot method at room temperature, and the schematic synthesis routes are shown in Scheme 1.

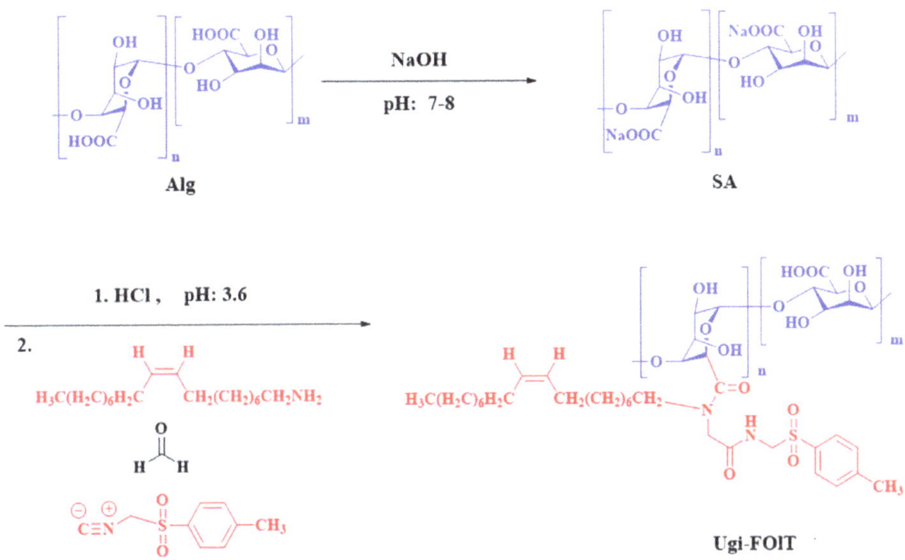

Scheme 1. Schematic illustration of synthesis routes of Ugi-FOlT.

In detail, 1.0 g (5.05 mmol) of SA was precisely weighed and completely dissolved in 80 mL deionized water under mechanical agitation in a three-orifice flask to prepare the SA aqueous solution with a mass fraction of 1.25%, and then the pH of this solution was adjusted to 3.6 by dropping a certain amount of 0.1 mol/L HCl aqueous solution. Subsequently, 1.01 mL (6.1 mmol) oleylamine and 0.46 mL (6.1 mmol) formaldehyde solutions were pre-added to a round-bottom flask, which were magnetically stirred at room temperature for 30 min before being added to the prepared aqueous SA solution under mechanical agitation. Then, 1.38 g (7.1 mmol) p-TI was incorporated into the reaction solution. After continuous stirring at room temperature for 24 h, 5 times the volume of absolute ethanol was introduced to the reaction solution to completely precipitate the reaction product. After centrifugation, the precipitate was loaded into a dialysis bag with MWCO 3500 and dialyzed against deionized water for 3 days, and the deionized water was replaced three times a day. The final dialysis-purified product was freeze-dried to obtain dried Ugi-FOlT.

The degree of substitution (DS) of Ugi-FOlT was determined by elemental analysis according to our previous method [24]. The same batch of SA was used to prepare Ugi-FOlT by reacting with oleylamine (OLA), formaldehyde (FA) and p-TI through Ugi-4CR. The effects of reaction parameters such as molar ratio of raw materials, reaction time, reaction temperature and pH value of solution on the DS of Ugi-FOIT were explored via controlling single-factor mode. The DS was used as the target factor to optimize the parameter conditions of the Ugi-4CR reaction in order to achieve simple and efficient synthesis of Ugi-FOIT.

3.3. Structural Characterization of Ugi-FOIT

The structure of Ugi-FOIT was characterized by FT-IR, ^{1}H NMR, TGA and XRD. FT-IR spectra of the sample was recorded on a Nicolet-6700 (Thermo Fisher Scientific, Carlsbad, CA, USA) with KBr pellets in the range of wavenumbers between 4000 and 400 cm^{-1} for 64 scans with a spectral resolution of 2.0 cm^{-1}. ^{1}H NMR spectra were recorded at 25 °C using a ULTRASHIELD 400 PLUS spectrometer (Bruker, Berne, Switzerland) operating at 400 MHz with deuterated water (D_2O) as solvent and tetramethylsilane (TMS) as an internal standard; the concentration of the sample was approximately 8.0~10.0 mg/mL. The thermal stability was studied by means of a 449F3 thermogravimetric analyzer (Netzsch, Selb, Germany) with aluminum crucibles as a sample holder; 8.0~12.0 mg sample was required for the measurement, and the temperature was increased from 25 to 800 °C at a heating rate of 20 °C/min under the N_2 atmosphere. The XRD pattern of the sample was determined on an AXS/D8 X-ray diffractometer (Bruker, Karlsruhe, Germany) equipped with graphite monochromatized high-intensity Cu-Kα radiation (λ = 0.154 nm, 40 kV, 100 mA), in the range 2θ = 5°~60° and with a scan rate of 5 °/min.

3.4. Determination of the Critical Micelle Concentration Value of Ugi-FOIT

According to previous reports, both the pyrene fluorescent probe method and surface tension method can measure the critical micelle concentration (CMC) value of Ugi-FOIT [37].

The fluorescence spectroscopy of Ugi-FOIT was recorded using pyrene as a fluorescence probe on a fluorescence spectrophotometer (F7000 spectrophotometer, Hitachi, Tokyo, Japan); 0.15 mol/L aqueous NaCl solution was used as the solvent to prepare the sample solution, which was beneficial to facilitating the self-assembly of polymer samples. The CMC value can be obtained by plotting the curves of the pyrene fluorescence intensity ratio of the first (373 nm) and the third (384 nm) (I_1/I_3) as a function of polymer concentration in the emission spectra of pyrene. This method is regarded as an effective way to determine the self-assembled behavior of amphiphilic polymers. The surface tension measurement of Ugi-FOIT solution with various concentrations was performed on a Dataphysics OCA15EC tension meter (Bruker, Karlsruhe, Germany) using the hanging drop method. The final data were the average value of three measurements.

3.5. Preparation and Colloidal Properties of Ugi-FOIT Micelles

The Ugi-FOIT micelles were prepared by directly dispersing Ugi-FOIT in deionized water followed by sonication to facilitate aqueous dispersion and self-assembly. Briefly, the Ugi-FOIT samples were dissolved in distilled water to achieve a micellar solution with the concentration of 0.5 mg/mL at 25 °C. The solution was then sonicated for 4 min by a probe sonicator at 120 W and repeated 3 times to guarantee optically clear dispersion. The sonication pulse was set for 2 s with a waiting time of 4 s between pulses. Finally, the micellar solution was further filtered through a 0.45 µm syringe filter and stored at 4 °C.

The morphology, size and zeta potential of the Ugi-FOIT micelles were examined by TEM and DLS measurement. The morphology of the Ugi-FOIT micelles were examined by a transmission electron microscope (JEM 2100 TEM, JEOL Co., Akishima, Japan) after the sample solution was evaporated on a carbon-coated copper grid. The hydrodynamic diameter and zeta potential of Ugi-FOIT micelles were examined by using a dynamic light scattering meter (Malvern Nano-ZS90 Zetasizer, Malvern, UK) at 25 °C. Additionally, to evaluate the storage stability of the prepared Ugi-FOIT micelles, they were dispersed in PBS (pH 7.4) and shaken at 100 rpm at 37 °C for 5, 10, 15, 20, 25, 30, 35 and 40 h. The stability of the Ugi-FOIT micelles was analyzed by measuring the changes in particle size and PDI with time.

3.6. Loading and In Vitro Release of Ibuprofen

The encapsulation of ibuprofen (IBU) into the Ugi-FOIT micelles was achieved by the ultrasound method [54]. A certain amount of IBU was dissolved in anhydrous ethanol to prepare an ethanol solution with a drug concentration of 3.0 mg/mL. Then, 1 mL of this ethanol solution and 2 mL of 2.5 mg/mL Ugi-FOIT micellar solution were mixed well under high speed stirring to form the drug-loaded micelles, followed by dialysis in water to remove the ethanol. In addition, IBU solutions at concentrations of 5.0, 10.0, 15.0, 20.0, 30.0 and 40.0 µg/mL were prepared with PBS at pH 7.4 and 5.0, respectively. The absorbance (Abs) at the maximum absorption wavelength of 222 nm was measured using a U-3900 UV–vis spectrophotometer (Hitachi, Tokyo, Japan). The concentration (C) of IBU was used as the abscissa, and the absorbance value (A) was used as the ordinate for data regression. After linear fitting, its standard curve was obtained, and the results are shown in Table 4. Subsequently, the amount of unencapsulated IBU was measured by the biphasic method [31]. After high-speed centrifugation (12,000 rpm) of the drug-loaded Ugi-FOIT micellar solution, the unencapsulated IBU in the supernatant aqueous solution could be extracted by chloroform. Furthermore, the amount of unencapsulated IBU could be determined by the abovementioned UV–vis spectrophotometer with the standard curve.

Table 4. Standard curve equations and correlation coefficients of IBU in PBS (pH 7.4 and 5.0).

Solution	λ_{max} (nm)	Standard Curve Equation	R^2
PBS (pH 7.4)	222	A = 0.04909C + 0.02814	0.99928
PBS (pH 5.0)	222	A = 0.01520C + 0.06280	0.99683

So, the Encapsulation Efficiency (EE) for the drug-loaded Ugi-FOIT micelles can be calculated using Equation (1) [31]:

$$Encapsulation\ efficiency\ (EE) = \frac{Total\ IBU - Unencapsulated\ IBU}{Total\ IBU} \times 100\% \quad (1)$$

The in vitro release behavior of *IBU* from drug-loaded Ugi-FOIT micelles was investigated by a commonly used dialysis method at 37 °C with a medium of PBS buffer (pH 5.0 and 7.4). Tween 80 (0.5%, *w*/*v*) used as a surfactant was added to the PBS media to maintain the sink conditions [55]. Briefly, 1.0 mL of drug-loaded Ugi-FOIT micelles was placed in dialysis bag (MWCO 3500 Da) and then immersed absolutely in 30 mL release medium. The whole release system was incubated in a thermostat water bath and shaken at 100 rpm at 37 °C. At the predetermined time point, 3 mL of the release medium was withdrawn, and the same volume of fresh release medium was then added to maintain the constant volume. The released amounts of *IBU* were measured by UV–vis spectrophotometer with the corresponding standard calibration curves. For comparison, similar release experiments were performed with free *IBU*, with the same amount as found in the drug-loaded Ugi-FOIT micelles. All experiments were performed in triplicate. The percentage of cumulative drug release was calculated from Equation (2):

$$E_r = \frac{V_e\sum_1^{n-1} C_i + V_b C_n}{m_{loaded\ IBU}} \times 100\% \quad (2)$$

where V_e is the volume of sample extracted from the release medium each time (1.0 mL); V_b is the total volume of release medium (30.0 mL); C_i and C_n refer to the concentration of IBU extracted from release medium on the *i*th and nth time, respectively; n is the number of times to withdraw the sample solution; $m_{loaded\ IBU}$ is the weight of *IBU* previously loaded in the Ugi-FOIT micelles.

3.7. Cytotoxicity of Ugi-FOIT

RAW 264.7 macrophages were used for an MTT assay to evaluate the cytotoxicity of Ugi-FOIT in vitro [56]. RAW 264.7 macrophages were first cultured in DMEM supplemented with 10% fetal bovine serum, 100 U/mL penicillin and 100 μg/mL streptomycin in an incubator at 37 °C and 5% CO_2 saturated humidity. Cells seeded in 96-well plates (4.0×10^3 cells/well) were incubated in an incubator for 24 h. Then, 100 μL of polymer solution (0, 100, 200, 300, 400 and 500 μg/mL) prepared entirely in DMEM was added to the 96-well plate (100 μL DMEM was added in the control group) and incubated for 48 h. Then, 20 μL of MTT solution in PBS (5.0 mg/mL) was added to each well for incubation for 4 h. The medium and MTT solution in the well were removed, and 100 μL of DMSO was added to dissolve the purple methyl chain crystals after washing with PBS. Finally, the 96-well plate was placed on an oscillator for 10 min at low speed, and the absorbance value of each well at a wavelength of 570 nm was measured with a Multiskan MK3 microplate reader (Thermo Fisher Scientific, Carlsbad, CA, USA). The cell viability was calculated by the ratio of the absorbance of the sample to that of the control group with Equation (3):

$$Cell\ viability = \frac{Abs_{570\ \mathrm{nm}}\ sample - Abs_{570\ \mathrm{nm}}\ blank}{Abs_{570\ \mathrm{nm}}\ control - Abs_{570\ \mathrm{nm}}\ blank} \times 100\% \tag{3}$$

where $Abs_{570\ \mathrm{nm}}$ blank was the absorbance of PBS at 570 nm without cells. $Abs_{570\ \mathrm{nm}}$ sample and $Abs_{570\ \mathrm{nm}}$ control were the absorbance at 570 nm in the presence and in the absence of sample treatment, respectively.

3.8. Statistical Analysis

SPSS 12.0 statistical software was applied to process and analyze the data, and the results are expressed as mean ± standard deviation. Single-factor ANOVA was used for comparison of variables, and $p < 0.05$ was considered statistically significant.

4. Conclusions

In conclusion, to achieve the optimal hydrophobic oral formulation for biomedical applications, we further optimized the synthesis process parameters of Ugi-FOIT and explored the effects of different DSs on the molecular self-assembly properties and self-assembled nanomicelles' colloidal properties of Ugi-FOIT, as well as the in vitro cytotoxicity and drug release behavior of Ugi-FOIT. The p-TI was used as the isonitrile component of the reaction to synthesize amphiphilic Ugi-FOIT with SA as the hydrophilic main chain, and oleylamine as the hydrophobic side chain. The synthesized Ugi-FOIT possessed good amphiphilic properties, spherical micellar aggregates with particle sizes in the range of 653 ± 25~710 ± 40 nm were successfully prepared by the ultrasonic-assisted dissolution method, and their zeta potentials ranged from −58.2 ± 1.92 to −48.9 ± 2.86 mV, exhibiting good stability. In addition, the in vitro release of Ugi-FOIT micellar aggregates encapsulated in IBU showed an initial release phase and a continuous slow release process. To note, the carboxyl groups on the molecular skeleton of Ugi-FOIT showed different hydrophilicity at different pH values, which also had a certain effect on the release of IBU in vitro. At the same time, Ugi-FOIT displayed good biocompatibility, with a cell survival rate above 90%. These results indicate that the self-assembled amphiphilic Ugi-FOIT micelles have potential value for the delivery of hydrophobic drugs for oral formulations.

Author Contributions: Writing—original draft preparation, Y.B.; data curation, X.C.; software, T.W.; conceptualization, R.Z.; writing—review and editing, H.Y.; supervision, Q.L. All authors have read and agreed to the published version of the manuscript.

Funding: This research was funded by the Key Research and Development Project of Hainan Province, ZDYF2023SHFZ124; the China Scholarship Council, 202208460007; the National Natural Science Foundation of China, 51963009; the Innovation and Scientific Research Projects for Graduates of Hainan Province, Qhyb2023-141 and Qhys2023-335; the Open Fund for Innovation and Entrepreneurship of College Students of Hainan Province; and the Open Fund for Innovation and

Entrepreneurship of College Students of Hainan Normal University. The APC was funded by the Key Research and Development Project of Hainan Province, ZDYF2023SHFZ121.

Institutional Review Board Statement: Not applicable.

Informed Consent Statement: Informed consent was obtained from all subjects involved in this study.

Data Availability Statement: The original contributions presented in this study are included in the article; further inquiries can be directed to the corresponding author.

Conflicts of Interest: There are no conflicts to declare.

References

1. Kubiak, T. Polymeric capsules and micelles as promising carriers of anticancer drugs. *Polym. Med.* **2022**, *52*, 37–50. [CrossRef] [PubMed]
2. Hu, S.; Yan, J.; Yang, G.; Ma, C.; Yin, J. Self-Assembled Polymeric Materials: Design, Morphology, and Functional-Oriented Applications. *Macromol. Rapid Commun.* **2022**, *43*, 2100791. [CrossRef] [PubMed]
3. Foroughi-Nia, B.; Barar, J.; Memar, M.Y.; Aghanejad, A.; Davaran, S. Progresses in polymeric nanoparticles for delivery of tyrosine kinase inhibitors. *Life Sci.* **2021**, *278*, 119642. [CrossRef] [PubMed]
4. Bhia, M.; Motallebi, M.; Abadi, B.; Zarepour, A.; Pereira-Silva, M.; Saremnejad, F.; Santos, A.C.; Zarrabi, A.; Melero, A.; Jafari, S.M.; et al. Naringenin Nano-Delivery Systems and Their Therapeutic Applications. *Pharmaceutics* **2021**, *13*, 291. [CrossRef] [PubMed]
5. Fan, Y.; Liu, Y.; Wu, Y.; Dai, F.; Yuan, M.; Wang, F.; Bai, Y.; Deng, H. Natural polysaccharides based self-assembled nanoparticles for biomedical applications—A review. *Int. J. Biol. Macromol.* **2021**, *192*, 1240–1255. [CrossRef] [PubMed]
6. Mizrahy, S.; Peer, D. Polysaccharides as building blocks for nanotherapeutics. *Chem. Soc. Rev.* **2012**, *41*, 2623–2640. [CrossRef] [PubMed]
7. Manivasagan, P.; Oh, J. Marine polysaccharide-based nanomaterials as a novel source of nanobiotechnological applications. *Int. J. Biol. Macromol.* **2016**, *82*, 315–327. [CrossRef] [PubMed]
8. Prajapati, S.K.; Jain, A.; Jain, A.; Jain, S. Biodegradable polymers and constructs: A novel approach in drug delivery. *Eur. Polym. J.* **2019**, *120*, 109191. [CrossRef]
9. Lee, K.Y.; Mooney, D.J. Alginate: Properties and biomedical applications. *Prog. Polym. Sci.* **2012**, *37*, 106–126. [CrossRef] [PubMed]
10. Yang, L.Q.; Zhang, B.F.; Wen, L.Q.; Liang, Q.Y.; Zhang, L.-M. Amphiphilic cholesteryl grafted sodium alginate derivative: Synthesis and self-assembly in aqueous solution. *Carbohydr. Polym.* **2007**, *68*, 218–225. [CrossRef]
11. Emami, Z.; Ehsani, M.; Zandi, M.; Foudazi, R. Controlling alginate oxidation conditions for making alginate-gelatin hydrogels. *Carbohydr. Polym.* **2018**, *198*, 509–517. [CrossRef] [PubMed]
12. Sun, J.; Tan, H. Alginate-Based Biomaterials for Regenerative Medicine Applications. *Materials* **2013**, *6*, 1285–1309. [CrossRef] [PubMed]
13. Zhang, H.T.; Cheng, J.Q.; Ao, Q. Preparation of alginate-based biomaterials and their applications in biomedicine. *Mar. Drugs* **2021**, *19*, 264. [CrossRef] [PubMed]
14. Gheorghita Puscaselu, R.; Lobiuc, A.; Dimian, M.; Covasa, M. Alginate: From Food Industry to Biomedical Applications and Management of Metabolic Disorders. *Polymers* **2020**, *12*, 2417. [CrossRef] [PubMed]
15. Ugi, I. The α-addition of immonium ions and anions to isonitriles accompanied by secondary reactions. *Angew. Chem. Int. Ed. Engl.* **1962**, *1*, 8–21. [CrossRef]
16. Xu, M.H. Asymmetric Ugi four-component reaction catalyzed by chiral phosphoric acid. *Chin. Sci. Bull.* **2018**, *63*, 3159–3160.
17. Sehlinger, A.; Ochsenreither, K.; Bartnick, N.; Meier, M.A.R. Potentially biocompatible polyacrylamides derived by the Ugi four-component reaction. *Eur. Polym. J.* **2015**, *65*, 313–324. [CrossRef]
18. Wang, S.; Fu, C.; Wei, Y.; Tao, L. Facile One-Pot Synthesis of New Functional Polymers through Multicomponent Systems. *Macromol. Chem. Phys.* **2014**, *215*, 486–492. [CrossRef]
19. Bu, H.; Kjøniksen, A.L.; Knudsen, K.D.; Nyström, B. Rheological and Structural Properties of Aqueous Alginate during Gelation via the Ugi Multicomponent Condensation Reaction. *Biomacromolecules* **2004**, *5*, 1470–1479. [CrossRef] [PubMed]
20. Patil, P.; Ahmadian-Moghaddam, M.; Dömling, A. Isocyanide 2.0. *Green Chem.* **2020**, *22*, 6902–6911. [CrossRef]
21. Wen, Y.; Chen, X.; Liu, Z.; Zhu, Q.; Li, Z.; He, G.; Yan, H.; Lin, Q. Hydrophobically Modified Alginate Derivatives via the Ugi Multicomponent Reaction for the Development of Hydrophobic Pharmaceutical Formulations. *ChemistrySelect* **2021**, *6*, 10965–10973. [CrossRef]
22. Liu, Z.; Chen, X.; Wen, Y.; Bao, C.; Liu, C.; Cao, S.; Yan, H.; Lin, Q. Chemical modification of alginate with tosylmethyl isocyanide, propionaldehyde and octylamine via the Ugi reaction for hydrophobic drug delivery. *Polym. Bull.* **2021**, *79*, 7809–7826. [CrossRef]
23. Liu, Z.; Chen, X.; Huang, Z.; Wang, H.; Cao, S.; Liu, C.; Yan, H.; Lin, Q. One-Pot Synthesis of Amphiphilic Biopolymers from Oxidized Alginate and Self-Assembly as a Carrier for Sustained Release of Hydrophobic Drugs. *Polymers* **2022**, *14*, 694. [CrossRef] [PubMed]
24. Liu, Z.; Chen, X.; Huang, Z.; Shi, J.; Liu, C.; Cao, S.; Yan, H.; Lin, Q. Self-assembled oleylamine grafted alginate aggregates for hydrophobic drugs loading and controlled release. *Int. J. Polym. Mater. Polym. Biomater.* **2021**, *72*, 212–223. [CrossRef]

25. Lin, W.J.; Juang, L.W.; Lin, C.C. Stability and release performance of a series of pegylated copolymeric micelles. *Pharm. Res.* **2003**, *20*, 668–673. [CrossRef] [PubMed]
26. Yang, J.S.; Ren, H.B.; Xie, Y.J. Synthesis of Amidic Alginate Derivatives and Their Application in Microencapsulation of λ-Cyhalothrin. *Biomacromolecules* **2011**, *12*, 2982–2987. [CrossRef] [PubMed]
27. Mourdikoudis, S.; Liz-Marzán, L.M. Oleylamine in Nanoparticle Synthesis. *Chem. Mater.* **2013**, *25*, 1465–1476. [CrossRef]
28. Galant, C.; Kjøniksen, A.L.; Nguyen, G.T.M.; Knudsen, K.D.; Nyström, B. Altering Associations in Aqueous Solutions of a Hydrophobically Modified Alginate in the Presence of β-Cyclodextrin Monomers. *J. Phys. Chem. B* **2005**, *110*, 190–195. [CrossRef] [PubMed]
29. Singh, B.; Sharma, D.K.; Kumar, R.; Gupta, A. Controlled release of the fungicide thiram from starch-alginate-clay based formulation. *Appl. Clay Sci.* **2009**, *45*, 76–82. [CrossRef]
30. Walvekar, P.; Gannimani, R.; Salih, M.; Makhathini, S.; Mocktar, C.; Govender, T. Self-assembled oleylamine grafted hyaluronic acid polymersomes for delivery of vancomycin against methicillin resistant Staphylococcus aureus (MRSA). *Colloids Surf. B Biointerfaces* **2019**, *182*, 110388. [CrossRef] [PubMed]
31. Yan, H.; Chen, X.; Feng, M.; Shi, Z.; Zhang, W.; Wang, Y.; Ke, C.; Lin, Q. Entrapment of bacterial cellulose nanocrystals stabilized Pickering emulsions droplets in alginate beads for hydrophobic drug delivery. *Colloids Surf. B Biointerfaces* **2019**, *177*, 112–120. [CrossRef] [PubMed]
32. Kang, H.; Shu, Y.; Li, Z.; Guan, B.; Peng, S.; Huang, Y.; Liu, R. An effect of alginate on the stability of LDH nanosheets in aqueous solution and preparation of alginate/LDH nanocomposites. *Carbohydr. Polym.* **2014**, *100*, 158–165. [CrossRef]
33. Yang, X.D.; Zhang, C.G.; Qiao, C.D.; Mu, X.L.; Li, T.D.; Xu, J.K.; Shi, L.; Zhang, D.J. A simple and convenient method to synthesize N-[(2-hydroxyl)-propyl-3-trimethylammonium] chitosan chloride in an ionic liquid. *Carbohydr. Polym.* **2015**, *130*, 325–332. [CrossRef] [PubMed]
34. Samanta, H.S.; Ray, S.K. Synthesis, characterization, swelling and drug release behavior of semi-interpenetrating network hydrogels of sodium alginate and polyacrylamide. *Carbohydr. Polym.* **2014**, *99*, 666–678. [CrossRef] [PubMed]
35. Ionita, M.; Pandele, M.A.; Iovu, H. Sodium alginate/graphene oxide composite films with enhanced thermal and mechanical properties. *Carbohydr. Polym.* **2013**, *94*, 339–344. [CrossRef] [PubMed]
36. Xue, Y.N.; Huang, Z.Z.; Zhang, J.T.; Liu, M.; Zhang, M.; Huang, S.W.; Zhuo, R.X. Synthesis and self-assembly of amphiphilic poly(acrylic acid-b-dl-lactide) to form micelles for pH-responsive drug delivery. *Polymer* **2009**, *50*, 3706–3713. [CrossRef]
37. Gaucher, G.; Dufresne, M.H.; Sant, V.P.; Kang, N.; Maysinger, D.; Leroux, J.C. Block copolymer micelles: Preparation, characterization and application in drug delivery. *J. Control. Release* **2005**, *109*, 169–188. [CrossRef] [PubMed]
38. Mai, Y.; Eisenberg, A. Self-assembly of block copolymers. *Chem. Soc. Rev.* **2012**, *41*, 5969–5985. [CrossRef] [PubMed]
39. Du, H.; Liu, M.; Yang, X.; Zhai, G. The role of glycyrrhetinic acid modification on preparation and evaluation of quercetin-loaded chitosan-based self-aggregates. *J. Colloid Interface Sci.* **2015**, *460*, 87–96. [CrossRef]
40. Jiang, C.; Wang, X.; Sun, P.; Yang, C. Synthesis and solution behavior of poly(ε-caprolactone) grafted hydroxyethyl cellulose copolymers. *Int. J. Biol. Macromol.* **2011**, *48*, 210–214. [CrossRef]
41. Chen, X.; Zhu, Q.; Liu, Z.; Zhang, T.; Gong, C.; Li, J.; Yan, H.; Lin, Q. Self-aggregate performance of hexyl alginate ester derivative synthesized via SN2 reaction for controlled release of λ-cyhalothrin. *Polym. Bull.* **2023**, *80*, 495–514. [CrossRef]
42. Li, Q.; Liu, C.G.; Huang, Z.H.; Xue, F.F. Preparation and characterization of nanoparticles based on hydrophobic alginate derivative as carriers for sustained release of vitamin D3. *J. Agric. Food Chem.* **2011**, *59*, 1962–1967. [CrossRef] [PubMed]
43. Ahmad, N.; Ramsch, R.; Llinàs, M.; Solans, C.; Hashim, R.; Tajuddin, H.A. Influence of nonionic branched-chain alkyl glycosides on a model nano-emulsion for drug delivery systems. *Colloids Surf. B Biointerfaces* **2014**, *115*, 267–274. [CrossRef] [PubMed]
44. Wang, K.; Luo, G.F.; Liu, Y.; Li, C.; Cheng, S.X.; Zhuo, R.X.; Zhang, X.Z. Redox-sensitive shell cross-linked PEG–polypeptide hybrid micelles for controlled drug release. *Polym. Chem.* **2012**, *3*, 1084. [CrossRef]
45. Fang, X.; Zhao, X.; Yu, G.; Zhang, L.; Feng, Y.; Zhou, Y.; Liu, Y.; Li, J. Effect of molecular weight and pH on the self-assembly microstructural and emulsification of amphiphilic sodium alginate colloid particles. *Food Hydrocoll.* **2020**, *103*, 105593. [CrossRef]
46. He, C.; Hu, Y.; Yin, L.; Tang, C.; Yin, C. Effects of particle size and surface charge on cellular uptake and biodistribution of polymeric nanoparticles. *Biomaterials* **2010**, *31*, 3657–3666. [CrossRef] [PubMed]
47. Astete, C.E.; Sabliov, C.M.; Watanabe, F.; Biris, A. Ca2+ cross-linked alginic acid nanoparticles for solubilization of lipophilic natural colorants. *J. Agric. Food Chem.* **2009**, *57*, 7505–7512. [CrossRef]
48. Kulkarni, S.A.; Feng, S.S. Effects of particle size and surface modification on cellular uptake and biodistribution of polymeric nanoparticles for drug delivery. *Pharm. Res.* **2013**, *30*, 2512–2522. [CrossRef] [PubMed]
49. Sukamporn, P.; Baek, S.J.; Gritsanapan, W.; Chirachanchai, S.; Nualsanit, T.; Rojanapanthu, P. Self-assembled nanomicelles of damnacanthal-loaded amphiphilic modified chitosan: Preparation, characterization and cytotoxicity study. *Mater. Sci. Eng. C* **2017**, *77*, 1068–1077. [CrossRef] [PubMed]
50. Ritsema, J.; Herschberg, E.; Borgos, S.; Løvmo, C.; Schmid, R.; Te Welscher, Y.; Storm, G.; van Nostrum, C.F. Relationship between polarities of antibiotic and polymer matrix on nanoparticle formulations based on aliphatic polyesters. *Int. J. Pharm.* **2018**, *548*, 730–739. [CrossRef] [PubMed]
51. Kietzmann, D.; Béduneau, A.; Pellequer, Y.; Lamprecht, A. pH-sensitive microparticles prepared by an oil/water emulsification method using n-butanol. *Int. J. Pharm.* **2009**, *375*, 61–66. [CrossRef] [PubMed]

52. Zhou, J.; Tai, G.X.; Liu, H.Y.; Ge, J.Y.; Feng, Y.; Chen, F.F.; Yu, F.; Liu, Z.H. Activin A down-regulates the phagocytosis of lipopolysaccharide-activated mouse peritoneal macrophages in vitro and in vivo. *Cell. Immunol* **2009**, *255*, 69–75. [CrossRef] [PubMed]
53. Cao, X.L.; Cheng, C.; Ma, Y.L.; Zhao, C.S. Preparation of silver nanoparticles with antimicrobial activities and the researches of their biocompatibilities. *J. Mater. Sci. Mater. Med.* **2010**, *21*, 2861–2868. [CrossRef] [PubMed]
54. Opanasopit, P.; Ngawhirunpat, T.; Chaidedgumjorn, A.; Rojanarata, T.; Apirakaramwong, A.; Phongying, S.; Choochottiros, C.; Chirachanchai, S. Incorporation of camptothecin into N-phthaloyl chitosan-g-mPEG self-assembly micellar system. *Eur. J. Pharm. Biopharm.* **2006**, *64*, 269–276. [CrossRef] [PubMed]
55. Naksuriya, O.; Shi, Y.; van Nostrum, C.F.; Anuchapreeda, S.; Hennink, W.E.; Okonogi, S. HPMA-based polymeric micelles for curcumin solubilization and inhibition of cancer cell growth. *Eur. J. Pharm. Biopharm.* **2015**, *94*, 501–512. [CrossRef] [PubMed]
56. Malarvizhi, G.L.; Retnakumari, A.P.; Nair, S.; Koyakutty, M. Transferrin targeted core-shell nanomedicine for combinatorial delivery of doxorubicin and sorafenib against hepatocellular carcinoma. *Nanomed. Nanotechnol. Biol. Med.* **2014**, *10*, 1649–1659. [CrossRef] [PubMed]

International Journal of
Molecular Sciences

MDPI

Article

Antioxidant Activity and the Therapeutic Effect of Sinomenine Hydrochloride-Loaded Liposomes-in-Hydrogel on Atopic Dermatitis

Xue Chen, Yang Wu †, Ruoyang Jia †, Yuqing Fang, Keang Cao, Xinying Yang, Xiaobo Qu and Hongmei Xia *

College of Pharmacy, Anhui University of Chinese Medicine, Hefei 230012, China; 2023205221004@stu.ahtcm.edu.cn (X.C.); 2023205227034@stu.ahtcm.edu.cn (Y.W.); 2022205221003@stu.ahtcm.edu.cn (R.J.); 2023205221005@stu.ahtcm.edu.cn (Y.F.); 2023205221003@stu.ahtcm.edu.cn (K.C.); 2020205227037@stu.ahtcm.edu.cn (X.Y.); xiaoboqu2022@163.com (X.Q.)

* Correspondence: sm_shine@ahtcm.edu.cn; Tel./Fax: +86-13965033210

† These authors should be co-first authors due to their contributed equally to this work.

Abstract: Sinomenine hydrochloride is an excellent drug with anti-inflammatory, antioxidant, immune-regulatory, and other functions. Atopic dermatitis is an inherited allergic inflammation that causes itchiness, redness, and swelling in the affected area, which can have a significant impact on the life of the patient. There are many therapeutic methods for atopic dermatitis, and sinomenine with immunomodulatory activity might be effective in the treatment of atopic dermatitis. In this study, the atopic dermatitis model was established in experimental mice, and physical experiments were carried out on the mice. In the experiment, sinomenine hydrochloride liposomes-in-hydrogel as a new preparation was selected for delivery. In this case, liposomes were dispersed in the colloidal hydrogel on a mesoscopic scale and could provide specific transfer properties. The results showed that the sinomenine hydrochloride-loaded liposomes-in-hydrogel system could effectively inhibit atopic dermatitis.

Keywords: sinomenine; liposomes-in-hydrogel; colloidal hydrogel; anti-inflammatory; atopic dermatitis

1. Introduction

Atopic dermatitis (AD) is a chronic, hereditary, recurrent, inflammatory skin disease that is clinically manifested by rashes, erythema, exudates, epidermal peeling, edema, and papules on the skin [1]. The disease occurs at all stages of life, with a higher prevalence among women than men in adulthood [2]. Modern medical research has shown that when various external factors (including exposure to hot and cold stimuli, chemicals, pollen and dust, infection) interact with various internal factors of the immune system (including family genetics, mental state, skin dysfunction), it triggers an abnormal response of the immune system, which ultimately leads to the development of AD [3,4]. The pathogenesis of AD is often accompanied by inflammation and oxidative stress responses, which produce a large number of free radicals, further aggravating skin damage [5]. The course of AD is divided into an acute phase and a chronic phase. The acute phase is the immune response mainly by T helper 2 cells (Th2), which manifests as Th2 cell infiltration, and the levels of interleukin (IL)-4, IL-5, and IL-13 were increased. The chronic phase is an immune response dominated by Th1 cells, manifested by delayed hypersensitivity through interferon-γ (IFN-γ) and IL-2 responses and excessive collagen accumulation, inducing skin thickening and tissue remodeling [6,7]. There is no good cure for this disease in Western medicine. Glucocorticoids and immunosuppressants are used to treat AD, which can temporarily alleviate the related symptoms, but the long-term effect is not ideal, and there are more adverse reactions [8,9]. Therefore, further development of drugs from traditional Chinese medicine is needed to effectively treat AD.

Sinomenine (SIN, $C_{19}H_{23}NO_4$) (Figure 1) is the main component of the Chinese herb Sinomenium acutum (Thunb.) Rehder & E.H. Wilson. Because of the poor water solubility of SIN, the pharmaceutical industry mostly uses its sinomenine hydrochloride (SINH) as the water-soluble salt form of SIN. Experiments have proved that SINH has many biological activities such as anti-oxidation, anti-inflammation, immunomodulation, and improving microcirculation [10–12]. The existing clinical formulations of SINH are mainly used for rheumatoid arthritis and cardiac arrhythmias [13,14]. The mechanisms of action of SINH include regulating the balance of Th17/Treg cells, increasing the level of IL-10 (mainly secreted by Treg cells), and decreasing the level of IL-17A mainly secreted by Th17 cells [15]; decreasing the activity of antigen-presenting cells and affecting the nuclear factor κB (NF-κB) pathway [16]; inhibiting the production of serum pro-inflammatory cytokines IL-1β and IL-6; and suppressing the expression of matrix metalloproteinases (MMP)-2 and MMP-9 proteins [17,18]. In addition, SINH is able to inhibit the inflammatory response by reducing oxidative stress [19]. Therefore, SINH has the potential to treat AD.

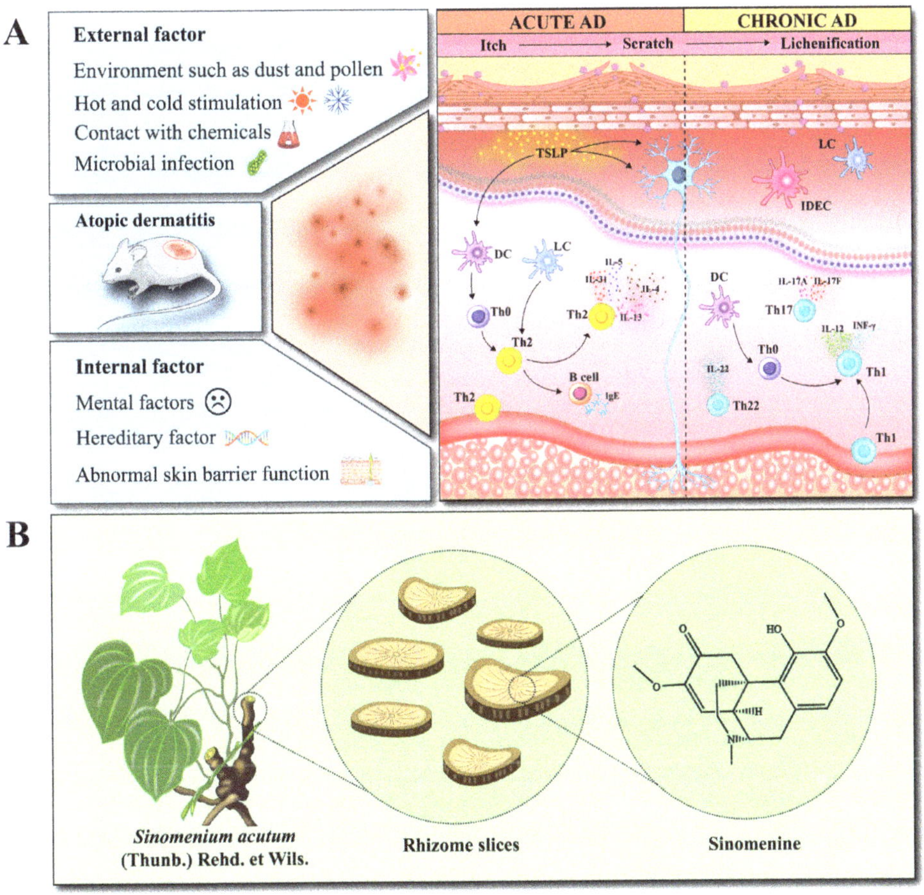

Figure 1. Pathogenesis of AD and source of SIN: (**A**) The sketch of the pathogenesis of atopic dermatitis, (**B**) Sinomenium acutum, rhizome slices, and the structure of sinomenine.

In view of the problems of SINH with a short half-life, low oral bioavailability, and gastrointestinal irritation caused by high blood concentrations [20,21], this study aimed to explore a new sinomenine hydrochloride delivery system—SINH-liposome-hydrogel

(SINH-L-H) system. SINH loaded into liposomes and evenly dispersed in hyaluronic acid colloidal hydrogel take advantage of their excellent biocompatibility and high hydration capacity, enabling the drug to remain on the surface of the skin for a longer period of time [22], improving the bioavailability of the drug and providing benefits for the treatment of atopic dermatitis. Due to their structural properties, the liposomes-in-hydrogel not only act as carriers but also achieve slow drug release and keep the affected area moist, which is crucial for the treatment of dry, inflammatory skin diseases such as AD. In addition, SINH-L incorporated into colloidal hydrogel provided the dual effect of enhanced drug permeability and sustained release to reduce skin lesions and provide antioxidant and anti-inflammatory effects [23]. Therefore, this paper will construct a mouse model of atopic dermatitis, observe and evaluate the therapeutic effect of the SINH-L-H system on atopic dermatitis, explore the mechanism of action of the SINH-L-H system in the treatment of atopic dermatitis, and verify the effectiveness of the SINH-L-H system in the treatment of atopic dermatitis.

2. Results

2.1. Molecular Docking of SIN as the Active Structure of SINH

The computer software Discovery Studio 4.5 was used to predict the interactions between the components of the formula. Through software screening, the three-dimensional structure of proteins IL4 and IL2 combined with SIN was selected, and the results showed that SIN had a good docking effect with IL4 and IL2. There are pi-alkyl and carbon-hydrogen bonds between SIN and IL4 (Figure 2A). There are van der Waals forces, pi-alkyl, and carbon-hydrogen bonds between SIN and IL2, and there is also a pi-pi stacked interaction between molecules (Figure 2B). In the same space, there are carbon-hydrogen bonds and bump interactions between SIN, cholesterol, and phospholipid in different conformations (Figure 2C,D). There is also a bump interaction between SIN and hydrogel (Figure 2E).

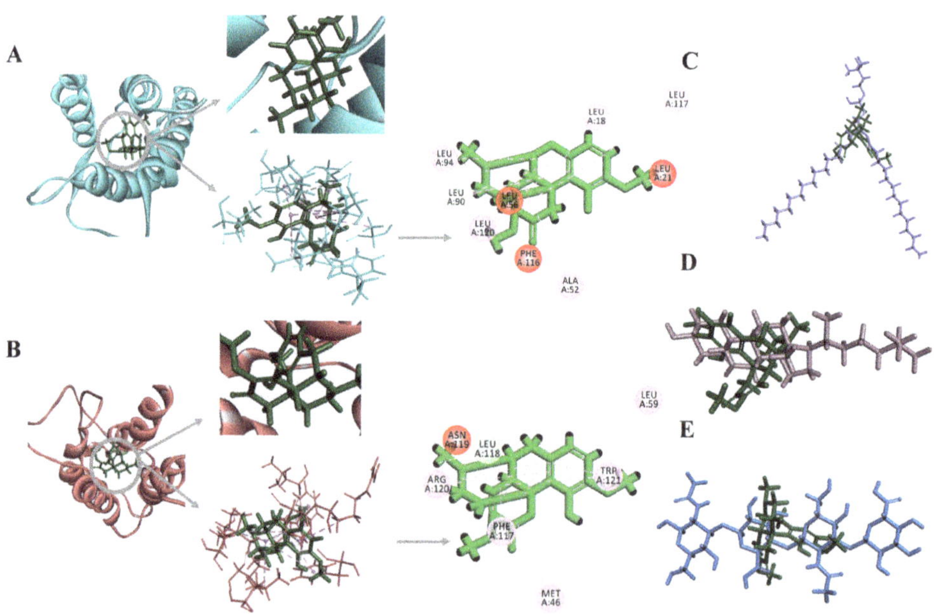

Figure 2. Molecular docking of SIN through Discovery Studio 4.5: (**A**) IL4 and SIN, (**B**) IL2 and SIN, (**C**) Cholesterol and SIN, (**D**) Phospholipid and SIN, (**E**) Hydrogel and SIN.

2.2. Basic Characteristics of SINH-L-H

The absorbance of SINH was measured by an ultraviolet spectrophotometer; SINH at 10–60 µg/mL concentration is proportional to their absorbance (R = 0.9992), and the linear relationship is good (Figure 3A). The SINH-loaded liposomes-in-hydrogel (SINH-L-H) is a creamy white, homogeneous, and fine hydrogel with no viscous lumps and low fluidity, remaining colloidal hydrogel at room temperature (Figure 3B). The liposomes in SINH-L-H were spherical and uniformly dispersed in the hyaluronic acid hydrogel (HA) via Cyro-EM (Figure 3B); the size of the SH-L is about 133.8 ± 14.5. Viscosity measurements showed the viscosity of SINH-L-H was 9.83 ± 0.11 mPa·s at 25 °C, and its pH was 7.27 ± 0.01. The encapsulation rate of SINH-L is 72.27%.

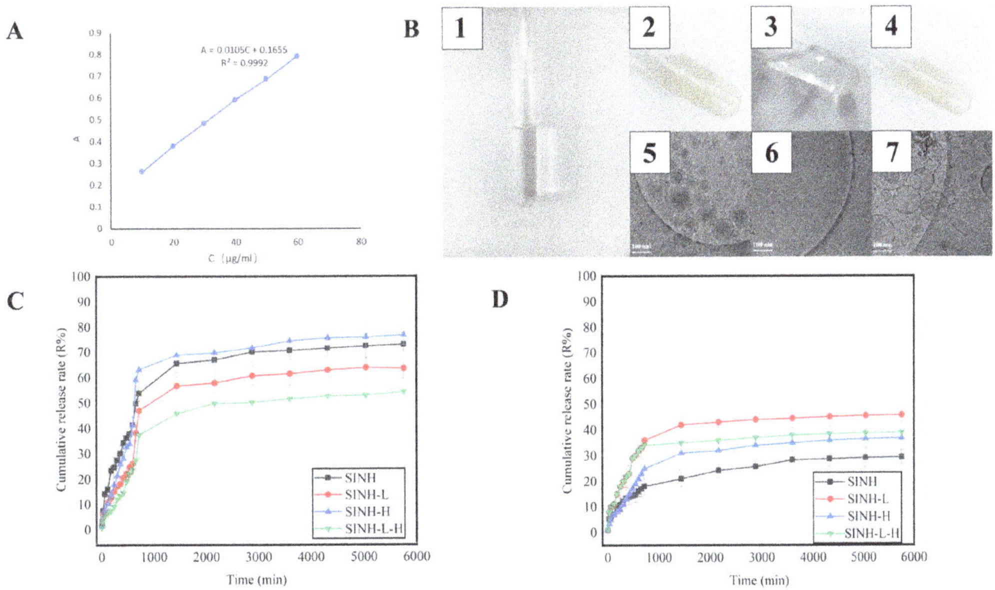

Figure 3. Characteristics of SINH. (**A**) Standard curve of determination of SINH. (**B**) The appearances of SINH, SINH-L, SINH-H, and SINH-L-H (1–4) and the images of SINH-L, SINH-H, and SINH-L-H under the cryo-EM (5–7). (**C**) Cumulative release curve of SINH and its three preparations through the dialysis membrane at different time points. (**D**) Cumulative release curve of SINH and its three preparations through mice skin at different time points.

Stability measurements showed that SH-L-H did not change significantly when stored for 30 days under refrigeration at 4 °C. The appearance and viscosity of SH-L-H did not differ from day 0, indicating that SH-L-H was stable under these conditions. SH-L-H is more variable when stored at ambient conditions: for the first 14 days, the SH-L-H was similar to the initial preparation, creamy white, homogeneous, and fine in texture; on the 21st day, the color of the SH-L-H changed from creamy white to yellowish and the viscosity decreased; on the 30th day, the color of the SH-L-H continued to deepen to yellow, and there was stratification and a decrease in viscosity. The above indicates that SH-L-H is unstable when stored under these conditions and that the drug is easily released from the liposomes and oxidized and discolored (Table 1).

Table 1. Determination results of SH-L-H 4 °C cold storage experiment and normal temperature storage experiment.

Storage Temperature	Storage Period (d)	Characters	Exterior	Viscosity (_mPa·s)
4 °C	0	Milky white, even and delicate		9.84 ± 0.14
	7	Milky white, even and delicate		9.76 ± 0.08
	14	Milky white, even and delicate		9.80 ± 0.03
	21	Milky white, even and delicate		9.79 ± 0.05
	30	Milky white, even and delicate		9.82 ± 0.03
Normal temperature	0	Milky white, even and delicate		9.84 ± 0.02
	7	Milky white, even and delicate		9.73 ± 0.01
	14	Milky white, even and delicate		9.70 ± 0.03
	21	Pale yellow, even and delicate		9.65 ± 0.04
	30	Yellow, layered		9.57 ± 0.06

2.3. *In Vitro/Ex Vivo Dialysis Diffusion Dialysis*

The in vitro and ex vivo release profiles of SINH, SINH-L, SINH-H, and SINH-L-H are shown below (Figure 3C,D). It can be seen that the sample from the upper diffusion pool gradually penetrates into the lower diffusion pool as the release time increases.

In the in vitro diffusion experiment, the release rate of the SH group was greater than the other three groups during the first 11 h. After 11 h, the release rate of SINH in SINH-H began to exceed that of the SINH group, which indicates that the hydrophilic porous structure of SINH-H promoted the release of SINH, so that SINH rapidly passed through the dialysis membrane into the lower diffusion pool, and the release rate of SINH in SINH-H did not show a large change with the increase in time, and the release curve showed that it entered the plateau period at about 12 h, and the final release degree could reach about 77%; the largest amount of SINH released by SINH-L-H occurs at 96 h, up to about 55%, this may be due to the fact that SINH-L-H possesses the three-dimensional mesh structure of HA hydrogels and the phospholipid bilayer of liposomes, and thus SINH is tightly bound in it and cannot be easily released.

In the ex vivo diffusion experiment, the total diffusion rate was SINH-L > SINH-L-H > SINH H > SINH, and SINH L reached the highest permeation efficiency of about 46% at 96 h, while the release rate of SINH only reached 30% at this time, and the liposomes contained phospholipids similar to skin tissues, which were "similarly compatible". Therefore, the release of SINH in SINH-L is the greatest, while SINH is in the free state and less compatible with the skin, so the amount and speed of penetration through the skin are smaller. SINH-L-H contains phospholipids and HA, which are also contained in skin tissues, and although it can contact a larger area of skin, it also takes time to cross the three-dimensional structure of the phospholipid bilayer and hydrogel, so the diffusion rate of SINH in SINH-L-H is lower than that of SINH-L; the amount and rate of release of SINH in SINH-H with the help of HA is slightly higher than that of SINH.

Comparing the release profiles of in vitro and ex vivo diffusion, we found that the speed of SINH through the dialysis membrane was faster and the amount of transmission was greater in each group, while the transdermal speed of the drug in each group showed irregularities and significant differences. Dialysis membrane is a homogeneous porous membrane of organic polymers that mainly relies on the osmotic pressure generated by the different concentration gradients on both sides of the membrane to promote the flow of solutes, so the speed and amount of SINH passage mainly depend on the osmotic pressure on both sides of the membrane. The skin includes the dermis and epidermis with abundant cells and blood vessels, and it will take more time for the drug to pass through the skin. The principle of similar compatibility between liposomes and cell membranes means that more SINH in SINH-L and SINH-L-H is deposited in the skin.

2.4. *Determination of Antioxidant Activity*

2.4.1. The Scavenging Rate on Hydroxyl Radicals

In the Fenton reaction, hydroxyl radicals react with salicylic acid to generate 2,3-dihydroxybenzoic acid, which exhibits a special absorption at 510 nm (Figure 4A). SINH had antioxidant activity in the reaction system and could reduce the production of hydroxyl radicals. Therefore, the generation of 2,3-dihydroxybenzoic acid in the reaction system decreased, and the color became lighter. It was indicated that the clearance rate of SINH for 2,3-dihydroxybenzoic acid was directly proportional to the concentration of SINH in the concentration range of 0.2 to 1.0 mg/mL. The IC_{50} of SINH for scavenging hydroxyl radicals was about 0.7034 mg/mL.

In the antioxidant reaction experiment, the drugs in SINH-L-H were surrounded by liposomes and hydrogels, and SINH was difficult to release. However, because of the synergistic effect of the antioxidant capacity of liposomal hydrogel and SINH itself, the clearance rate of SINH-L-H was higher than that of SINH. At the concentration of 0.2 mg/mL, the scavenging rates of SINH, SINH-L, SINH-H, and SINH-L-H on hydroxyl radicals were about 0.42 $\pm$ 0.1%, 6.73 $\pm$ 1.7%, 7.98 $\pm$ 3.4%, and 3.36 $\pm$ 1.5%, respectively. The clearance rate of SINH was lower than that of the same concentration formulation (Figure 4B). In the Fenton reaction, the removal effect of SINH-L-H was lower than that of SINH-L and SINH-H but higher than that of SINH. This showed that SINH-L-H had better stability and durability while maintaining the sinomenine's effect so that the drug could work in the body for a longer time.

2.4.2. The Inhibitory Rate on ABTS Radicals

The 1:1 mixture reaction of ABTS and $K_2S_2O_8$ produced stable cationic free radicals $ABTS^+$. The maximum absorption wavelength of $ABTS^+$ is 734 nm, and the reaction between antioxidants and $ABTS^+$ will cause the solution to fade (Figure 4C). The stronger the antioxidant capacity, the lighter the color of the solution, and the higher the free radical scavenging rate.

The scavenging ability of SINH on $ABTS^+$ was directly proportional to the concentration between 0.001 and 0.02 mg/mL, and the higher the concentration, the stronger the scavenging ability. The IC_{50} value of SINH was about 0.0115 mg/mL. The reaction steps of

SINH, SINH-L, SINH-H, and SINH-L-H with the same concentration of 0.001 mg/mL have clearance rates of 11.33 ± 4.21%, 5.76 ± 0.32%, 6.96 ± 0.41%, and 5.74 ± 0.36% for $ABTS^+$, respectively (Figure 4D). The clearance rate of $ABTS^+$ in the preparations was lower than that of SINH. The clearance of $ABTS^+$ was lower than that of SINH. This was because SINH was encapsulated in the liposomal hydrogel, and the release rate of SINH was slow. For a certain period of time, the free radical scavenging ability of SINH-L-H was weaker than that of SINH. But SINH-L-H had a slow release effect, which made it have a long-term effect. Therefore, SINH-L-H had the ability for long-term, slow release.

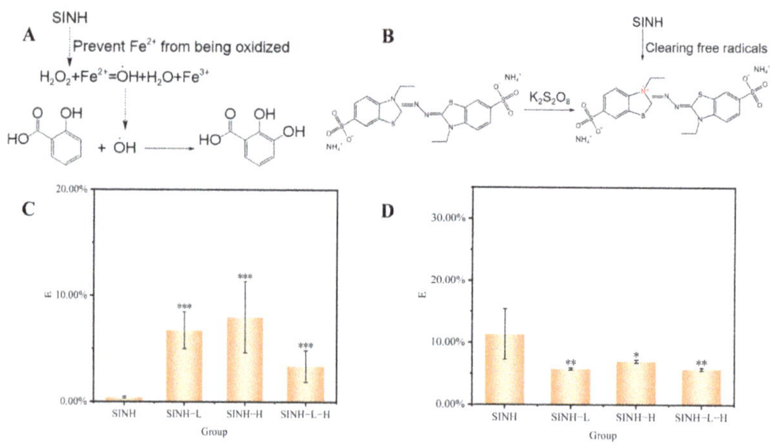

Figure 4. Characteristics and antioxidant activity of SINH and its formulations. (**A**) Fenton reaction mechanism of SINH, (**B**) SINH scavenging $ABTS^+$ reaction, (**C**) The effect of SINH, SINH-L, SINH-H, and SINH-L-H on scavenging hydroxyl radicals, (**D**) The effect of SINH, SINH-L, SINH-H, and SINH-L-H on scavenging ABTS radicals (* $p < 0.05$, ** $p < 0.01$, *** $p < 0.001$: Pharmaceutical preparation group vs. SINH group).

2.5. The Therapeutic Effects on AD Mice

2.5.1. Measurements Results of Skin Irritation of SINH-L-H

In the single-dose skin irritation test, no mice in the B-L-H and SINH-L-H groups showed any skin changes within 72 h after administration (Table 2). In multiple skin irritation experiments, the mice in the B-L-H (Blank-loaded liposomes-in-hydrogel) and SINH-L-H groups did not show any erythema or edema within 7 days after administration, indicating that the formula matrix we selected was safe and non-toxic (Table 3). In terms of scoring results, B-L-H and SINH-L-H were safe because of no irritation to intact mouse skin.

Table 2. Skin irritation score of a single administration.

Time	Group				Irritation Intensity
	B-L-H Skin Status	B-L-H Skin Score	SINH-L-H Skin Status	SINH-L-H Skin Score	
1 h		0		0	None
24 h		0		0.2	None

Table 2. *Cont.*

Time	Group				Irritation Intensity
	B-L-H Skin Status	B-L-H Skin Score	SINH-L-H Skin Status	SINH-L-H Skin Score	
48 h		0		0	None
72 h		0		0	None

Table 3. Skin irritation score of multiple administration.

Time	Group				Irritation Intensity
	B-L-H Skin Status	B-L-H Skin Score	SINH-L-H Skin Status	SINH-L-H Skin Score	
1 d		0		0	None
2 d		0		0.2	None
3 d		0		0	None
4 d		0		0	None
5 d		0		0	None
6 d		0		0	None
7 d		0		0	None

2.5.2. General Condition of Mice and Manifestation of Skin Lesions

The establishment of the AD mouse model and the time point of drug administration according to Figure 5A. The body weight and skin thickness of each group of mice are shown in Figure 5B, and the skin healing is shown in Figure 5C. The mice in the blank group showed an increasing trend in body weight and skin thickness, with smooth hair, good condition, and no abnormal behavior; the ears were red and transparent, with an intact appearance and clearly visible blood vessels. The body weight of mice in the other groups peaked on the fourth day and decreased thereafter, among which the body weight of mice in the positive control group with ointment decreased more rapidly, presumably due to the side effects caused by dexamethasone as a kind of glucocorticoids. After 2,4-dinitrochlorobenzene (DNCB) excitation twice, erythema appeared on the dorsal skin of mice in each group and gradually thickened and dried, with scabs appearing, and mice scratched frequently and became agitated. On the 4th day, the erythema and thickening of the skin of the mice in each group were gradually reduced, and the skin of the mice in the positive control group was obviously thinner, with slight folds and erythema and no hair growth at the excitation site. As shown in Figure 5D, compared with the model group, the skin injury scores of mice in the model group were significantly higher than those in the normal control group, and the skin scores of mice in the positive control group were significantly lower than those in the SINH-L-H group.

2.5.3. Ear Swelling in Mice

Except for the blank group, the ears of mice in the other groups showed swelling and crusting after excitation, and the symptoms significantly improved after SINH-L-H administration. Compared with the blank group, the swelling and crusting status of the remaining groups increased to varying degrees, with the most significant change in mice's status in the model group. Compared with the model group, the mice were administered to the SINH-L-H group with the most significant ear changes (Figure 6A,B).

2.5.4. Mouse Organ Index

The effect of SINH-L-H on AD was initially evaluated by organ indices, including the spleen index and thymus index, and the results are shown in Figure 6C. Compared with the blank group, the spleen index and thymus index of mice in the model group increased, and the differences were statistically significant; compared with the model group, the spleen index and thymus index of mice using SINH, SINH-L, SINH-H, and SINH-L-H decreased, and the effect of using SINH-L-H was almost similar to that of using dexamethasone, indicating that SINH and its three preparations all have good efficacy on AD.

2.5.5. Malondialdehyde (MDA) in Mouse Organ

MDA, one of the products of lipid peroxidation, is one of the important indexes to measure the degree of oxidative damage. The content of MDA in each mouse tissue is shown in Figure 6D,E. Compared with the blank group, the content of MDA in the skin and liver of mice in the model group increased with significant differences, and the content of MDA in the kidney also increased. All MDA levels in the skin, liver, and kidney of mice in the SINH, SINH-L, SINH-H, and SINH-L-H groups also decreased, with the most pronounced decrease in mice in the SINH-L-H group, indicating that SH-L-H was the most effective in AD among these three groups.

2.5.6. HE Staining

The results of HE staining of the dorsal skin of mice in each group are shown in Figure 6F. Compared with the blank group, the dorsal skin of the model mice showed obvious hyperplasia, significant thickening of the epidermal layer, intracellular and intercellular edema, and a large amount of inflammatory infiltration around the blood vessels in the dermis, showing a severe inflammatory state overall. In the control group, the structure of each layer was more complete with clear borders; the epidermis was not obviously keratinized; the spiny layer was mildly hyperplastic; there was mild intracellular and intercellular edema; and the inflammatory infiltrate around the blood vessels in the dermis was mild. The rest of the administered groups were less symptomatic than the model group, showing varying degrees of epidermal layer thickening, intracellular and intercellular edema, and vascular inflammatory infiltration, with the SINH-L-H group producing the mildest manifestations (Figure 6).

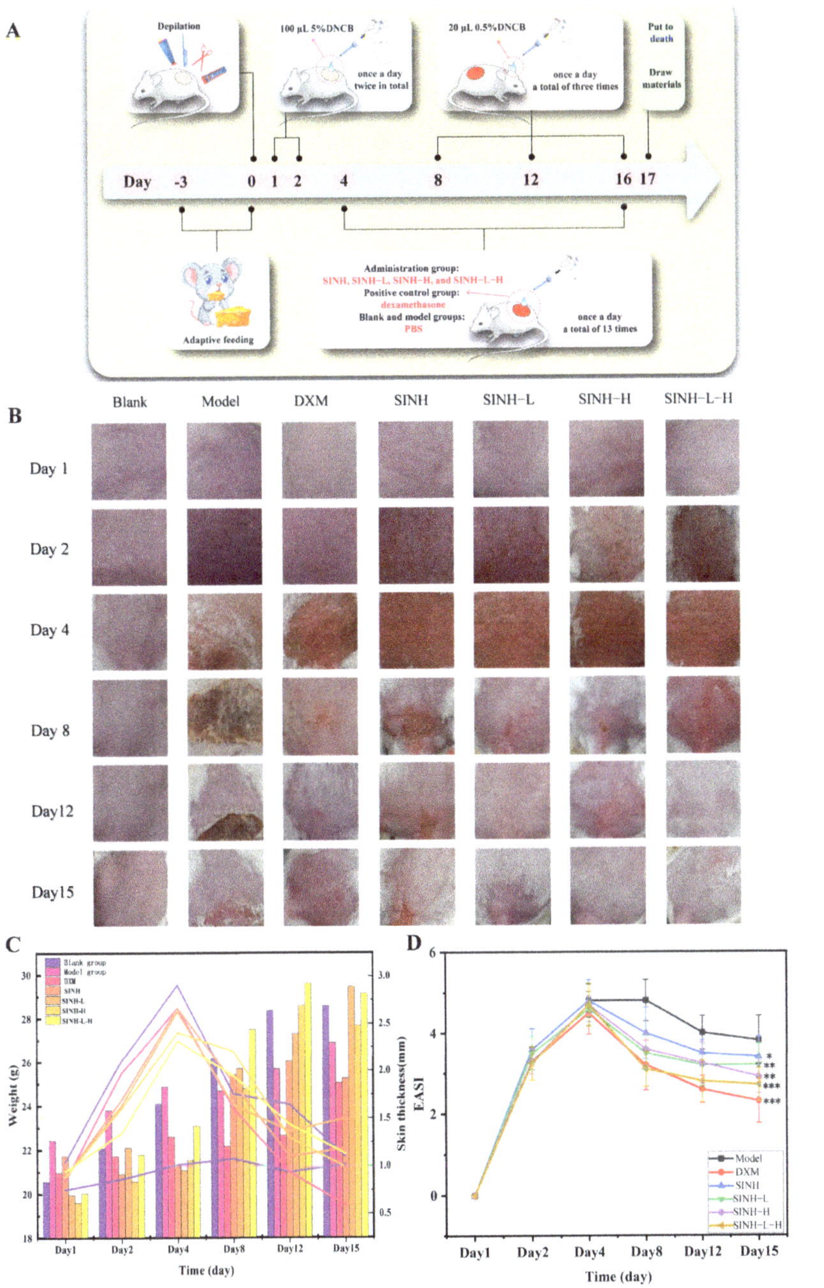

Figure 5. Pharmacological activity of SINH, SINH-L, SINH-H, and SINH-L-H. (**A**)The establishment of an eczema mouse model and the time point of drug administration, (**B**) Skin lesions in mice, (**C**) Changes in body weight and skin thickness of mice (n = 10), (**D**) Degree of skin lesion score (compared with the model group: * $p < 0.05$, ** $p < 0.01$, *** $p < 0.001$).

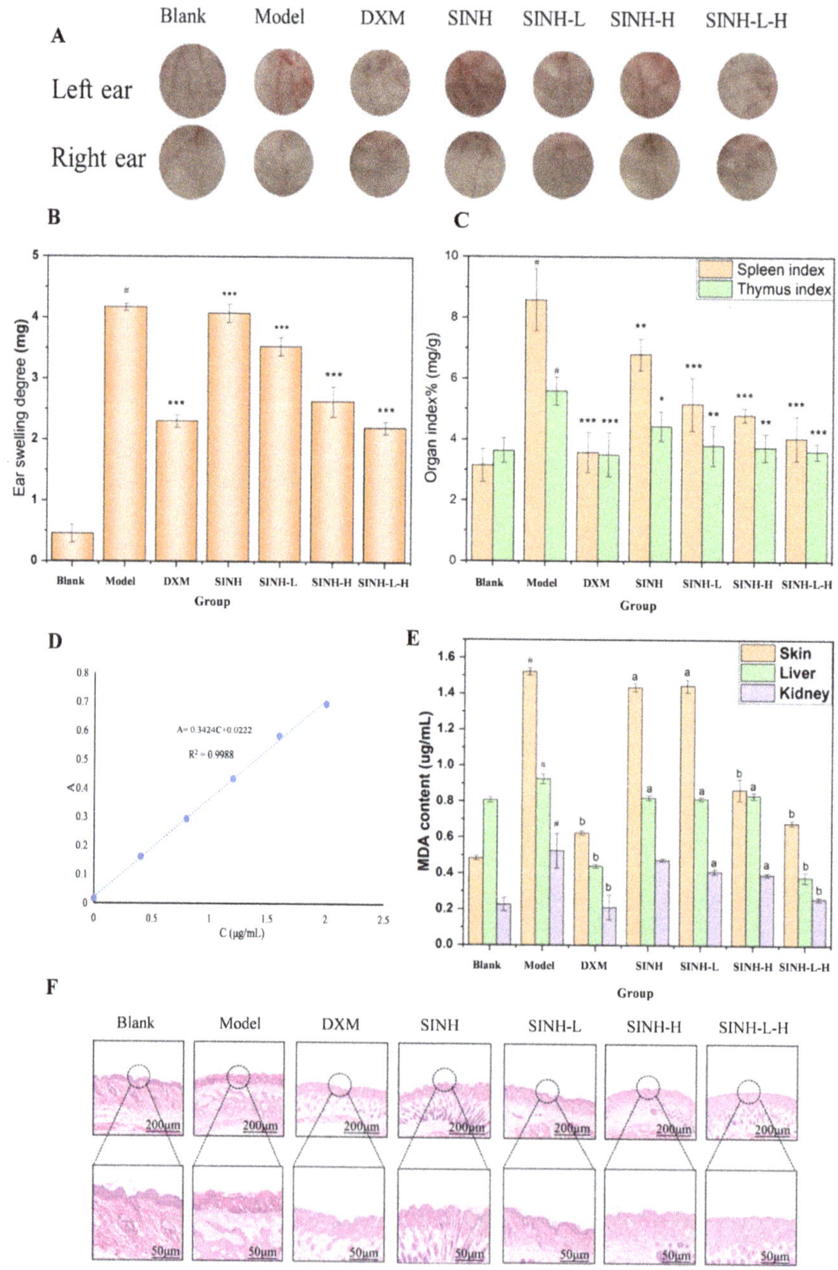

Figure 6. Effect of SINH-L-H on eczema mice. (**A**) Effect of SINH-L-H on mice ears in eczema mice; (**B**) Effect of SINH-L-H on ear swelling in AD mice (Compared with the blank group: # $p < 0.001$; compared with the model group: *** $p < 0.001$); (**C**) Effect of SINH-L-H on the organ index of AD mice (# $p < 0.001$: model group vs. blank group; * $p < 0.05$, ** $p < 0.01$, *** $p < 0.001$: drug group vs. model group); (**D**) MDA standard curve; (**E**) Effect of DNCB induction on MDA content in various tissues of mice ($\overline{X} \pm SD$, $n = 10$) Note: Compared with blank group: # $p < 0.001$; compared with model group: a $p < 0.01$, b $p < 0.001$; (**F**) Pathological skin changes on the back of each group of mice.

3. Discussion

AD is a common skin disease that can have a serious impact on a patient's life. It is currently considered to have a strong correlation with susceptibility genes, epidermal barrier dysfunction, and immune dysfunction. The AD model was established by DNCB, leading to epidermal barrier dysfunction, and the therapeutic effect of SINH-L-H on AD was studied.

Liposomes were reported to be an excellent drug carrier, which could load drugs for the treatment of skin injuries, and hydrogels could clear wound exudates, provide a moist environment, and stimulate and guide tissue regeneration, which could be used as external materials for wounds [24,25]. In this study, hyaluronic acid was selected for the colloidal hydrogel matrix. Hyaluronic acid itself is a polysaccharide and has additional characteristics that make it very suitable for drug delivery systems, such as biocompatibility, antiviral, antibacterial, and anti-tumor properties. The combination of liposomes and hyaluronic acid to load drugs could form an excellent delivery system to enable the drug to be delivered to its target. The successful preparation of SINH-L-H had the characteristics of a slow and controlled release. In addition, SINH-L-H prepared in the study had stable properties, good dispersion, and an excellent dermal penetration effect, which was in line with the expected effect.

The onset of AD could be caused by multiple factors. Abnormal Notch signaling pathways can induce the development of severe AD, while normal Notch signaling pathways can inhibit the production of thymic stromal lymphopoietin (TSLP) in keratinocytes and alleviate the pathogenesis of AD [26]. Pro-inflammatory cytokines can promote the occurrence of AD, including TSLP, IL-25, and IL-33 [27]. Th2 cells are derived from the immune differentiation of T cells and serve as pro-inflammatory mediators in AD. Activate Th2 cells, promote specific cytokine-related inflammation (IL-4, IL-5, IL-13), an increase in eosinophils, and the production of immunoglobulin E (IgE) [28]. B cells participate in T cell activation and Th2 production and differentiate into cytokine-producing cells that may alter the differentiation of effector T cells [29]. One of the main signs of AD is an increase in serum total IgE levels. Activated Th2 cells release IL-4 and IL-13, which promote B cells to produce IgE antibodies through the signal transducer and transcriptional activator (STAT) pathway [30]. The various pathogenic factors mentioned above can serve as targets for the treatment of AD.

In order to better predict the target of action between AD and SIN, we used Genecards (GeneCards—Human Genes | Gene Database | Gene Search (https://www.genecards.org/), retrieved from 15 January 2023) to screen out the target of disease and drug action and created a Venn diagram to find the intersection of the target of action (Figure 7A). DIVID (DAVID Functional Annotation Bioinformatics Microarray Analysis (ncifcrf.gov), retrieved from 16 January 2023) was used to find the relevant pathways of the intersection genes, and bioinformatics (bioinformatics.com.cn) was used to enrich the pathways and find the best pathway (Figure 7C,D). At the same time, use STRING (STRING: functional protein association networks (string-db.org), retrieved from 17 January 2023) to create a Protein-Protein interaction network (PPI) graph of the intersecting genes and identify the genes that have the most significant impact (Figure 7B). Finally, the most relevant gene targets were extracted from the screened pathways and rescreened with significant genes in STRING. IL2 and IL4 were selected as the predicted targets for SIN treatment of AD.

Studies have shown that there is an excess of reactive oxygen species in the skin of atopic dermatitis patients, and these reactive oxygen species can cause high oxidative stress, resulting in oxidative damage to skin cell DNA and proteins, further aggravating skin inflammation and damage [31,32]. SINH, SINH-L, SINH-H, and SINH-L-H are able to remove reactive oxygen species, which helps reduce oxidative damage to cells due to the accumulation of reactive oxygen species. By reducing the accumulation of reactive oxygen species, SINH-L-H may help reduce oxidative damage to skin cells and improve skin barrier function, thereby relieving the symptoms of atopic dermatitis. Therefore, SINH-L-H is a potential drug for the treatment of AD.

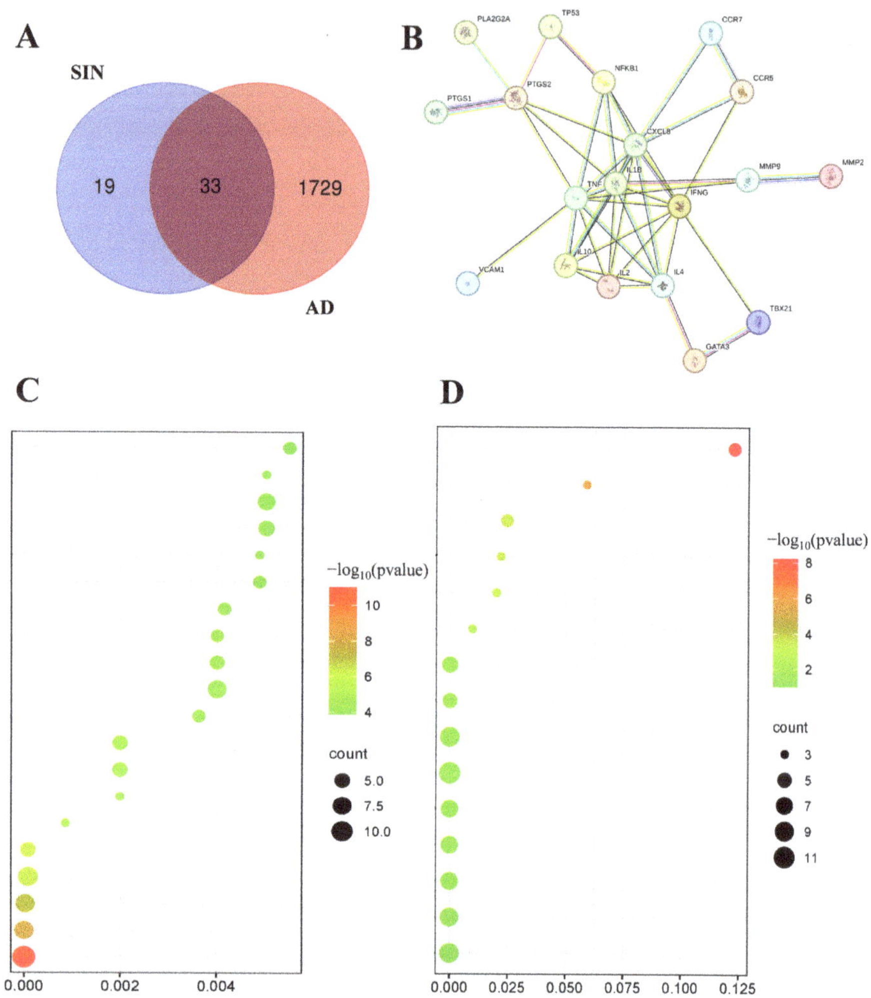

Figure 7. Molecular formation and target screening of drug systems: (**A**) Venn diagram of AD and SIN targets, (**B**) Bubble diagram of target PPI, (**C**,**D**) GO and KEGG enrichment.

4. Materials and Methods

4.1. Materials

Sinomenine hydrochloric (≥98%, China), cholesterol (AR, Sinopharm Chemical Reagent Co., Ltd., Shanghai, China), soy lecithin (Shanghai Jinsong Industry Co., Ltd., Shanghai, China), absolute ethanol (AR, Shanghai RichJoint Chemical Reagents Co., Ltd., Shanghai, China), hyaluronic acid (HA, 99%, Shandong Xiya Reagents Co., Ltd., Linyi, China), 1,1-diphenyl-2-picrylhydrazyl (DPPH, 98%, Shanghai Yuanye Biology Science and Technology Co., Ltd., Shanghai, China), H_2O_2 (AR, Shanghai SuYi Chemical Reagent Co., Ltd., Shanghai, China), $FeSO_4{\cdot}7H_2O$ (AR, Sinopharm Chemical Reagent Co., Ltd., Shanghai, China), thiobarbituric acid (TBA, 98%, Shanghai Yuanye Biology Science and Technology Co., Ltd., Shanghai, China), trichloroacetic acid (TCA, AR, Tianjin Damao Chemical Reagent Factory, Tianjin, China), 2,4-dinitrochlorobenzene (DNCB, Shandong Xiya Reagents Co., Ltd., Linyi, China), 2,2′-Azinobis-(3-ethylbenzthiazoline-6-sulphonate) (ABTS, 98%, Shanghai Macklin Biochemical Technology Co., Ltd., Shanghai, China), $K_2S_2O_8$

(99.5%, Macklin Biochemical Technology Co., Ltd., Shanghai, China), phosphate buffered saline (PBS).

4.2. Animals

Healthy Kunming mice (female, 20 ± 2 g) were provided by the Animal Experimental Center of the Anhui University of Traditional Chinese Medicine (Hefei, China). All animal experiments conform to the guidelines approved by the Ethics Committee of the Anhui University of Traditional Chinese Medicine (Hefei, China). Animals are kept under constant environmental conditions (25 ± 2 °C, 40–70% relative humidity), and provided with unlimited supplies of food and water.

4.3. Preparation SINH-L, SINH-H, and SINH-L-H

The synthesis of SINH-L, SINH-H, and SINH-L-H was adopted from our previous work as cited [23] (Figure 8). Took 0.3 g of phospholipids and 0.1 g of cholesterol in a rotary evaporation flask, added the appropriate amount of absolute ethanol to dissolve, fixed the flask on a rotary evaporator to rotated and evaporated to form a thin film, set the temperature at 70 °C, vacuum pressure at 70 kPa, speed at 100 r/min until the absolute ethanol evaporated completely. Then, 10 mL of 2 mg/mL of SINH solution was added to the film and rotated and hydrated for 30 min to form a milky white suspension. The suspension was sonicated for 10 min and filtered across the 0.22 μm microporous membrane three times to obtain SINH-L. 0.1 g HA was added to the prepared 10 mL SINH-L suspension and stirred at the speed of 100 rpm for 12 h with a magnetic stirrer till HA naturally swelled to form a uniform hydrogel (SINH-L-H). 0.1 g of HA was also added to 10 mL of SINH solution, and the same was used to form SINH-H. In addition, blank liposomes (B-L) with PBS instead of SINH solution, blank liposome hydrogels (B-L-H) with B-L instead of SINH-L, and blank HA hydrogels (B-H) with PBS instead of SINH solution were prepared.

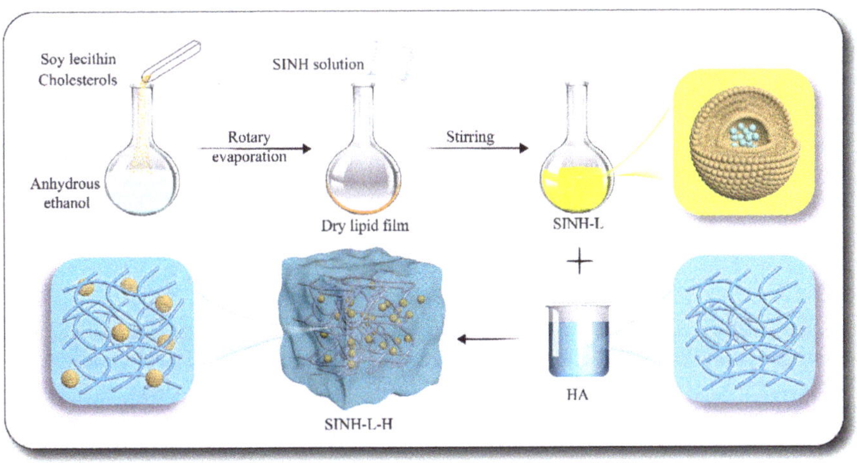

Figure 8. Preparation process of SINH-L-H.

4.4. Quality Evaluation of SINH-L

4.4.1. Appearance and Morphology of Cryo-EM

The prepared SINH-L suspension was placed in a transparent glass test tube to observe its appearance. An appropriate amount of prepared SINH-L was mixed with a turbine apparatus, and the morphological characteristics of SINH-L were observed with cryo-EM.

4.4.2. Particle Size and Zeta Potential

First, we applied the Malvern potential particle simeter (n = 3) to measure the particle size of SINH-L and the electrophoretic mobility of SINH-L in electric fields. The electrophoretic mobility was then converted to the Zeta potential value according to the Smoluchowski equation.

4.4.3. Encapsulation Rate and Drug Load

In this experiment, the drug amount of free sinomenine hydrochloride in SH-L suspension was obtained by the high-speed centrifugal method, and the drug amount of total sinomenine hydrochloride in SINH-L suspension was obtained by the ethanol demulsification method: 1 mL SINH-L suspension was precisely measured and diluted by adding an appropriate amount of PBS solution. After centrifugation at 15,000 RPM for 10 min, the supernatant was taken to measure the dosage of free SINH (W_1) at 262 nm. In addition, 1 mL SINH-L suspension was precisely measured and placed in a 10 mL volumetric bottle; anhydrous ethanol was added to the scale line, and the volumetric bottle was placed in a CNC ultrasonic cleaner for 30 min for ultrasonic demulsification. An appropriate amount of demulsified solution was taken and measured at 262 nm to measure the dosage of total SINH (W_2). Calculate the *EE* of SINH-L according to the following Equation:

$$EE = \frac{(W_2 - W_1) \times 100\%}{W_2}$$

Among them, W_1 represents the amount of free SINH in liposome suspension, and W_2 represents the amount of total SINH in liposome suspension.

4.5. Molecular Docking

The 2D molecular structures of SIN, hyaluronic acid hydrogel, phospholipid, and cholesterol were obtained from PubChem. They were imported into Discovery Studio 4.5 and exported as small-molecule compound files in SDF format. Then, we downloaded the core target proteins IL4 and IL6 from the UniProt database and selected the appropriate protein structure. Finally, after ligand extraction, hydrogenation, and water extraction, the docking program of Discovery Studio 4.5 was used for docking between small molecules and therapeutic targets. A molecular docking pattern diagram was achieved.

4.6. Ex Vivo/Ex Vivo Dialysis Diffusion Dialysis of SH-L-H

We performed this experiment using the Franz diffusion cell. For in vitro diffusion: first fill each diffusion cell with diffusion solution (PBS) and add a stirrer of appropriate size. A dialysis membrane of appropriate size was fixed in the middle of the upper and lower cells of the diffusion cell, and 1 mL of SH, SH-L, SH-L-H, and their corresponding blanks were added to the upper cell, respectively. At 5 min, 30 min, 60 min, 120 min, . . . 720 min, 1440 min, 2160 min, and 2880 min, 2 mL of samples were aspirated and immediately supplemented with the same volume of PBS solution. The absorbance was measured at 262 nm using a UV spectrophotometer and the corresponding cumulative drug release rates were calculated according to the Equation.

For the ex vivo skin dialysis experiments, we used the dorsal skin of Kunming mice, and the other operations were the same as for dialysis membrane diffusion.

$$R(\%) = \frac{C_n \times V_n + \sum_{i=1}^{n-1} C_i \times V}{Q_t} \times 100\%$$

where "R" is the cumulative drug release rate of SH at the different sampling points, "C_n" is the mass concentration of SH at the nth time point (μg/mL), "V_n" is the volume of the solution sampled at the nth time point (mL), "C_i" is the mass concentration of SH in the receiving solution at the sampling point ($i \leq n - 1$) (μg/mL), "V" is the total volume of the lower cells (mL), and "Q_t" is the theoretical drug content (2000 μg/mL).

4.7. Determination of Antioxidant Activity of SINH-L-H

4.7.1. The Scavenging Rate on Hydroxyl Radicals

SINH (0.2, 0.4, 0.6, 0.8, and 1.0 mg/mL) was added to the Fenton reaction system with salicylic acid. The experimental operation was as follows: in the sample group, add 1 mL $FeSO_4$ solution, 1 mL salicylic acid solution, and 1 mL H_2O_2 solution into 1 mL SINH successively to measure the absorbance (A_1); in the control group, add 1 mL of salicylic acid and 2 mL water to 1 mL SINH to measure the absorbance (A_2); in the blank group, add 1 mL $FeSO_4$ solution, 1 mL salicylic acid solution and 1 mL H_2O_2 solution into 1 mL water to measure the absorbance (A_3). The solution of each group was mixed evenly and reacted in a 37 °C water bath for 30 min. The absorbance at a wavelength of 510 nm was measured. The same operation was used for SINH-L-H, SINH-H, and SINH-L of 0.2 mg/mL as SINH mentioned above.

$$E\% = \frac{A_3 - (A_1 - A_2)}{A_3} \times 100\%$$

Calculate the clearance rate and IC50 of SINH and its preparations on hydroxyl radicals based on changes in the absorbance.

4.7.2. The Inhibitory Rate on ABTS Radicals

Mix the ABTS aqueous solution and $K_2S_2O_8$ solution (1:1) evenly, and react in a dark place for 12 h. Dilute the ABTS mixture 30 times so that the absorbance of the solution at 734 nm is within the range of 0.7–0.8. Take 0.5 mL of SINH with different concentrations (0.001, 0.005, 0.01, 0.02, 0.03 mg/mL) and place it in a reaction vessel. Add 2 mL of the 30 times diluted ABTS mixture as the sample group (A_S) and react in the dark at room temperature for 5 min. Then measure the absorbance at 734 nm. The control group (A_C) added 2 mL of PBS to SINH, while the blank group (A_0) added 2 mL of ABTS mixture to 0.5 mL of PBS, and then measured the absorbance using the same procedure.

$$E\% = \frac{A_0 - (A_S - A_C)}{A_0} \times 100\%$$

SINH-L, SINH-H, and SINH-L-H with a concentration of 0.001 mg/mL, replace the SINH position with formulations and perform the same operation. Finally, based on the measurement results, calculate IC_{50} and the clearance rate of the samples.

4.8. The Therapeutic Effect of SINH-L-H on AD

4.8.1. Skin Irritation of SINH-L-H

Forty healthy Kunming mice (20 ± 2 g) were provided by the Experimental Animal Center of Anhui University of Chinese Medicine under license No. SYXK (SU): 2020-0009. This research protocol has been approved by the Animal Care and Use Committee of Anhui University of Traditional Chinese Medicine (AHUCM mouse-2023136). The mice were randomly divided into a single-dosing test group and a multiple-dosing test group.

The back hair of the mice was removed before the experiment began. The mice were divided into three groups, one group was evenly applied with 0.3 mL of B-L-H, one group was evenly applied with 0.3 mL of SINH-L-H, and the other group was not treated as a control group. After application, cover the affected area with gauze and secure it with medical pressure-sensitive tape. The erythema and edema, pigmentation, bleeding, and skin roughness were observed at 1 h, 24 h, 48 h, and 72 h after removal of the gauze.

The skin irritation test after multiple doses was administered in a similar manner to the single dose, with the exception that it required 1 week of continuous dosing at the same site for the same duration and amount of dose each time. The gauze and medical pressure-sensitive tape were also removed 6 h after administration and the site was washed with gauze soaked in warm water. Observe the skin condition of the mice 1 h after each removal of the gauze. Score according to Tables 4 and 5, (at the same time do the individual project score and then do the comprehensive score).

Table 4. Skin irritation response score.

Skin Reaction	Degree of Response	Score
Erythema and eschar	Erythema free	0
	Slight erythema (barely visible)	1
	Moderate erythema	2
	Erythema severa	3
	Purplish red erythema to slight eschar formation	4
Edema condition	No edema	0
	Slight edema (barely visible)	1
	Mild edema (edema bulge clearly defined)	2
	Moderate edema (edema bulge about 1 mm)	3
	Severe edema (swelling of more than 1 mm, extended range)	4
Highest score		8

Table 5. Evaluation criteria of skin irritation intensity.

Average Score	Irritation Intensity
0–0.49	Non-irritant
0.5–2.99	Mild irritation
3–5.99	Moderate irritation
6–8	Strong irritation

4.8.2. Establishment of AD Model and Administration

The healthy Kunming mice were domesticated in a specific pathogen-free environment for 3 days. One day before the experiment, shave skin hair with an area of about 2 cm × 2 cm on the back. The shaved back skin was treated with 100 μL 5% DNCB solution (a mixture of acetone:olive oil = 3:1 as solvent) on the first and second days, and the blank group was replaced with the same volume of acetone/olive oil mixture after sensitization. After the first induction, the mice were fed normally with an interval of 5 days. The second sensitization, the skin of the ear was applied with a 0.5% DNCB solution every 3 days, for a total of 3 times. The treatment group was treated with SINH, SINH-L, SINH-H, and SINH-L-H, respectively, starting on the 4th day (if sensitized on the same day, dosed 2 h before sensitization) for 12 days, and the mice of the blank and model groups were treated with equal amounts of PBS, while DXM group was treated with dexamethasone (DXM) as a positive control. The mice were fasted for 12 h after the final administration. The ear pieces, spleen, serum, and dorsal skin of the mouse were collected for subsequent tests after execution (Figure 4A).

4.8.3. Degree of Skin Lesion Score (EASI Score) in AD Mice

During the modeling process, the mice were observed for changes in their physiological status, including activity, body weight, changes in skin thickness, fur color, diet, and water intake. Skin thickness is measured using a sebum thickness meter (also known as a skin fold caliper) to measure the thickness of the skin and underlying adipose tissue by picking up skin folds. The mice were scored 12 h after the last administration according to the Eczema Severity Index (EASI) scale. The scoring criteria included erythema, papules, desquamation, crusting, and exudate, each of which was divided into three degrees of mild, moderate, and severe and expressed in different data. The scoring criteria were as follows, and the total score was the mouse skin lesion score (Table 6). Based on this scale, the establishment of the AD model and the intervention of SINH, SINH-L, SINH-H, and SINH-L-H on this model were obtained.

Table 6. Score of eczema skin lesion in mice (EASI Score).

Skin Condition	Degree of Response	Points
No obvious skin injury was observed by naked eye	None	0
Visual inspection requires careful examination to see the lesions	Mild	1
The lesions are clearly visible to the naked eye	Medium	2
The lesions were severe and very visible	Severe	3
The score between the various symptoms can be recorded as a half grade (0.5)		

4.8.4. Determination of Ear Swelling Rate in Mice

After the mice were executed, the right and left ear pieces (about 8 mm) were immediately removed with a punch and weighed on an analytical balance, and the ear swelling rate of the mice was calculated according to the following formula.

$$\text{Ear swelling rate\%} = \frac{\text{Right ear mass} - \text{Left ear mass}}{\text{Left ear mass}} \times 100\%$$

4.8.5. Measurement of Organ Index in Mice

After the mice were executed, the thymus and spleen tissues were separated and quickly weighed, and the thymus (spleen) index was calculated according to the formula as follows:

$$\text{Spleen (thymus) index} = \frac{\text{Spleen (thymus) mass (mg)}}{\text{Body mass (mg)}}$$

4.8.6. Determination of MDA Content in Organ

Mouse liver and kidney were taken to prepare 10% organ homogenate. The content of MDA in mouse organs was determined according to the following steps: (1) The experiment was divided into the blank, model, DXM, SINH, SINH-L, SINH-H, and SINH-L-H groups. (2) 1 mL of skin, liver, and kidney homogenate from each group was added to 3 mL of TBA working solution and placed in a water bath at 95 °C for 40 min. (3) After removal and cooling to room temperature with running water to stop the reaction, the supernatant was centrifuged at 4000 r/min for 8 min and the absorbance value was measured at 532 nm. (4) The absorbance values obtained were substituted into the MDA content standard curve to calculate the MDA content in the tissue and determine the degree of oxidation of the tissue. The MDA content in the skin, liver, and kidney were determined the same as above.

4.8.7. Hematoxylin and Eosin Staining

After removing the skin tissue from the back of the mice, the normal saline was rinsed and quickly fixed in 4% formalin buffer and embedded in paraffin 24 h later. Skin sections with a thickness of 5 μm made by the section mechanism and stained with hematoxylin-eosin (HE) were observed in the skin histopathological characteristics of mice.

4.8.8. Statistical Methods

All data were expressed as mean ± standard deviation (x ± s), one-way ANOVA was used, and the LSD method was used for two-way comparison between groups, and the test level was $\alpha = 0.05$, and the difference was considered statistically significant at $p < 0.05$.

5. Conclusions

To sum up, SINH was encapsulated in liposomes and then evenly dispersed into the HA hydrogel matrix to form SINH-L-H, which could increase the residence time on the skin surface to improve the bioavailability of SINH to treat AD. The experimental results showed that SINH-L-H had uniform texture, good stability, and a slow release effect. SINH-L-H was a safe preparation based on its non-irritating effect on the skin. In the antioxidant test, SINH-L-H showed strong scavenging ability on hydroxyl free radicals and ABTS free radicals, and the scavenging ability of SINH was positively correlated with the concentration of SINH, indicating that SINH had a good antioxidant effect. For AD in mouse models, the use of SINH-L-H significantly alleviated the symptoms of AD and reduced the amount of MDA in the skin. SINH penetrated into the skin, depending on the liposomes-in-gel delivery. SINH-L-H has a good transdermal effect. SINH-L-H was a safe and effective new dosage form for the treatment of AD and provided a novel carrier for the clinical application of SINH.

Author Contributions: X.C.: Data curation, Formal analysis, Investigation, Methodology, Writing—original draft. Y.W.: Formal analysis, Investigation, Software, Validation. Writing—original draft. R.J.: Formal analysis, Investigation, Software. Y.F.: Data curation, Formal analysis, Resources. K.C.: Data curation, Formal analysis, Resources. X.Y.: Investigation, Writing—original draft. X.Q.: Investigation, Software, Validation. H.X.: Conceptualization, Funding acquisition, Project administration, Supervision, Writing—review & editing. All authors have read and agreed to the published version of the manuscript.

Funding: This study was supported by the Department of Education of Anhui Province of China (No. 2023AH050754 and No. KJ2019A0470), Anhui Provincial Quality Engineering Project for Education in the New Era—Research on the Reform of Graduate Education and Teaching (No. 2023jyjxggyjY168), the Quality Engineering Project of the Anhui Provincial Department of Educa-tion in 2021 (No. 2021jyxm0824), the Natural Science Foundation of Anhui Province of China (No. 1608085MH227), the Research and Development Project Entrusted by the Enterprise: Research on Antioxidant Biological Activity of Quercetin (No. 2022HZ049), the Quality Engineering Project of Anhui University of Chinese Medicine in 2021 (No. 2021zlgc042) and the Anhui University of Chinese Medicine Annual Innovation Training Program for College Students (No. 2017142 and No. 2017171).

Institutional Review Board Statement: This study involves animal experiments, and its research program has been approved by the Experimental Animal Ethics Committee of Anhui University of Chinese Medicine. The approval number is AHUCM-mouse-2023136 and the approval date is 6 November 2023. We strictly comply with all applicable animal welfare regulations and guidelines to ensure the reasonable use and minimum suffering of laboratory animals.

Informed Consent Statement: Not applicable.

Data Availability Statement: Data will be made available on request.

Conflicts of Interest: The authors declare no conflict of interest.

References

1. Fang, Z.; Lu, W.; Zhao, J.; Zhang, H.; Qian, L.; Wang, Q.; Chen, W. Probiotics modulate the gut microbiota composition and immune responses in patients with atopic dermatitis: A pilot study. *Eur. J. Nutr.* **2020**, *59*, 2119–2130. [CrossRef]
2. Nowicka, D.; Grywalska, E. The Role of Immune Defects and Colonization of Staphylococcus aureus in the Pathogenesis of Atopic Dermatitis. *Anal. Cell. Pathol.* **2018**, *2018*, 1956403. [CrossRef]
3. Dizon, M.P.; Yu, A.M.; Singh, R.K.; Wan, J.; Chren, M.M.; Flohr, C.; Silverberg, J.I.; Margolis, D.J.; Langan, S.M.; Abuabara, K. Systematic review of atopic dermatitis disease definition in studies using routinely collected health data. *Br. J. Dermatol.* **2018**, *178*, 1280–1287. [CrossRef]
4. Puar, N.; Chovatiya, R.; Paller, A.S. New treatments in atopic dermatitis. *Ann. Allergy Asthma Immunol.* **2021**, *126*, 21–31. [CrossRef]
5. Jia, Y.; Hu, J.; An, K.; Zhao, Q.; Dang, Y.; Liu, H.; Wei, Z.; Geng, S.; Xu, F. Hydrogel dressing integrating FAK inhibition and ROS scavenging for mechano-chemical treatment of atopic dermatitis. *Nat. Commun.* **2023**, *14*, 2478. [CrossRef]
6. Kim, J.; Kim, B.E.; Leung, D.Y.M. Pathophysiology of atopic dermatitis: Clinical implications. *Allergy Asthma Proc.* **2019**, *40*, 84–92. [CrossRef]
7. Kim, G.D.; Lee, S.E.; Park, Y.S.; Shin, D.H.; Park, G.G.; Park, C.S. Immunosuppressive effects of fisetin against dinitrofluorobenzene-induced atopic dermatitis-like symptoms in NC/Nga mice. *Food Chem. Toxicol.* **2014**, *66*, 341–349. [CrossRef]
8. Frazier, W.; Bhardwaj, N. Atopic Dermatitis: Diagnosis and Treatment. *Am. Fam. Physician* **2020**, *101*, 590–598.
9. Sidbury, R.; Kodama, S. Atopic dermatitis guidelines: Diagnosis, systemic therapy, and adjunctive care. *Clin. Dermatol.* **2018**, *36*, 648–652. [CrossRef]
10. Yin, N.; Xiong, Y.; Tao, W.; Chen, J.; Wang, Z. Sinomenine alleviates lipopolysaccharide-induced inflammatory responses in RAW264.7 macrophages. *Immunopharmacol. Immunotoxicol.* **2020**, *42*, 147–155. [CrossRef]

11. Kiasalari, Z.; Afshin-Majd, S.; Baluchnejadmojarad, T.; Azadi-Ahmadabadi, E.; Fakour, M.; Ghasemi-Tarie, R.; Jalalzade-Ogvar, S.; Khodashenas, V.; Tashakori-Miyanroudi, M.; Roghani, M. Sinomenine Alleviates Murine Experimental Autoimmune Encephalomyelitis Model of Multiple Sclerosis through Inhibiting NLRP3 Inflammasome. *J. Mol. Neurosci.* **2021**, *71*, 215–224. [CrossRef]
12. Liu, W.; Zhang, Y.; Zhu, W.; Ma, C.; Ruan, J.; Long, H.; Wang, Y. Sinomenine Inhibits the Progression of Rheumatoid Arthritis by Regulating the Secretion of Inflammatory Cytokines and Monocyte/Macrophage Subsets. *Front. Immunol.* **2018**, *9*, 2228. [CrossRef]
13. Jutley, G.; Raza, K.; Buckley, C.D. New pathogenic insights into rheumatoid arthritis. *Curr. Opin. Rheumatol.* **2015**, *27*, 249–255. [CrossRef]
14. Zhang, M.W.; Wang, X.H.; Shi, J.; Yu, J.G. Sinomenine in Cardio-Cerebrovascular Diseases: Potential Therapeutic Effects and Pharmacological Evidences. *Front. Cardiovasc. Med.* **2021**, *8*, 749113. [CrossRef]
15. Tong, B.; Yu, J.; Wang, T.; Dou, Y.; Wu, X.; Kong, L.; Dai, Y.; Xia, Y. Sinomenine suppresses collagen-induced arthritis by reciprocal modulation of regulatory T cells and Th17 cells in gut-associated lymphoid tissues. *Mol. Immunol.* **2015**, *65*, 94–103. [CrossRef]
16. Zhao, Y.; Li, J.; Yu, K.; Liu, Y.; Chen, X. Sinomenine inhibits maturation of monocyte-derived dendritic cells through blocking activation of NF-kappa B. *Int. Immunopharmacol.* **2007**, *7*, 637–645. [CrossRef]
17. Zhou, H.; Wong, Y.F.; Wang, J.; Cai, X.; Liu, L. Sinomenine ameliorates arthritis via MMPs, TIMPs, and cytokines in rats. *Biochem. Biophys. Res. Commun.* **2008**, *376*, 352–357. [CrossRef]
18. Li, R.Z.; Guan, X.X.; Wang, X.R.; Bao, W.-Q.; Lian, L.-R.; Choi, S.W.; Zhang, F.Y.; Yan, P.-Y.; Leung, E.L.H.; Pan, H.-D.; et al. Sinomenine hydrochloride bidirectionally inhibits progression of tumor and autoimmune diseases by regulating AMPK pathway. *Phytomedicine* **2023**, *114*, 154751. [CrossRef]
19. Xia, B.; Li, Q.; Wu, J.; Yuan, X.; Wang, F.; Lu, X.; Huang, C.; Zheng, K.; Yang, R.; Yin, L.; et al. Sinomenine Confers Protection Against Myocardial Ischemia Reperfusion Injury by Preventing Oxidative Stress, Cellular Apoptosis, and Inflammation. *Front. Pharmacol.* **2022**, *13*, 922484. [CrossRef]
20. Song, H.; Wen, J.; Li, H.; Meng, Y.; Zhang, Y.; Zhang, N.; Zheng, W. Enhanced transdermal permeability and drug deposition of rheumatoid arthritis via sinomenine hydrochloride-loaded antioxidant surface transethosome. *Int. J. Nanomedicine* **2019**, *14*, 3177–3188. [CrossRef]
21. Shen, Q.; Zhang, X.; Qi, J.; Shu, G.; Du, Y.; Ying, X. Sinomenine hydrochloride loaded thermosensitive liposomes combined with microwave hyperthermia for the treatment of rheumatoid arthritis. *Int. J. Pharm.* **2020**, *576*, 119001. [CrossRef]
22. Freitas, B.; Lavrador, P.; Almeida, R.J.; Gaspar, V.M.; Mano, J.F. Self-Assembled Bioactive Colloidal Gels as Injectable Multiparticle Shedding Platforms. *ACS Appl. Mater. Interfaces* **2020**, *12*, 31282–31291. [CrossRef]
23. Yang, X.; Xia, H.; Li, Y.; Cheng, Y.; Wang, Y.; Xia, Y.; Yue, Y.; Cheng, X.; Chu, Z. In vitro and Ex vivo Antioxidant Activity and Sustained Release Properties of Sinomenine-Loaded Liposomes-in-Hydrogel Biomaterials Simulating Cells-in-Extracellular Matrix. *Nat. Prod. Commun.* **2022**, *17*, 1934578X221130699. [CrossRef]
24. Ho, T.-C.; Chang, C.-C.; Chan, H.-P.; Chung, T.-W.; Shu, C.-W.; Chuang, K.-P.; Duh, T.-H.; Yang, M.-H.; Tyan, Y.-C. Hydrogels: Properties and Applications in Biomedicine. *Molecules* **2022**, *27*, 2902. [CrossRef]
25. Large, D.E.; Abdelmessih, R.G.; Fink, E.A.; Auguste, D.T. Liposome composition in drug delivery design, synthesis, characterization, and clinical application. *Adv. Drug Deliv. Rev.* **2021**, *176*, 113851. [CrossRef]
26. Dumortier, A.; Durham, A.D.; Di Piazza, M.; Vauclair, S.; Koch, U.; Ferrand, G.; Ferrero, I.; Demehri, S.; Song, L.L.; Farr, A.G.; et al. Atopic dermatitis-like disease and associated lethal myeloproliferative disorder arise from loss of Notch signaling in the murine skin. *PLoS ONE* **2010**, *5*, e9258. [CrossRef]
27. Ziegler, S.F. The role of thymic stromal lymphopoietin (TSLP) in allergic disorders. *Curr. Opin. Immunol.* **2010**, *22*, 795–799. [CrossRef]
28. Tidwell, W.J.; Fowler, J.F., Jr. T-cell inhibitors for atopic dermatitis. *J. Am. Acad. Dermatol.* **2018**, *78*, S67–S70. [CrossRef]
29. Czarnowicki, T.; Gonzalez, J.; Bonifacio, K.M.; Shemer, A.; Xiangyu, P.; Kunjravia, N.; Malajian, D.; Fuentes-Duculan, J.; Esaki, H.; Noda, S.; et al. Diverse activation and differentiation of multiple B-cell subsets in patients with atopic dermatitis but not in patients with psoriasis. *J. Allergy Clin. Immunol.* **2016**, *137*, 118–129 e115. [CrossRef] [PubMed]
30. Kasperkiewicz, M.; Schmidt, E.; Ludwig, R.J.; Zillikens, D. Targeting IgE Antibodies by Immunoadsorption in Atopic Dermatitis. *Front. Immunol.* **2018**, *9*, 254. [CrossRef] [PubMed]
31. Wang, T.; Jian, Z.; Baskys, A.; Yang, J.; Li, J.; Guo, H.; Hei, Y.; Xian, P.; He, Z.; Li, Z.; et al. MSC-derived exosomes protect against oxidative stress-induced skin injury via adaptive regulation of the NRF2 defense system. *Biomaterials* **2020**, *257*, 120264. [CrossRef] [PubMed]
32. Mews, J.; Tomaszewska, A.; Siewiera, J.; Lewicki, S.; Kuczborska, K.; Lipińska-Opałka, A.; Kalicki, B. Effects of Hyperbaric Oxygen Therapy in Children with Severe Atopic Dermatitis. *J. Clin. Med.* **2021**, *10*, 1157. [CrossRef] [PubMed]

International Journal of
Molecular Sciences

MDPI

Article

Fabrication and Characterization of Pectin Films Containing Solid Lipid Nanoparticles for Buccal Delivery of Fluconazole

Namon Hirun [1], Jongjan Mahadlek [2], Sontaya Limmatvapirat [3], Pornsak Sriamornsak [3], Etsuo Yonemochi [4], Takayuki Furuishi [4] and Pakorn Kraisit [1,*]

1 Thammasat University Research Unit in Smart Materials and Innovative Technology for Pharmaceutical Applications (SMIT-Pharm), Faculty of Pharmacy, Thammasat University, Pathumthani 12120, Thailand; namon.hi@tu.ac.th

2 Pharmaceutical Intellectual Center "Prachote Plengwittaya", Faculty of Pharmacy, Silpakorn University, Nakhon Pathom 73000, Thailand; mahadlek_j@su.ac.th

3 Pharmaceutical Biopolymer Group (PBiG), Faculty of Pharmacy, Silpakorn University, Nakhon Pathom 73000, Thailand; limmatvapirat_s@su.ac.th (S.L.); sriamornsak_p@su.ac.th (P.S.)

4 Department of Physical Chemistry, School of Pharmacy and Pharmaceutical Sciences, Hoshi University, Tokyo 142-8501, Japan; e-yonemochi@hoshi.ac.jp (E.Y.); t-furuishi@hoshi.ac.jp (T.F.)

* Correspondence: pakorn54@tu.ac.th

Abstract: Fluconazole (FZ) is a potential antifungal compound for treating superficial and systemic candidiasis. However, the use of conventional oral drug products has some limitations. The development of buccal film may be a potential alternative to oral formulations for FZ delivery. The present study involved the development of novel FZ-loaded solid lipid nanoparticles (FZ-SLNs) in pectin solutions and the investigation of their particle characteristics. The particle sizes of the obtained FZ-SLNs were in the nanoscale range. To produce pectin films with FZ-SLNs, four formulations were selected based on the small particle size of FZ-SLNs and their suitable polydispersity index. The mean particle sizes of all chosen FZ-SLNs formulations did not exceed 131.7 nm, and the mean polydispersity index of each formulation was less than 0.5. The properties of films containing FZ-SLNs were then assessed. The preparation of all FZ-SLN-loaded pectin films provided the mucoadhesive matrices. The evaluation of mechanical properties unveiled the influence of particle size variation in FZ-SLNs on the integrity of the film. The Fourier-transform infrared spectra indicated that hydrogen bonds could potentially form between the pectin-based matrix and the constituents of FZ-SLNs. The differential scanning calorimetry thermogram of each pectin film with FZ-SLNs revealed that the formulation was thermally stable and behaved in a solid state at 37 °C. According to a drug release study, a sustained drug release pattern with a burst in the initial stage for all films may be advantageous for reducing the lag period of drug release. All prepared films with FZ-SLNs provided a sustained release of FZ over 6 h. The films containing FZ-SLNs with a small particle size provided good permeability across the porcine mucosa. All film samples demonstrated antifungal properties. These results suggest the potential utility of pectin films incorporating FZ-SLNs for buccal administration.

Keywords: fluconazole; pectin; solid lipid nanoparticles (SLNs); buccal film; antifungal

1. Introduction

Fluconazole (FZ) is a synthetic triazole agent belonging to the therapeutic compound class of azole antifungals. The antifungal action of FZ arises from the depletion of fungal ergosterol synthesis and impairment of the fungal cell membrane [1]. FZ is a potential antifungal used to treat superficial and systemic candidiasis caused by a fungus of the *Candida* genus. Although oral tablets and suspensions of FZ are commercially available, the oral administration of FZ has some limitations associated with first-pass metabolism and side effects, such as stomach upset, diarrhea, nausea, and vomiting [2–4]. In addition, it has been suggested that the rapid dissolution of FZ from solid oral dosage forms may contribute to the increased chance of side effects [5]. Lipid-based nanocarriers represent a potential strategy for addressing limitations associated with traditional oral therapy [6]. In recent years, the fabrication of novel lipid-based nanocarriers has attracted increasing attention and has been considered a promising vehicle for delivering antifungals [3,7]. To ensure efficient drug delivery and avoid the risk of side effects, the incorporation of nanocarriers into dosage forms has been proposed for buccal delivery instead of conventional oral drug administration [8]. Therefore, the development of a suitable dosage form containing lipid-based nanocarriers for buccal administration is a promising approach for FZ delivery.

Solid lipid nanoparticles (SLNs), which typically range in size from 10 to 1000 nm in diameter, are colloidal lipid-based nanocarriers made up of a solid lipid core surrounded by a surfactant layer [7,9]. The structural characteristics and types of SLNs depend on the preparation techniques, the preparation parameters, the chemical nature of formulation components, and the composition of the formulation [6,10]. The main advantages of using SLNs to deliver active compounds are their ability to enhance absorption and increase safety [10]. Furthermore, the ability of the SLNs to sustain the release of the active compound and promote the relevant biological activity makes them promising lipid-based nanocarriers for a variety of applications [11]. Over the years, the fabrication of drug-loaded SLNs has gained attention to accomplish enhanced permeability for transdermal drug delivery [12]. In addition, SLNs have been considered as a potential platform to improve drug permeability through the buccal mucosa [13]. The buccal delivery of drugs from the site of application to the bloodstream can bypass first-pass metabolism, avoid pre-systemic inactivation in the gastrointestinal tract, and minimize gastrointestinal side effects [8,14]. Therefore, the buccal administration of SLNs offers a viable, non-invasive strategy for local and systemic therapy. However, the use of plain drug-loaded SLNs for buccal delivery faces a challenge due to their short residence time on the buccal mucosa. The clearance of SLNs from the buccal mucosa is a consequence of tongue and masticatory movements, dilution by saliva, and swallowing [14,15]. Recently, the strategy of incorporating SLNs into the mucoadhesive matrix, which can achieve prolonged adhesion on the mucosa surface, has become an attractive approach to overcome the limitation [15,16].

Mucoadhesion, the phenomenon of adhesion between material and mucous membrane, attracts considerable research attention that has been devoted to extending the residence time of pharmaceutical drug formulations at the mucosal surface [17,18]. Mucoadhesive preparations are capable of mucoadhesion to offer drug release at the target site and/or to facilitate drug permeation across the mucosa [19]. Several mucoadhesive dosage forms, such as tablets, discs, and films, have been explored for buccal administration [20]. Mucoadhesive films are often preferred over other dosage forms as buccal delivery systems. This is because of the high flexibility, appropriate thickness, and suitable softness of the films, which provide better patient compliance [20,21]. To achieve the mucoadhesive capability, the buccal mucoadhesive films are fabricated with mucoadhesive polymers used as major film formers [22]. The existing literature emphasizes that the mucoadhesion produced by the mucoadhesive polymers is derived from the interconnection between the mucus and the polymers through physical association (chain interpenetration and entanglement) and chemical interactions, mainly H-bonding and van der Waals forces [20,22,23]. The presence of H-bonding functional groups, including hydroxyl, carboxyl, amine, and amide groups, in the polymer structure contributes to adhesion to the mucosa [23]. In addition, the chain flexibility of the mucoadhesive polymeric component is anticipated to encourage mucoadhesion of the dosage forms [19,22].

Pectin is a natural heteropolysaccharide, the structure of which is rich in carboxylic groups. Although the structure of pectin is complex and possesses interlinked distinct domains, the main component of pectin is homogalacturonan residue, which is partially methyl-esterified. According to the degree of esterification (DE), pectin can be classified into two major groups: low DE pectins (DE < 50%) and high DE pectins (DE > 50%). High DE pectins with a high molecular weight have high mucoadhesive performance on the mucosa of the gastrointestinal tract, especially the tissues of the buccal, stomach, small intestine, and large intestine, because their molecular features enable the formation of H-bonds as well as great entanglements between polymer and mucin chains [23,24]. To constitute a film layer, pectins with high DE values have the potential to be utilized as film-forming material [25,26]. As the DE of pectins increased, the films became more hydrophobic [27]. The formulations made from pectin with high DE are superior to the more hydrophilic pectin-derived matrices in terms of the sustained release of entrapped molecules [27]. Several previous studies have looked into the fabrication of pectin-based films enriched with nano-compositions, such as sulfur nanoparticles, silver nanoparticles, and nano clays, for food and biomedical applications [28–30]. A distribution of nanomaterials within the film may enhance interface contact between film constituents and refine film performance [31]. However, there is limited research investigating pectin-based film incorporated with drug-loaded SLNs.

The main objective of the present study was to fabricate novel pectin mucoadhesive films incorporated with FZ-loaded SLNs (FZ-SLNs) for drug delivery through the buccal mucosa. High-DE pectin with a 70% DE and molecular weight of 200 kDa was employed as a mucoadhesive film-forming polymer. According to previous work conducted by our research group on the preparation of FZ-SLNs, Tween® 80 and glyceryl monostearate (GMS) are potential ingredients that can form the FZ-loaded solid lipid particle at the nanoscale level [13]. Hence, Tween® 80 and GMS were suitable for use as stabilizers and solid lipids, respectively. The dispersion of FZ-SLNs in pectin solutions was characterized. After the film preparation, the FZ-SLN-incorporated films were characterized in terms of thickness, mucoadhesion, mechanical characteristics, Fourier-transform infrared (FTIR) spectrum, drug release profile, and permeability study. In addition, the antifungal activity of the film samples against *Candida albicans* was also determined.

2. Results and Discussion

2.1. The Particle Characteristic of FZ-SLNs in Pectin Solutions

To assess the effect of the amount of film-forming polymer on the particle characteristics of FZ-SLNs in dispersions, the amounts of pectin were varied as described in formulations C1–C3 (Table 1), while the content of the stabilizer and solid lipid in the formulations was fixed.

Table 1. Formulation composition of FZ-SLNs in pectin solutions.

Formulation Code	Formulation Composition (g) *		
	Pectin	GMS	Tween® 80
C1	3.8	2.0	2.0
C2	3.9	2.0	2.0
C3	4.0	2.0	2.0
C4	4.0	1.9	2.0
C5	4.0	1.8	2.0
C6	4.0	2.0	2.1
C7	4.0	2.0	2.2

* The amounts of FZ and glycerol used for each formulation are 0.075 g and 2 g, respectively.

Figure 1a shows the particle size and polydispersity index (PDI) for FZ-SLNs in pectin solutions with different pectin content. As can be seen, the mean values of the particle size of FZ-SLNs were in the range between 131.7 and 136.6 nm, which were in the nanosize range. The small mean PDI values of FZ-SLNs in C2 and C3 were 0.444 and 0.430, respectively, while C1 represented a high mean PDI value of 0.615. The PDI values found in C2 and C3 were significantly lower than that of C1 ($p < 0.05$). The PDI values decline with increasing pectin content, indicating that the addition of pectin narrowed the particle size distribution. The small PDI value of 0.5 and below is acceptable and a representation of uniform SLN dispersion with a narrow particle size distribution [32,33]. The homogeneous size distribution leads to the desired effect on the physical stability of the nanoparticle dispersion and is a favorable particle characteristic [31]. Because the formulation C3 contained high mucoadhesive polymer content and possessed nanoparticles of a small size and a small PDI value, the pectin content as described in the formulation C3 was selected, and the content of solid lipid and stabilizer was varied further to investigate the effect of solid lipid and stabilizer contents.

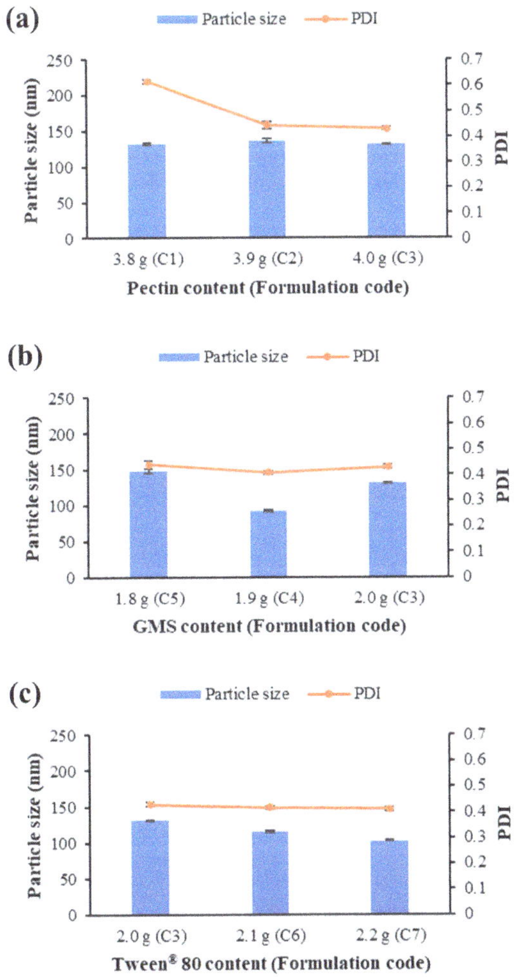

Figure 1. Particle size and polydispersity index (PDI) of FZ-SLNs in pectin solutions containing different content of (**a**) pectin, (**b**) GMS, and (**c**) Tween® 80.

As described in Table 1, the content of GMS was varied in formulations C3, C4, and C5. Figure 1b shows the particle size and PDI for FZ-SLNs in pectin solutions with different GMS content. The differences in the particle sizes among formulations C3, C4, and C5 were statistically significant. The formulation C4 contained the smallest particles compared to other formulations with distinct lipid contents (C3 and C5). This indicated that a decrease in the particle size of the SLNs was observed as the lipid content increased to a certain value, after which an increase in the particle size of the SLNs was observed. Although the increase in lipid concentration generally produces the size increment of SLNs, the reduction in particle size caused by an increase in lipid content has also been observed [35]. It was reported that a concentration-dependent increase in particle size began as the lipid content of the SLN dispersion exceeded a critical value because of the insufficient amount of the coating stabilizer layer for high lipid content above a certain level [35,36]. At lipid contents lower than the critical level, the increase in lipid concentration exerts a negligible influence on the size increment of particles [35].

As described in Table 1, the content of the stabilizer, Tween® 80, was varied and increased in the formulations C3, C6, and C7. From Figure 1c, it can be seen that the mean values of the particle size decreased with increasing the content of Tween® 80. The differences in the particle size values among these sample formulations were statistically significant ($p < 0.05$). The PDI values of the samples tended to decrease when the amount of stabilizer was increased. The tendency of the stabilizer to form a stabilizer layer at the interface and decrease the interfacial tension may prevent the agglomeration of lipid particles, leading to the formation of small particles with a uniform size distribution [32].

Formulations C3, C4, C6, and C7, which possessed an acceptable PDI and were the top four formulations with the smallest particle size, were selected for film preparation and subsequent film testing.

2.2. The Mucoadhesion of Pectin Films Containing FZ-SLNs

The films containing FZ-SLNs were successfully prepared using the film casting method for formulations C3, C4, C6, and C7. The FZ-SLN-incorporated film types obtained from formulations C3, C4, C6, and C7 were referred to as FC3, FC4, FC6, and FC7, respectively. The thickness values of the obtained films for FC3, FC4, FC6, and FC7 were 0.194 ± 0.012, 0.193 ± 0.013, 0.186 ± 0.005, and 0.183 ± 0.037 mm, respectively. There was no statistical significance in the thickness values among these obtained films.

The mucoadhesion of FC3, FC4, FC6, and FC7 is shown in Figure 2. All films possessed the mucoadhesive ability. FC3 showed the greatest mean values for both work of adhesion (W_{ad}) and maximum force (F_{max}). However, there was no statistically significant difference in F_{max} values among FC3, FC4, FC6, and FC7. Also, the W_{ad} values of FC6 and FC7 were comparable. Although FC3, FC6, and FC7 contained distinct Tween® 80 contents, there was no statistically significant difference in the W_{ad} values among these films. This may imply that the amount of Tween® 80 had no discernible effect on the mucoadhesive property of the studied films. However, the W_{ad} values of the film formulations with high GMS content (FC3) were significantly greater than those of the film with low GMS content (FC4) ($p < 0.05$). The W_{ad} value reflects the adhesion energy between the mucoadhesive film and buccal mucosa. The structure of GMS contains hydroxyl groups, and pectin chains possess abundant carboxyl groups [15,23]. These functional groups play a crucial role in their interactions with mucin. Furthermore, the mucoadhesive characteristic of pectin may arise from their adsorption onto the mucin layer [37]. Therefore, using pectin as the film-forming polymer provided the FZ-SLN-incorporated film with the mucoadhesive ability. In addition, high GMS content in the FZ-SLN-incorporated film enriched the capacity for mucoadhesion compared with that of the film containing low GMS content.

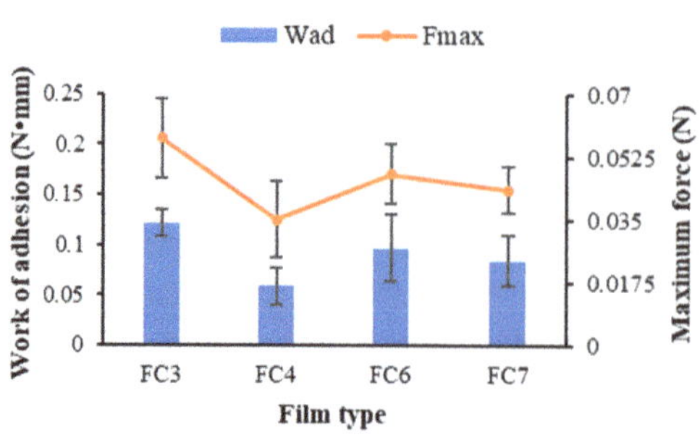

Figure 2. Mucoadhesive properties of prepared pectin films containing FZ-SLNs.

2.3. The Mechanical Properties of Pectin Films Containing FZ-SLNs

Tensile strength is a quantitative mechanical parameter representing film integrity, and the percent elongation at break is a measurement of film flexibility [38,39]. Figure 3 represents the tensile strength and percent elongation at the break of FC3, FC4, FC6, and FC7. The tensile strength of FC4 was significantly greater than that of other films, while the values of tensile strength and percent elongation at the break of FC3 were significantly lower than those of others ($p < 0.05$). Based on the particle size of the FZ-SLNs dispersed in the studied film types, FC4 had the smallest nanoparticles, while FC3 contained the largest nanoparticles. The impact of the particle size variation on the mechanical characteristics of polysaccharide-based films containing nanoparticles has been reported by de Moura et al. [40]. The high surface area of small nanoparticles accelerated their interaction with the polysaccharide matrix during film formation, while the large nanoparticles caused the hindrance effect on film formation. Therefore, the presence of small nanoparticles caused higher film integrity and tenacity compared with film containing nanoparticles of a large size [40].

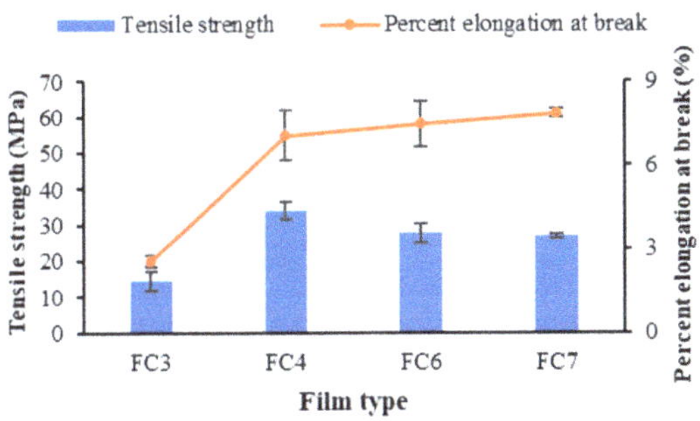

Figure 3. Mechanical properties of prepared pectin films containing FZ-SLNs.

2.4. FTIR Spectra

The FTIR spectra of pectin, FZ-loaded pectin film without SLNs (FZ film), and pectin films with FZ-SLNs are presented in Figure 4. As shown in the FTIR spectrum of pectin (Figure 4a), the two broad bands in the region between 3000 and 3700 cm^{-1} correspond to the stretching vibrations of hydroxyl groups, which are the abundant functional groups found in the chain structure of pectin [41,42]. In this OH stretching range, the band observed in the higher wavenumber region of 3450–3700 cm^{-1} reflects the presence of free OH groups, while the band appearing in the lower wavenumber region of 3000–3450 cm^{-1} corresponds to the existence of hydrogen-bonded OH groups [42]. The appearance of C=O peaks at 1629 cm^{-1} and 1739 cm^{-1} may be attributed to the presence of free and esterified carboxyl groups in the pectin chain structure [43]. The main characteristic peaks of pectin were also observed in FZ film (Figure 4a). According to the FTIR spectrum of FC3 (Figure 4a), two distinct peaks near 2848 cm^{-1} and 2914 cm^{-1}, which can be assigned to CH stretching of GMS, revealed the presence of GMS in the film [15]. Although there was no noticeable change in the other main characteristic peaks of pectin in the spectrum of FC3, the band of the free OH groups seemed to have disappeared, and a band that broadened in the OH stretching range (3000–3700 cm^{-1}) was observed in the FTIR pattern of FC3. The decrease in the sharpness and the broader band of OH stretching could relate to the formation of hydrogen bonds between the film matrix and nanocomponent [44]. Accordingly, the interaction between the pectin-based matrix and the composition of SLNs may cause a reduction in the number of free OH groups [44,45]. No apparent pattern alteration in the FTIR pattern of pectin films incorporating FZ-SLNs was observed when comparing the FTIR spectrum of FC3 with the spectra of various pectin films containing FZ-SLNs (Figure 4b).

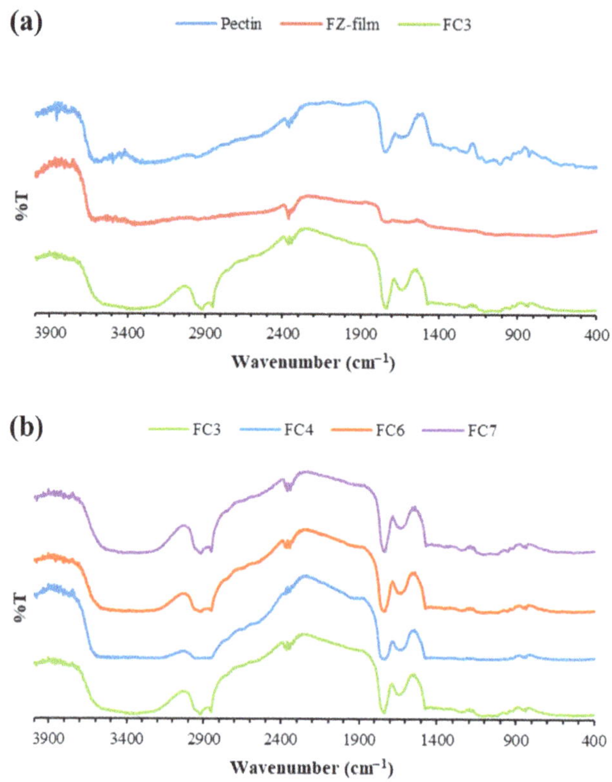

Figure 4. FTIR spectra of (**a**) pectin, FZ film, and FC3 and (**b**) distinct pectin films containing FZ-SLNs.

2.5. Differential Scanning Calorimetry

Figure 5 depicts the thermograms of pectin and pectin films with FZ-SLNs, which were characterized by differential scanning calorimetry (DSC). For pectin, the broad endothermic peak detected within the temperature range of 50 to 125 °C may reflect the release of water associated with the hydrophilic residues of the polymer. This thermal event has also been observed for pectin powder, as reported in the literature [46,47]. All pectin films with FZ-SLNs also exhibited this broad endothermic signal, but their thermograms displayed an additional endothermic peak at the low-temperature side in the range of 50 to 70 °C. The melting transition of GMS has been reported to be approximately 61.8 °C [13]. The presence of the endothermic signal in the range of 50 to 70 °C revealed the melting of GMS in the formulations upon heating. Although it has been reported that the plain FZ powder exhibited a sharp characteristic melting peak around 141 °C [13], the DSC thermograms of all pectin films with FZ-SLNs do not show the sharp melting peak of the drug. The integration of drug molecules into the carrier and the resulting conversion of the drug from a crystalline to an amorphous state may account for the lack of a drug melting peak in FC3, FC4, FC6, and FC7 [13]. All thermal events observed in pectin films with FZ-SLNs occurred at temperatures higher than 37 °C, suggesting that all formulations are thermally stable and maintain their solid state at body temperature.

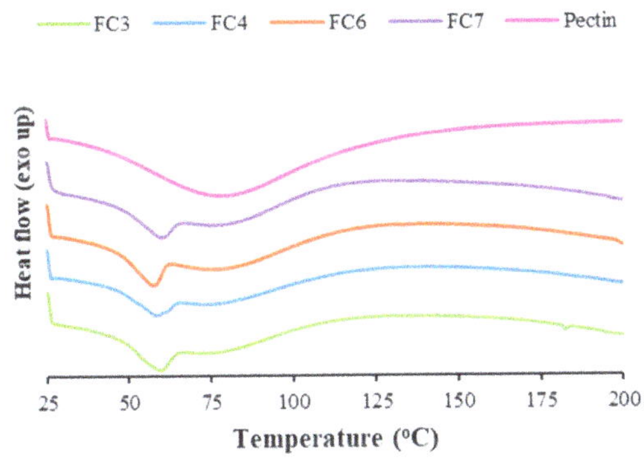

Figure 5. DSC thermograms of pectin and pectin films containing FZ-SLNs.

2.6. In Vitro Release of FZ

The in vitro release profiles of FZ from pectin films containing FZ-SLNs are depicted in Figure 6. For all FZ-SLN-incorporated film types, the drug release patterns represented the burst release of FZ that occurred in the early stage and was after that followed by a sustained release over time. The lipophilicity of FZ is moderate, and FZ demonstrates viability in both aqueous and lipophilic environments [48,49]. Therefore, some FZ molecules may deposit at the surface of SLNs and the polymer matrix during preparation. The presence of free drug molecules in the outer layers of SLNs may account for the initial burst release of the drug from the SLN-loaded matrix, while the subsequent sustained release of the drug could be attributed to the liberation of the encapsulated drug molecules from the internal lipid phase [50,51]. The SLN structure as well as the polymeric matrix act as barriers to drug diffusion and transportation, resulting in sustained drug release [51]. The sustained drug release pattern with the burst in the initial stage could be beneficial in reducing the lag period of drug release [50,52]. The release patterns of FZ from FC6 and FC7 were comparable and showed a more sustained drug release than those from FC3 and FC4. Considering the presence of a surfactant in each FZ-SLN-loaded film type, the amount of Tween® 80 in the films FC6 and FC7 was higher than in FC3 and FC4. It has been proposed that an inverse relationship between the Tween® 80 content and the order of drug release occurs because a higher surfactant content stabilizes the SLNs and retards the release of the encapsulated drug from the lipid core to the outside of the SLNs [53].

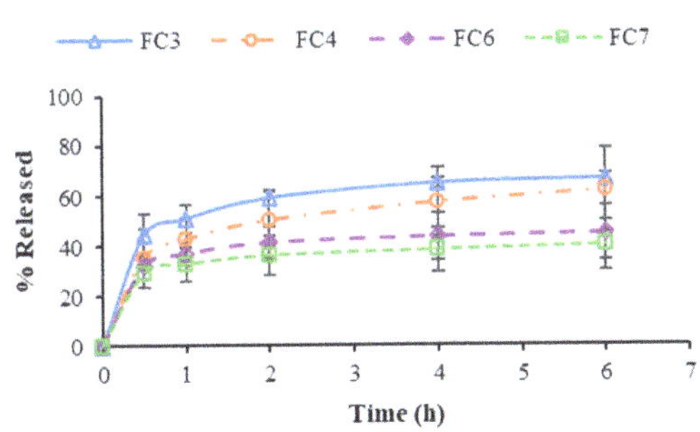

Figure 6. In vitro release of FZ from prepared pectin films containing FZ-SLNs.

2.7. Permeation Study

The permeability parameters, like the flux and apparent permeability coefficient for FC3, FC4, FC6, and FC7, are represented in Figure 7. Although there is no statistical difference among different film types for both parameters, the mean values of both parameters were found to be the lowest for FC3, which contained the largest size of FZ-SLNs. The mean values of the permeability parameters were found to slightly increase for FC4, FC6, and FC7, which possessed smaller FZ-SLNs. The variation in the permeability of various samples could be related to the size of the nanoparticles. The flux of drug-loaded SLNs across the mucosal membrane likely varies in relation to particle size, and drug carriers with smaller particle sizes provide good permeation ability [54].

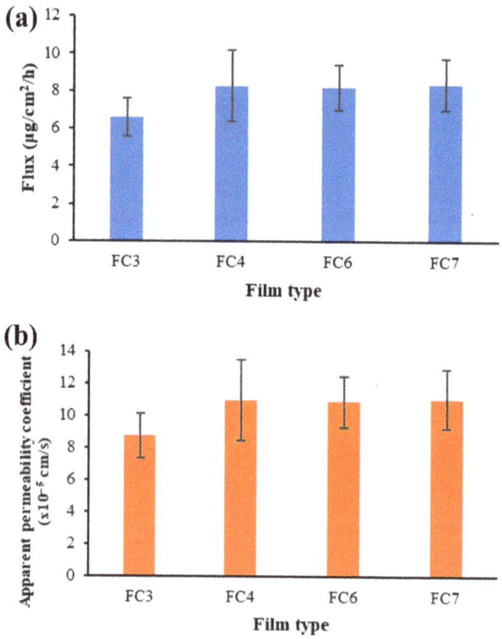

Figure 7. Permeability parameters, (**a**) flux and (**b**) apparent permeability coefficient, for prepared pectin films containing FZ-SLNs.

2.8. In Vitro Antifungal Activity

The antifungal activity of the prepared FZ-SLN-incorporated films was investigated. Ethanol solution and the solution of FZ in ethanol were utilized as the negative control and positive control, respectively. The diameters of the inhibition zones are shown in Figure 8. The positive control showed antifungal activity, while no antifungal activity was observed for the negative control due to the lack of FZ in the negative control. All prepared FZ-SLN-incorporated films exhibited antifungal activity against *Candida albicans*. The diameter of the inhibition zone was virtually greater for the positive control compared to all prepared films. This seems to be attributed to the sustained release behavior of the prepared FZ-SLN-incorporated films, as described in the previous section, while the positive control contained FZ in the already-dissolved form in ethanol [15]. It has been reported that antifungal-loaded vehicles have similar but slightly lower in vitro antifungal activity when drugs are released from sustained drug delivery systems compared to the pure drug in media, and the antifungal activity could account for the release of drug molecules from carriers [15,55].

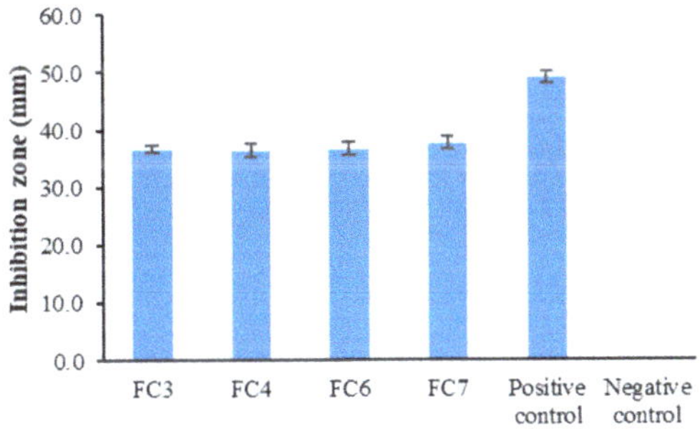

Figure 8. Diameters of the inhibition zone of prepared pectin films containing FZ-SLNs and controls.

3. Materials and Methods

3.1. Materials

FZ was purchased from TCI (Tokyo, Japan). High-DE pectin with a 70% DE and molecular weight of 200 kDa was kindly provided by Herbstreith & Fox GmbH (Werder, Germany). GMS and Tween® 80 (polysorbate) were purchased from PC Drug Co., Ltd. (Bangkok, Thailand). *Candida albicans* ATCC 10231 was purchased from the Department of Medical Sciences (Bangkok, Thailand). All additional chemicals employed in this investigation were of analytical grade and were utilized without any modifications.

3.2. Preparation of FZ-SLNs in Pectin Solutions and Pectin Films Containing FZ-SLNs

Table 1 shows the amounts of pectin, Tween® 80, and GMS in each formulation of FZ-SLNs in pectin solution. The amounts of the drug (FZ) and glycerol used as a plasticizer were 0.075 g and 2 g, respectively, for all formulations. The remaining weight of the composition after subtracting the quantities of the ingredients mentioned above from the total formulation weight was the weight of the water used to prepare the formulation, with a total weight of 150 g.

The polymer solution was composed of pectin and glycerol. The required amount of pectin was dispersed in 3/4 of the water content of the formulation. After the homogenous pectin solution was obtained, 2 g of glycerol was dissolved in the pectin solution.

The FZ-SLNs were prepared using the hot homogenization method. To prepare the water phase, Tween® 80 was dissolved in the remaining 1/4 of the water content, and the water phase was heated to 80 °C. The oil phase consisted of 0.075 g of FZ and GMS regarding the content described in Table 1 for each formulation. GMS and FZ were both melted and homogenized to prepare the oil phase at 80 °C. To prepare the primary emulsion, the oil phase was poured into the water phase, and both phases were mixed using a high-shear mixer at 7000 rpm for 5 min. The size of the internal phase was then reduced under ultrasonic amplitude at 100% by using an ultrasonic probe sonicator. During ultrasonication for 15 min, the sample was kept in an ice bath to acerate the formation of FZ-SLNs.

The FZ-SLNs in pectin solutions were fabricated by dispersing the prepared FZ-SLNs into the polymer solution with continuous mechanical stirring at 800 rpm for 30 min.

For preparing the pectin films containing FZ-SLNs, the FZ-SLNs in the pectin solution were transferred into the glass plate and dried according to the film casting method as previously described [15].

The FZ in the polymer solution for preparing FZ-loaded pectin film without SLNs (FZ film) was prepared by dissolving the same amount of FZ in 7.5 mL of ethanol and then adding the ethanol solution containing FZ to the polymer solution. FZ film was also obtained by using the film casting method.

3.3. The Particle Size and the Nanoparticle Size Distribution Characterization

The particle size and the nanoparticle size distribution reported as PDI were evaluated using dynamic light scattering. Each sample of FZ-SLNs in the pectin solution was diluted with distilled water before investigation using a Malvern Zetasizer Nano ZS (Malvern Instruments, Malvern, Worcestershire, UK) to prevent multi-scattering effects [15]. The experiments were conducted in triplicate.

3.4. Mechanical Characterization and Mucoadhesion Studies

A Mini Test 600 thickness gauge was employed to determine the thickness of the obtained films. Each prepared film was cut into a square shape with a size of 2.0 cm × 2.0 cm. The mechanical properties, including tensile strength and percent elongation at break, of the films were determined by using a TA-XT plus texture analyzer (Stable Micro Systems, Godalming, Surrey, UK). Porcine buccal mucosa was utilized as a biological membrane in the present work, owing to its resemblance to human buccal tissue [21]. The TA-XT plus texture analyzer fitted with a mucoadhesive holder was also used to measure the mucoadhesive characteristics according to the method of Kraisit et al. [21].

3.5. FTIR Spectroscopy

The FTIR spectra of plain pectin, FZ film, and pectin films containing FZ-SLNs were obtained using an FTIR spectrometer (IRTracer-100, Shimadzu, Kyoto, Japan). The samples were prepared using the KBr method. Each sample was ground, mixed with KBr powder, and compressed using a hydraulic press. The sample disc was placed in the sample holder and scanned from 4000 to 400 cm^{-1} at a resolution of 4 cm^{-1}, with 20 scans.

3.6. DSC Measurements

DSC thermograms of plain pectin and pectin films containing FZ-SLNs were obtained in the temperature range of 25–200 °C with a heating rate of 10 °C/min, using a differential scanning calorimeter (DSC 3+, Mettler Toledo, Viroflay, France). The DSC scans were performed under a nitrogen atmosphere at a gas flow rate of 50 mL/min.

3.7. In Vitro Drug Release Study

The release patterns of FZ from the pectin films containing FZ-SLNs were determined in triplicate utilizing a modified Franz-type diffusion cell. The films with a circular shape and a diameter of 1.5 cm were placed on a fiber mesh before being applied to the receptor cell. A mixture of simulated salivary fluid (pH 6.8) and ethanol at a volume ratio of 1:1 was used as the release medium. The receptor chamber was filled with 15 mL of the release medium, and the temperature of the diffusion cell was controlled using a water jacket at 37 °C. At each time point, 0.5 mL of the release medium was taken and replaced with an equivalent volume of the new medium. The quantity of FZ in the collected release medium was analyzed using high-performance liquid chromatography (HPLC) coupled with a C18 column at 260 nm UV detection. The mobile phase used for HPLC separation was methanol/water (40/60, v/v), and the flow rate was 0.8 mL/min.

3.8. In Vitro Permeation Study

The in vitro permeation study was performed in triplicate using a modified Franz-type diffusion cell. The experimental setup closely resembled that outlined in the section on in vitro drug release study, with the exception that porcine buccal mucosa was utilized as the barrier between the donor compartment and the receptor chamber for the permeation study. The flux (μg/cm^2/h), and the apparent permeability coefficient ($\times 10^{-5}$ cm/s) were estimated from the permeation data.

3.9. In Vitro Antifungal Activity

The agar diffusion method was employed to assess the in vitro antifungal activity of the chosen films against *Candida albicans* ATCC 10231, as Phaechamud and Mahadlek described [56]. The fungi were cultivated and preserved at 37 °C for 48 h using Sabouraud dextrose broth (Difco™, Franklin Lakes, NJ, USA). Subsequently, the broth was diluted to achieve a turbidity level of approximately 10^8 colony-forming units (CFUs) per milliliter. The inoculated medium was swabbed and spread on a Sabouraud dextrose agar plate (Difco™). Every film was precisely sliced into a circular shape with a diameter of approximately 1.0 cm and carefully positioned on the surface of the inoculated medium. In this experiment, sterilized cylinder cups were meticulously placed on the surface of the inoculated medium. Each cylinder cup was then filled with the positive control, which consisted of FZ dissolved in ethanol, and the negative control, which consisted of an ethanol solution. The films and controls were incubated at 37 °C for 48 h. The diameter of the inhibition zone for antifungal activity was measured in millimeters (mm).

3.10. Statistical Analysis

The means and standard deviations of the results from the experiments conducted in triplicate were computed and reported. The statistical comparison was performed using the analysis of variance (ANOVA). Statistical significance was accepted at the 0.05 level.

4. Conclusions

In this study, FZ-SLNs in pectin solutions were formulated, and their particle characteristics were investigated. The high amount of pectin resulted in a decrease in the size of the obtained FZ-SLNs. Similarly, increasing the amount of Tween® 80 also reduced the size of the FZ-SLNs when pectin was present in high quantities. A decrease in the particle size of the FZ-SLNs as the GMS content increased was observed, followed by an increase in the particle size when the GMS content exceeded a certain value. Four formulations with small particle sizes and an acceptable PDI were chosen to fabricate novel pectin films with FZ-SLNs. Each of the selected formulations had a mean particle size not greater than 131.7 nm and a mean PDI value below 0.5. Despite variations in the excipient composition within the film, comparable film thicknesses were achieved. All prepared films showed mucoadhesive ability; the presence of high GMS content in FZ-SLN-incorporated films tended to enrich the mucoadhesiveness. The mechanical characteristic investigation revealed the impact of the particle size variation on film integrity. Based on the FTIR spectra, hydrogen bonds might form between the pectin-based matrix and the SLNs' components. The DSC thermogram of each pectin film containing FZ-SLNs did not show any thermal events below 37 °C, indicating that the formulation stayed solid at body temperature. According to the drug release patterns, a sustained drug release pattern with a burst in the initial stage for all films could be beneficial in reducing the lag period of drug release. The film with a smaller particle size of FZ-SLNs showed a slight increase in permeability across the porcine mucosa. All film samples possessed antifungal activity. These results reveal the potential application of pectin films containing FZ-SLNs for buccal drug delivery.

Author Contributions: Conceptualization, N.H., S.L., E.Y., T.F. and P.K.; methodology, N.H., J.M. and P.K.; formal analysis, N.H.; investigation, J.M. and P.K.; resources, J.M., P.S., E.Y., T.F. and P.K.; writing—original draft preparation, N.H.; writing—review and editing, P.K.; visualization, N.H. and P.K.; supervision, S.L., P.S., E.Y. and P.K.; project administration, P.K.; funding acquisition, P.K. All authors have read and agreed to the published version of the manuscript.

Funding: The authors wish to express their gratitude to the Thailand Research Funds and the Office of the Higher Education Commission for financial support under Contract No. MRG6180269. This work was supported by Thammasat University Research Unit in Smart Materials and Innovative Technology for Pharmaceutical Applications (SMIT-Pharm) [Project ID 6305016].

Institutional Review Board Statement: Not applicable.

Informed Consent Statement: Not applicable.

Data Availability Statement: Data will be made available on request.

Acknowledgments: The authors wish to express their gratitude to the Thailand Research Funds and the Office of the Higher Education Commission for financial support under Contract No. MRG6180269. This work was supported by Thammasat University Research Unit in Smart Materials and Innovative Technology for Pharmaceutical Applications (SMIT-Pharm) [Project ID 6305016]. The authors would like to thank Herbstreith & Fox GmbH (Werder, Germany) for contributing pectin samples. Additionally, Premjit Limpamanoch's laboratory assistance is greatly appreciated. Correspondingly, we would like to express our gratitude to Hoshi University's School of Pharmacy and Pharmaceutical Sciences and Silpakorn University's Faculty of Pharmacy for providing supportive facilities and equipment.

Conflicts of Interest: The authors declare no conflicts of interest.

References

1. Sawant, B.; Khan, T. Recent advances in delivery of antifungal agents for therapeutic management of candidiasis. *Biomed. Pharmacother.* **2017**, *96*, 1478–1490. [CrossRef] [PubMed]
2. Naz, K.; Shahnaz, G.; Ahmed, N.; Qureshi, N.A.; Sarwar, H.S.; Imran, M.; Khan, G.M. Formulation and in vitro characterization of thiolated buccoadhesive film of fluconazole. *AAPS PharmSciTech* **2017**, *18*, 1043–1055. [CrossRef] [PubMed]
3. Fernandes, A.V.; Pydi, C.R.; Verma, R.; Jose, J.; Kumar, L. Design, preparation and in vitro characterizations of fluconazole loaded nanostructured lipid carriers. *Braz. J. Pharm. Sci.* **2020**, *56*, e18069. [CrossRef]
4. Hatamiazar, M.; Mohammadnejad, J.; Khaleghi, S. Chitosan-albumin nanocomposite as a promising nanocarrier for efficient delivery of fluconazole against vaginal candidiasis. *Appl. Biochem. Biotechnol.* **2024**, *196*, 701–716. [CrossRef] [PubMed]
5. Mushtaq, A.; Baseer, A.; Zaidi, S.S.; Waseem Khan, M.; Batool, S.; Elahi, E.; Aman, W.; Naeem, M.; Din, F.U. Fluconazole-loaded thermosensitive system: In vitro release, pharmacokinetics and safety study. *J. Drug Deliv. Sci. Technol.* **2022**, *67*, 102972. [CrossRef]
6. Shirodkar, R.K.; Kumar, L.; Mutalik, S.; Lewis, S. Solid lipid nanoparticles and nanostructured lipid carriers: Emerging lipid based drug delivery systems. *Pharm. Chem. J.* **2019**, *53*, 440–453. [CrossRef]
7. Almawash, S. Solid lipid nanoparticles, an effective carrier for classical antifungal drugs. *Saudi Pharm. J.* **2023**, *31*, 1167–1180. [CrossRef]
8. Tran, P.H.L.; Duan, W.; Tran, T.T.D. Recent developments of nanoparticle-delivered dosage forms for buccal delivery. *Int. J. Pharm.* **2019**, *571*, 118697. [CrossRef]
9. Son, G.H.; Lee, B.J.; Cho, C.W. Mechanisms of drug release from advanced drug formulations such as polymeric-based drug-delivery systems and lipid nanoparticles. *J. Pharm. Investig.* **2017**, *47*, 287–296. [CrossRef]
10. Lacatusu, I.; Badea, N.; Murariu, A.; Oprea, O.; Bojin, D.; Meghea, A. Antioxidant activity of solid lipid nanoparticles loaded with umbelliferone. *Soft Mater.* **2013**, *11*, 75–84. [CrossRef]
11. Lacatusu, I.; Badea, N.; Stan, R.; Meghea, A. Novel bio-active lipid nanocarriers for the stabilization and sustained release of sitosterol. *Nanotechnology* **2012**, *23*, 455702. [CrossRef] [PubMed]
12. Souto, E.B.; Baldim, I.; Oliveira, W.P.; Rao, R.; Yadav, N.; Gama, F.M.; Mahant, S. SLN and NLC for topical, dermal, and transdermal drug delivery. *Expert Opin. Drug Deliv.* **2020**, *17*, 357–377. [CrossRef] [PubMed]
13. Kraisit, P.; Hirun, N.; Mahadlek, J.; Limmatvapirat, S. Fluconazole-loaded solid lipid nanoparticles (SLNs) as a potential carrier for buccal drug delivery of oral candidiasis treatment using the Box-Behnken design. *J. Drug Deliv. Sci. Technol.* **2021**, *63*, 102437. [CrossRef]
14. Nair, V.V.; Cabrera, P.; Ramírez-Lecaros, C.; Jara, M.O.; Brayden, D.J.; Morales, J.O. Buccal delivery of small molecules and biologics: Of mucoadhesive polymers, films, and nanoparticles—An update. *Int. J. Pharm.* **2023**, *636*, 122789. [CrossRef] [PubMed]
15. Kraisit, P.; Yonemochi, E.; Furuishi, T.; Mahadlek, J.; Limmatvapirat, S. Chitosan film containing antifungal agent-loaded SLNs for the treatment of candidiasis using a Box-Behnken design. *Carbohydr. Polym.* **2022**, *283*, 119178. [CrossRef] [PubMed]
16. Tzanova, M.M.; Hagesaether, E.; Tho, I. Solid lipid nanoparticle-loaded mucoadhesive buccal films—Critical quality attributes and in vitro safety & efficacy. *Int. J. Pharm.* **2021**, *592*, 120100. [PubMed]
17. Pham, Q.D.; Nöjd, S.; Edman, M.; Lindell, K.; Topgaard, D.; Wahlgren, M. Mucoadhesion: Mucin-polymer molecular interactions. *Int. J. Pharm.* **2021**, *610*, 121245. [CrossRef] [PubMed]
18. Bayer, I.S. Recent advances in mucoadhesive interface materials, mucoadhesion characterization, and technologies. *Adv. Mater. Interfaces* **2022**, *9*, 2200211. [CrossRef]
19. Sandri, G.; Rossi, S.; Ferrari, F.; Bonferoni, M.C.; Caramella, C.M. Mucoadhesive polymers as enabling excipients for oral mucosal drug delivery. In *Oral Mucosal Drug Delivery and Therapy*; Rathbone, M.J., Senel, S., Pather, I., Eds.; Springer: Boston, MA, USA, 2015; pp. 53–88.
20. Rohani Shirvan, A.; Hemmatinejad, N.; Bahrami, S.H.; Bashari, A. A comparison between solvent casting and electrospinning methods for the fabrication of neem extract-containing buccal films. *J. Ind. Text.* **2021**, *51*, 311S–335S. [CrossRef]
21. Kraisit, P.; Limmatvapirat, S.; Luangtana-Anan, M.; Sriamornsak, P. Buccal administration of mucoadhesive blend films saturated with propranolol loaded nanoparticles. *Asian J. Pharm. Sci.* **2018**, *13*, 34–43. [CrossRef]
22. Soe, M.T.; Pongjanyakul, T.; Limpongsa, E.; Jaipakdee, N. Films fabricated with native and ball-milled modified glutinous rice starch: Physicochemical and mucoadhesive properties. *Starch* **2021**, *73*, 2000012. [CrossRef]

23. Thirawong, N.; Nunthanid, J.; Puttipipatkhachorn, S.; Sriamornsak, P. Mucoadhesive properties of various pectins on gastrointestinal mucosa: An in vitro evaluation using texture analyzer. *Eur. J. Pharm. Biopharm.* **2007**, *67*, 132–140. [CrossRef] [PubMed]
24. Laurén, P.; Paukkonen, H.; Lipiäinen, T.; Dong, Y.; Oksanen, T.; Räikkönen, H.; Ehlers, H.; Laaksonen, P.; Yliperttula, M.; Laaksonen, T. Pectin and mucin enhance the bioadhesion of drug loaded nanofibrillated cellulose films. *Pharm. Res.* **2018**, *35*, 145. [CrossRef]
25. Maftoonazad, N.; Ramaswamy, H.S.; Marcotte, M. Evaluation of factors affecting barrier, mechanical and optical properties of pectin-based films using response surface methodology. *J. Food Process Eng.* **2007**, *30*, 539–563. [CrossRef]
26. Pérez, C.D.; Flores, S.K.; Marangoni, A.G.; Gerschenson, L.N.; Rojas, A.M. Development of a high methoxyl pectin edible film for retention of l-(+)-ascorbic acid. *J. Agric. Food Chem.* **2009**, *57*, 6844–6855. [CrossRef] [PubMed]
27. Liu, L.; Chen, G.; Fishman, M.L.; Hicks, K.B. Pectin gel vehicles for controlled fragrance delivery. *Drug Deliv.* **2005**, *12*, 149–157. [CrossRef] [PubMed]
28. Vartiainen, J.; Tammelin, T.; Pere, J.; Tapper, U.; Harlin, A. Biohybrid barrier films from fluidized pectin and nanoclay. *Carbohydr. Polym.* **2010**, *82*, 989–996. [CrossRef]
29. Shankar, S.; Tanomrod, N.; Rawdkuen, S.; Rhim, J.W. Preparation of pectin/silver nanoparticles composite films with UV-light barrier and properties. *Int. J. Biol. Macromol.* **2016**, *92*, 842–849. [CrossRef] [PubMed]
30. Ezati, P.; Rhim, J.W. pH-responsive pectin-based multifunctional films incorporated with curcumin and sulfur nanoparticles. *Carbohydr. Polym.* **2020**, *230*, 115638. [CrossRef]
31. Villa-Rojas, R.; Valdez-Fragoso, A.; Mújica-Paz, H. Manufacturing methods and engineering properties of pectin-based nanobiocomposite films. *Food Eng. Rev.* **2018**, *10*, 46–56. [CrossRef]
32. Alnusaire, T.S. Olive leaves (*Olea europaea* L) extract loaded lipid nanoparticles: Optimization of processing parameters by Box-Behnken statistical design, in-vitro characterization, and evaluation of anti-oxidant and anti-microbial activity. *J. Oleo Sci.* **2021**, *70*, 1403–1416. [CrossRef] [PubMed]
33. Pineda-Hernández, M.T.; Pérez-Urizar, J.T.; Ganem-Rondero, A. Thermo-reversible in situ forming implant with nanostructured lipid carriers (NLC) as a delivery system for the administration of estradiol valerate. *Drug Deliv. Transl. Res.* **2020**, *10*, 1393–1402. [CrossRef] [PubMed]
34. Tamjidi, F.; Shahedi, M.; Varshosaz, J.; Nasirpour, A. Nanostructured lipid carriers (NLC): A potential delivery system for bioactive food molecules. *Innov. Food Sci. Emerg. Technol.* **2013**, *19*, 29–43. [CrossRef]
35. Vitorino, C.; Carvalho, F.A.; Almeida, A.J.; Sousa, J.J.; Pais, A.A.C.C. The size of solid lipid nanoparticles: An interpretation from experimental design. *Colloids Surf. B Biointerfaces* **2011**, *84*, 117–130. [CrossRef]
36. Kim, B.D.; Na, K.; Choi, H.K. Preparation and characterization of solid lipid nanoparticles (SLN) made of cacao butter and curdlan. *Eur. J. Pharm. Sci.* **2005**, *24*, 199–205. [CrossRef] [PubMed]
37. Szekalska, M.; Czajkowska-Kośnik, A.; Maciejewski, B.; Misztalewska-Turkowicz, I.; Wilczewska, A.Z.; Bernatoniene, J.; Winnicka, K. Mucoadhesive alginate/pectin films crosslinked by calcium carbonate as carriers of a model antifungal drug—Posaconazole. *Pharmaceutics* **2023**, *15*, 2415. [CrossRef] [PubMed]
38. Nascimento, T.A.; Calado, V.; Carvalho, C.W.P. Development and characterization of flexible film based on starch and passion fruit mesocarp flour with nanoparticles. *Food Res. Int.* **2012**, *49*, 588–595. [CrossRef]
39. Gonçalves, M.M.; Carneiro, J.; Justus, B.; Espinoza, J.T.; Budel, J.M.; Farago, P.V.; Paula, J.P.D. Preparation and characterization of a novel antimicrobial film dressing for wound healing application. *Braz. J. Pharm. Sci.* **2020**, *56*, e18784. [CrossRef]
40. de Moura, M.R.; Aouada, F.A.; Avena-Bustillos, R.J.; McHugh, T.H.; Krochta, J.M.; Mattoso, L.H.C. Improved barrier and mechanical properties of novel hydroxypropyl methylcellulose edible films with chitosan/tripolyphosphate nanoparticles. *J. Food Eng.* **2009**, *92*, 448–453. [CrossRef]
41. Aburto, J.; Moran, M.; Galano, A.; Torres-García, E. Non-isothermal pyrolysis of pectin: A thermochemical and kinetic approach. *J. Anal. Appl. Pyrolysis* **2015**, *112*, 94–104. [CrossRef]
42. Rangelova, N.; Aleksandrov, L.; Nenkova, S. Synthesis and characterization of pectin/SiO_2 hybrid materials. *J. Solgel Sci. Technol.* **2018**, *85*, 330–339. [CrossRef]
43. Liu, L.; Cao, J.; Huang, J.; Cai, Y.; Yao, J. Extraction of pectins with different degrees of esterification from mulberry branch bark. *Bioresour. Technol.* **2010**, *101*, 3268–3273. [CrossRef]
44. Ghadiri Alamdari, N.; Salmasi, S.; Almasi, H. Tomato seed mucilage as a new source of biodegradable film-forming material: Effect of glycerol and cellulose nanofibers on the characteristics of resultant films. *Food Bioproc. Tech.* **2021**, *14*, 2380–2400. [CrossRef]
45. Mujtaba, M.; Akyuz, L.; Koc, B.; Kaya, M.; Ilk, S.; Cansaran-Duman, D.; Martinez, A.S.; Cakmak, Y.S.; Labidi, J.; Boufi, S. Novel, multifunctional mucilage composite films incorporated with cellulose nanofibers. *Food Hydrocoll.* **2019**, *89*, 20–28. [CrossRef]
46. Lorevice, M.V.; Otoni, C.G.; Moura, M.R.d.; Mattoso, L.H.C. Chitosan nanoparticles on the improvement of thermal, barrier, and mechanical properties of high- and low-methyl pectin films. *Food Hydrocoll.* **2016**, *52*, 732–740. [CrossRef]
47. Karaki, N.; Aljawish, A.; Muniglia, L.; Humeau, C.; Jasniewski, J. Physicochemical characterization of pectin grafted with exogenous phenols. *Food Hydrocoll.* **2016**, *60*, 486–493. [CrossRef]
48. Orgován, G.; Kelemen, H.; Noszál, B. Protonation and β-cyclodextrin complex formation equilibria of fluconazole. *J. Incl. Phenom. Macrocycl. Chem.* **2016**, *84*, 189–196. [CrossRef]

49. Vlaia, L.; Olariu, I.; Muţ, A.M.; Coneac, G.; Vlaia, V.; Anghel, D.F.; Maxim, M.E.; Stângă, G.; Dobrescu, A.; Suciu, M.; et al. New, biocompatible, chitosan-gelled microemulsions based on essential oils and sucrose esters as nanocarriers for topical delivery of fluconazole. *Pharmaceutics* **2022**, *14*, 75. [CrossRef] [PubMed]
50. Kaur, R.; Sharma, N.; Tikoo, K.; Sinha, V.R. Development of mirtazapine loaded solid lipid nanoparticles for topical delivery: Optimization, characterization and cytotoxicity evaluation. *Int. J. Pharm.* **2020**, *586*, 119439. [CrossRef]
51. Nirbhavane, P.; Sharma, G.; Singh, B.; Khuller, G.K.; Goni, V.G.; Patil, A.B.; Katare, O.P. Preclinical explorative assessment of celecoxib-based biocompatible lipidic nanocarriers for the management of CFA-induced rheumatoid arthritis in Wistar rats. *AAPS PharmSciTech* **2018**, *19*, 3187–3198. [CrossRef]
52. Li, H.; Dai, F.; Liu, H.; Tao, Q.; Hu, J.; Zhang, Y.; Xiao, Z.; Rupenthal, I.D.; Li, H.; Yang, F.; et al. Physicochemical properties and micro-interaction between micro-nanoparticles and anterior corneal multilayer biological interface film for improving drug delivery efficacy: The transformation of tear film turnover mode. *Drug Deliv.* **2023**, *30*, 2184312. [CrossRef] [PubMed]
53. Gaur, P.K.; Mishra, S.; Bajpai, M.; Mishra, A. Enhanced oral bioavailability of efavirenz by solid lipid nanoparticles: In vitro drug release and pharmacokinetics studies. *Biomed. Res. Int.* **2014**, *2014*, 363404. [CrossRef] [PubMed]
54. Eid, H.M.; Elkomy, M.H.; El Menshawe, S.F.; Salem, H.F. Development, optimization, and in vitro/in vivo characterization of enhanced lipid nanoparticles for ocular delivery of ofloxacin: The influence of pegylation and chitosan coating. *AAPS PharmSciTech* **2019**, *20*, 183. [CrossRef] [PubMed]
55. Guo, Y.; Karimi, F.; Fu, Q.; Qiao, G.; Zhang, H. Reduced administration frequency for the treatment of fungal keratitis: A sustained natamycin release from a micellar solution. *Expert Opin. Drug Deliv.* **2020**, *17*, 407–421. [CrossRef]
56. Phaechamud, T.; Mahadlek, J. Solvent exchange-induced in situ forming gel comprising ethyl cellulose-antimicrobial drugs. *Int. J. Pharm.* **2015**, *494*, 381–392. [CrossRef]

International Journal of
Molecular Sciences

MDPI

Article

Starch-Based Polysaccharide Systems with Bioactive Substances: Physicochemical and Wettability Characteristics

Agnieszka Ewa Wiącek * and Anna Furmaniuk

Department of Interfacial Phenomena, Faculty of Chemistry, Maria Curie-Skłodowska University, Maria Curie-Skłodowska Sq. 3, 20-031 Lublin, Poland; anna.furmaniuk@poczta.umcs.lublin.pl

* Correspondence: agnieszka.wiacek@mail.umcs.pl

Abstract: Polysaccharide-based systems have very good emulsifying and stabilizing properties, and starch plays a leading role. Their modifications should add new quality features to the product to such an extent that preserves the structure-forming properties of native starch. The aim of this manuscript was to examine the physicochemical characteristics of the combinations of starch with phospholipids or lysozymes and determine the effect of starch modification (surface hydrophobization or biological additives) and preparation temperature (before and after gelatinization). Changes in electrokinetic potential (zeta), effective diameter, and size distribution as a function of time were analyzed using the dynamic light scattering and microelectrophoresis techniques. The wettability of starch-coated glass plates before and after modification was checked by the advancing and receding contact angle measurements, as well as the angle hysteresis, using the settle drop method as a complement to profilometry and FTIR. It can be generalized that starch dispersions are more stable than analogous *n*-alkane/starch emulsions at room and physiological temperatures. On the other hand, the contact angle hysteresis values usually decrease with temperature increase, pointing to a more homogeneous surface, and the hydrophobization effect decreases vs. the thickness of the substrate. Surface hydrophobization of starch carried out using an *n*-alkane film does not change its bulk properties and leads to improvement of its mechanical and functional properties. The obtained specific starch-based hybrid systems, characterized in detail by switchable wettability, give the possibility to determine the energetic state of the starch surface and understand the strength and specificity of interactions with substances of different polarities in biological processes and their applicability for multidirectional use.

Keywords: starch; phosphatidylcholine; lysozyme; effective diameter; zeta potential; dynamic light scattering; electrophoresis; wettability; contact angle; hysteresis

1. Introduction

Polysaccharides are used in many areas of human life and industries, e.g., in food, pharmaceutical, cosmetics, medicine productions (as dressing materials, medications for use on the skin, for construction of contact lenses, surgical implants, catheters, diaphragms for artificial kidneys), and the production of paints and cleaning agents. One of the most common polysaccharides is starch, the most important in the food industry, which fulfills various functions, e.g., gelling, stabilizing, and clarifying, as well as the inhibition of the crystallization of sucrose as fat substitute or emulsifier [1–10].

Thanks to the possibility of forming specific ordered spatial structures by appropriate polymer chains, polysaccharides, including starches of various origins, have the ability to form hydrocolloids, which are associated with the presence of functional groups capable of interacting with other chains or molecules. Such interactions through hydrogen and/or coordination bonds can occur between fragments of the same chain as well as between two or more different chains. Starch is a very important, widely available biopolymer and polysaccharide of great industrial importance. It is obtained primarily from corn, potatoes, wheat, and rice [2–5]. Starch is often used in its natural form. It is a sugar composed of two polysaccharides, i.e., amylose and amylopectin. The ratio of amylose to amylopectin ranges from 1:3.5 to 1:5, and there are also starches that contain up to 98% of amylopectin, e.g., starch derived from corn, rice, and barley varieties [2–7]. What is important from a technological point of view is that amylose forms viscous solutions at 90 °C, while amylopectin forms a hydrogel with water at 70 °C. In addition, starch contains some amounts of non-saccharide components, such as protein (0.06–0.5%), fats (up to 0.9%), phosphorus (0.05–0.25%), and other minerals (including K, Na, Ca, Mg, Al, Fe). Starch also has very important biological functions in the human body. It is a source of energy for the central nervous system and red blood cells, and it is also necessary for the synthesis of amino acids and the oxidation of fatty acids in the body. In order to improve its functional properties or give it new features, it is mainly needed to decrease unfavorable properties, including high sensitivity to moisture and rather poor mechanical properties compared to synthetic polymers. Starch can be processed using traditional technologies, e.g., extrusion, embossing, and high-pressure injection. Modifications of native starch may also be helpful in increasing the functional properties [1–12].

The most commonly used physical modification is the gelatinization of starch accompanied by the destruction of the granular structure. During heating, most hydrogen bonds are broken, and a reduction in the melting point and glass transition is observable. Depending on the degree of destruction of the granular structure and the share of water, different products are obtained. On the other hand, chemical modifications may lead to a reduction in molecular weight, a disruption of glycosidic bonds in the chain, and consequently, a significant deterioration of the mechanical properties. The preparation of modified starches, mainly through esterification, allows them to obtain derivatives with the appropriate, expected hydrophobicity. However, a strategy based on chemical modification may be seriously limited by the toxicity, the high costs, and the removal of intermediate products. Therefore, scientists are looking for new, economical ways of modification to more biocompatible and multidirectional applications [8–12].

One of the applications for starch is its use as a substitute for fat and sugar. These types of substitutes perform numerous functions in food, e.g., they are fillers, stabilizers, and gelling and thickening agents, they affect the texture, maintain moisture, ensure proper crunchiness, and are also responsible for the taste sensation [5–14]. Texture is one of the most important features of food products. Thanks to the ability to bind water, starches that come from potatoes, waxy corn, and tapioca are used as thickeners in bakery products, fruit and vegetable products, meat products, frozen foods, and food concentrates. It is well known that starch properties are strictly dependent on the source of origin and the related gelatinization temperature. According to the literature, in the case of rice starch, the beginning and end of gelatinization are about 65 °C and 73 °C, respectively [11–18].

Another currently explored novel application of starch is the immobilization of enzymes, proteins, or drugs to improve catalytic activity or to create more efficient drug delivery systems [19]. On the other hand, S. Li and co-workers showed that the cross-linker-free chitosan/carboxymethylstarch–lysozyme microspheres could be a promising antibacterial additive, especially for *E. coli* for enteric infection treatment due to effectiveness and fast release at the intestinal tract. The development of an immobilized lysozyme is an urgent need in medicine, the food industry, and biotechnology. A higher antibacterial activity is possible owing to the synergistic effect between the membrane-attacking ability of chitosan and the cell wall-attacking ability of a lysozyme [8].

The aim of this manuscript was to characterize the combinations of starch with phospholipids and lysozymes (globular protein) to address two main issues: (a) enhancing starch's functional properties or adding new features due to its influence on taste and texture and (b) exploring novel applications, like enzyme immobilization, for improved catalysis or drug delivery systems. Starch contains hydroxyl groups and simple glycosidic bonds that can be easily modified physically, chemically, or enzymatically to incorporate specific features, enhancing cooking properties, gel clarity, texture, adhesion, and film formation, increasing freeze–thaw stability, or reducing syneresis, retrogradation, and gelling tendencies [19]. Starch's amphiphilicity, hydrophobicity, mechanical strength, and thermal stability are among the properties that can be improved as a result of starch modification. In the research described in this paper, surface hydrophobization was carried out using an *n*-alkane film without changing the bulk properties of starch, and it was checked how it would affect the properties of the entire system and the interactions between individual components. Starch modification usually leads to improvement of its mechanical properties and, above all, functional properties, and improves its processing parameters, which increases the possibilities of its use.

When making modifications, it is important to add new quality features to the product and also preserve the structure-forming properties of native starch—thickening and gelling. Starch-based systems with phospholipids (e.g., phosphatidylcholine, PC main component of biological membranes that self-assemble at both hydrophilic and hydrophobic surfaces in aqueous media) can be used to obtain films suitable in pharmacy and medicine because their structure resembles cell membranes, making them biocompatible; they do not cause allergies and have a higher stability due to steric and electrostatic repulsions. Regarding phosphatidylcholine, PC is one of the main ingredients of the popular lecithin widely used in the food, cosmetics, and pharmaceutical industries [20]. Additionally, there may be a scope to obtain starch-based products (e.g., rice) with a lipid composition to reach the desired starch digestibility [21], whereas combinations with lysozymes, especially in the treatment of bacterial, fungal, and viral infections are useful [16–19]. A lysozyme is an antibacterial substrate, and proteins, glycoside hydrolase, and enzymes damage bacterial cell walls by catalyzing the hydrolysis of 1,4-beta-linkages between N-acetylmuramic acid and N-acetyl-D-glucosamine residues in a peptidoglycan and between N-acetyl-D-glucosamine residues in chitodextrins. A lysozyme is a part of the innate immune system and is a natural form of protection from Gram-positive pathogens, like *Bacillus* and *Streptococcus*. So, the incorporation of a lysozyme enhances the antimicrobial activity of starch-based systems. Modification by antibacterial substrate's film may give new applications. Additionally, as was mentioned earlier, preparations of modified starch (e.g., acetylated annealed) may be deployed for a slow release of the therapeutic substances [22].

Regarding the above aspects, the physicochemical characterization of dispersion systems was aimed at determining the effect of starch modification and temperature of starch preparation, as well as checking the wettability parameters of starch-coated glass plates before and after modification [22–25]. Changes in effective diameter, multimodal size distribution, and electrokinetic potential (zeta) as a function of time were analyzed using the dynamic light scattering and microelectrophoresis technique, as well as the advancing and receding contact angle (and angle hysteresis), by the sessile drop method [26–29]. The obtained starch-based systems with strictly controlled parameters give the possibility of their multidirectional use [30–36].

Starch, because of its unique functional properties, is used in a number of novel foods, such as in the process of encapsulation of oils, aromas, and vitamins, as a pharmaceutical excipient, in a large number of agricultural products, and as a coating. However, such coatings are generally less permeable to gases and more permeable to water vapor due to the presence of significant amounts of hydrophilic groups, particularly hydroxyl groups. These properties can be modified by employing an appropriate procedure of biological alteration. In this type of product, water permeability is often not desired, hence the need to hydrophobize the surface. Hydrophobization using *n*-tetradecane film is one of the cheaper modification methods. These activities combined with modification with a lysozyme and phosphatidylcholine can lead to obtaining biodegradable starch with a wider range of applications without major costs. Starch is considered a biodegradable substance, and modifications with natural substances do not change this desirable property compared to other artificial polymers that do not usually have this feature [37–46]. Another scope of this manuscript was to find how the internal starch structure can influence its surface properties before and after biological modification. Such systems have been tested to obtain films applicable in food packaging or a suitable surface coating, e.g., pharmacy and medicine as an effect of potential biocompatibility with antiallergic, antibacterial, antifungal, and antiviral properties. To the best of our knowledge, these are the first reported studies of measuring advancing and receding contact angles on starch and modified starch films with additional hydrophobization and parallely studies of the physicochemical properties, including the stability of their analogous dispersions.

2. Results and Discussion

2.1. *Effective Diameter and Zeta Potential of Starch-Based Systems vs. Time*

During thermal technological processes, the stability of starch is improved by its cross-linking, especially in an acidic environment and during high shear. Unfortunately, such a modification also has negative effects, e.g., it reduces the transparency and clarity of gruels and resistance to cold storage [5–12,25–29]. Taking into account the gelatinization temperature for rice starch, three different temperatures were selected for testing: the temperature before the gelation point, the physiological temperature (useful for food and cosmetic products), and the temperature after the end of gelatinization.

In Figure 1, the relationships of the effective diameter and the zeta potential as a function of time for starch dispersion in a 10^{-3} M NaCl solution at three temperatures, 25 °C, 37 °C, and 70 °C, are presented. As was mentioned earlier, during gelatinization, firstly, the amorphous parts of the starch granule are hydrated, which induces swelling of the granules, leading to a distortion of the crystalline area. Secondly, the crystalline regions become more accessible to water. Thus, the size of starch granules increased with temperature increase from 25 °C to 70 °C (Figure 1a).

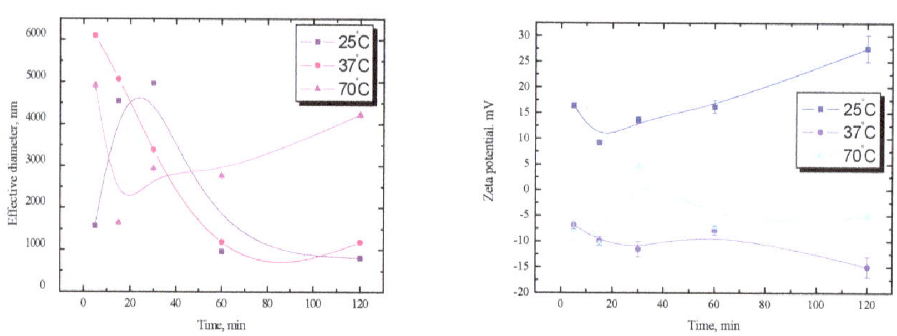

(a) (b)

Figure 1. Dependence of the (**a**) effective diameter and (**b**) zeta potential of starch dispersion in a 10^{-3} M NaCl solution as a function of time.

The morphological changes associated with gelatinization are swelling and the rupture of the granular crystalline structure. At room temperature, a rapid increase in the effective diameter is visible within 30 min, while at 37 °C, the equilibrium state is reached faster, and the effective diameter values are more stable. After an hour at both temperatures, 25 °C and 37 °C, the diameters are noticeably more stable, and after 2 h, they are even similar. At 70 °C after 30 min, a monotonic increase in the diameter value was observed, which lasts up to 2 h of measurements and proves that the tested system at this temperature is the least stable. Analyzing the dependence of the zeta potential (Figure 1b), there are also clear differences between the samples, both in terms of the value of the potential and the nature of changes during the measurements. The obtained dependencies of the zeta potential as a function of temperature may be a further indication that starch gelatinization involves a change of the ordered state from the crystalline state to the disordered state. The zeta potential describes the forces occurring in the system and is indirectly a parameter of stability.

Starch granules are held together by hydrogen bonds and interchain hydrophobic bonds. As the temperature increases, the main process may be the breaking of hydrogen bonds between poly-(1 → 4)-glucan chains in crystallites and subsequent hydrophobic bonds. During these processes occurring in starch granules, after breaking hydrogen bonds, the polar groups probably hide inside and the hydrophobic parts remain outside; so as the temperature increases, a decrease in zeta potential is noticeable (Figure 1b). At temperatures of 37 °C and 70 °C, small negative zeta potential values were obtained, while at room temperature, these values are positive and the highest, which may indirectly confirm this conclusion and also indicate better stability.

It is well known that the gelatinization process of starch is multi-stage, and its course strictly depends on the temperature. Typically, above the temperature of 50 °C, the hydrogen bonds between the amylose chains forming the crystalline fragments of the starch grain are broken. Water penetrates between them, creating new hydrogen bonds with released starch chains [18–20]. The gelatinization temperature is the temperature above which the starch does not show a granular structure in an aqueous medium and gives a paste characterized by high viscosity. Large grains lose their structure at the beginning. On the other hand, small grains are more resistant, which results in a large range of gelatinization temperature. This is why particle size measurements are so important in starch studies.

As mentioned, the gelatinization temperature depends on the starch source and for rice starch, it is 65 degrees. Differences between rice varieties can be found in the ratio of amylose/amylopectin, internal ordering, and side chain length distribution. Rice starches have a better resistance to processing stress, are much more stable in time (very slow retrogradation), and have smooth gels with low structure. Starches with 25–30% amylose and 70–75% amylopectin are called "normal" starches; starches with very high levels of amylopectin (98–99%) are named "waxy" starches; and a third group contains starches with high amylose content (50–70%) and has no specific name [19]. The ratio of amylose to amylopectin in starch profoundly impacts its physicochemical properties or starch-based industrial products' features, which, in turn, affect their functionality and potential applications. Higher amylose content implies increased film strength, while the presence of highly branched amylopectin leads to the formation of films with poor mechanical strength [19].

Remylgel belongs to waxy starches with a very high content of amylopectin. It is most probable that the branched chain of amylopectin is directed into the air in a gel film. Changing the parameters of starch during the gelification process (e.g., type of modification solution, temperature, time of modification), the wettability properties of the starch surface, and also the possibility of starch application, can be substantially changed. During gelatinization, the presence of a large amount of water is forming, and then the layered structure of the starch grains disappears and turns into bubbles. As a result of water entering the vesicles, amylose is completely dissolved, amylopectin swells more strongly, and peptization occurs [18–25]. These changes can also be seen in the example of measurements of size diameter and values of electrokinetic potential. For starch dispersion in a NaCl solution at 37 °C, slight fluctuations of this parameter were observed during 2 h of measurements. At 70 °C after 1 h, the zeta potential remained at the same level, about −7.0 mV. However, such a low potential does not guarantee the stability of the system; hence, large fluctuations of the effective diameter at this temperature are visible (Figure 1). At room temperature, higher potential values were noted, which resulted in more stable values of the effective diameter of starch dispersion in the NaCl solution. On this basis, a conclusion can be drawn about the correlation of these two parameters.

Due to the potential medical and pharmaceutical applications, further studies were carried out for an analogous dispersion of starch with the addition of a phospholipid (PC). As already mentioned, the phospholipid increases the biocompatibility of such a system due to the fact that it is one of the main components of cell membranes. The obtained dependencies of the effective diameter and zeta potential as a function of time for the starch/phospholipid dispersions at three temperatures are presented in Figure 2. At room temperature and 70° during the first hour, large fluctuations are visible; however, after 1 h, in both cases, a monotonic decrease in the value of the effective diameter for up to 2 h of measurements is observable.

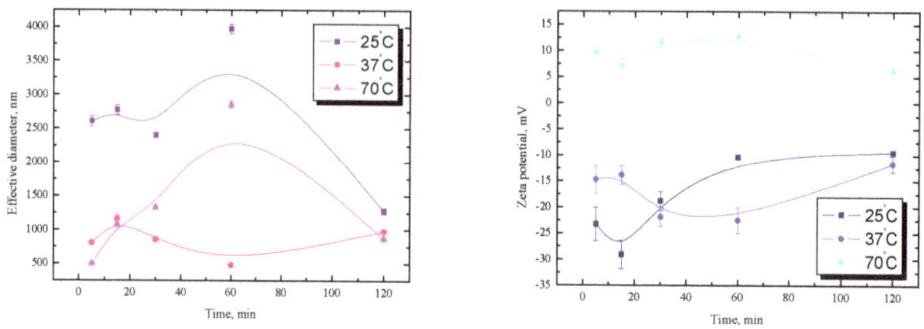

(**a**) (**b**)

Figure 2. Dependence of the (**a**) effective diameter and (**b**) zeta potential of starch/phosphatidylcholine dispersion in a 10^{-3} M NaCl solution as a function of time.

However, at a physiological temperature, the starch/PC dispersion in a 10^{-3} M NaCl solution is the most stable. The diameter values remained almost all the time at the same level of about 990 nm. Figure 2b shows the relationship of the zeta potential as a function of time. At room temperature and 37 °C, only negative values of the electrokinetic potential were recorded, unlike the measurements for the dispersion at 70 °C, which are only positive. This is directly related to the changes in starch structure below and above temperature gelatinization, as previously described in detail. The hydrophilic–hydrophobic character of starches depends on temperature, as well as their substitution type and starch pretreatment. At room temperature, the zeta potential values ranged from −29 mV to −9.4 mV, initially showing a decrease and then a monotonic increase. The reverse trend of changes was noted for the dispersion at 37 °C. The values of zeta potentials for room and physiological temperatures after 2 h are similar. On the other hand, the values of the electrokinetic potential of the starch/PC dispersion at 70 °C range from 6.2 mV to 9.7 mV, and fluctuations in time are visible. Such a low potential of approx. 10 mV does not guarantee the stability of the system; hence, large fluctuations of the effective diameter at this temperature are visible in Figure 2a.

In the next stage of the research, measurements were carried out for the starch/lysozyme dispersion in a NaCl solution. The obtained effective diameter versus time relationships are shown in Figure 3a. The stability of dispersions depends on several factors, e.g., the type and concentration of the emulsifier used and its surface activity or adsorption rate at the interface. A lysozyme is a protein, and due to its surface properties, it can be used to create and stabilize different kinds of dispersions. Regarding emulsions, it is known that a protein film is formed around the oil droplets, which has a positive charge and prevents coalescence or flocculation. It has a high emulsifying capacity, especially after unfolding [15–20]. A rapid increase in the effective diameter is visible within 5–30 min and then a gradual decrease during 2 h at 25 °C. For the dispersion prepared at the physiological temperature (about 37 °C), nearly a double increase in diameter was observed during the first hour. The further course of the curve for 37 °C looks like a mirror image of the dependence obtained for the dispersion at 25 °C (an increase in the effective diameter was noted in the 60th minute, and then there was a decrease at 2 h in the opposite trend for physiological temperature).

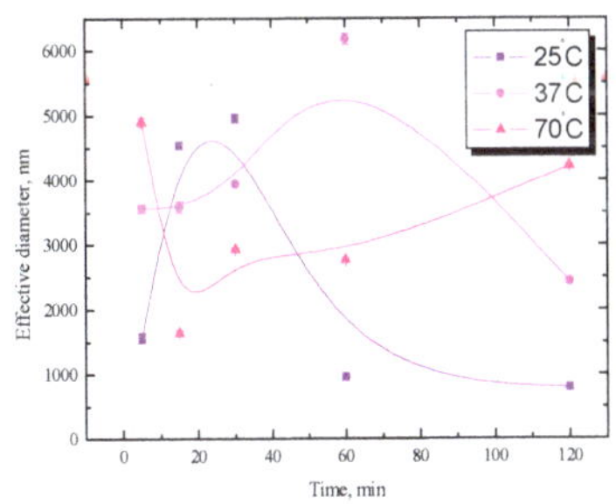

(a)

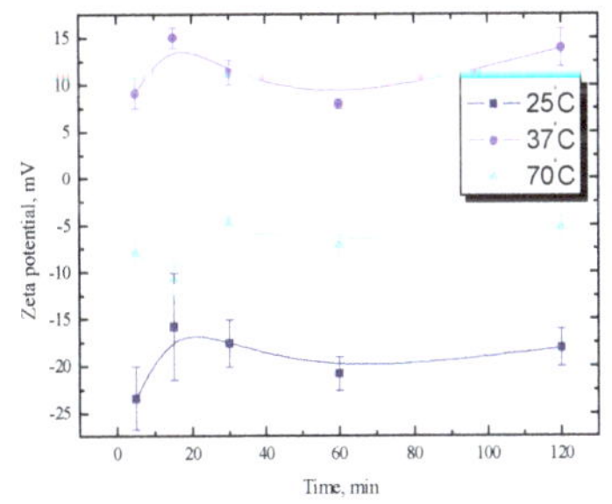

(b)

Figure 3. Dependence of the (**a**) effective diameter and (**b**) zeta potential of the starch/lysozyme dispersion in a 10^{-3} M NaCl solution as a function of time.

At a temperature of 70 °C, a rapid decrease in the effective diameter is visible in the first 15 min, followed by a gradual increase. The obtained values of the zeta electrokinetic potential of the starch/lysozyme dispersion in the NaCl solution are shown in Figure 3b. It is easy to see that the course of the curves at 25 °C and 37 °C is very similar, but the electrokinetic potential values differ primarily in sign. The electrokinetic potential values at room temperature range from −23.4 mV to −18.0 mV. The positive charge from the protein film is noticeable at the physiological temperature. At this temperature, positive zeta potential values were obtained. This behavior may be related indirectly to the mechanism proposed lately [22]. The authors stated that the rearrangement of starch chains during heating, especially in the excess of water, leads to the strengthening of bonds between amylose and amylopectin, increasing starch granule stability, which can inhibit enzymatic hydrolysis. Moreover, this process can lead to the leaching of amylose chains from amorphous regions of starch granules, which probably reinforces their crystalline structure. The amorphous regions of semicrystalline starch granules represented regions most susceptible to the initial water absorption. The hydration of starch granules leads to increased mobility, especially in the amorphous regions. During progress in heating/hydration, the initially weaker or imperfect crystallites successively disappear, whereas the other ones become more perfect upon fusion and recrystallization.

At a temperature of 70 °C, within 15 min from the creation of the dispersion, a decrease in potential from −7.9 mV to −10.7 mV was noted, and then the potential remained at a constant level and reached a value of approx. −5.5 mV. Such a low potential does not guarantee the stability of the system; hence, large fluctuations of the effective diameter for this dispersion are visible in Figure 3a. In the case of the *n*-tetradecane/starch emulsion, fluctuations in the diameter value are visible at all temperatures, and the largest was at 37 °C; therefore, for better visibility, some results have been reduced twice (Figure 4a). On this basis, it can be concluded that the analogous starch dispersions are more stable than the tested *n*-tetradecane/starch emulsions.

It is known that an O/W emulsion or its powdered form (dry emulsion) can be a carrier of oil-soluble compounds with health-promoting properties or biological activity (e.g., polyunsaturated fatty acids, sterols, vitamins, or antioxidants). A properly selected recipe and technology for the production of emulsion (dispersion) means that the structure of the particles must ensure the long-term stability of biologically active substances. For a high efficiency of such systems, it is necessary to disperse the oil phase to particles with a diameter of less than 1 mm using an appropriate homogenization technique, as the particle size also affects the stability of the emulsions produced [15–20].

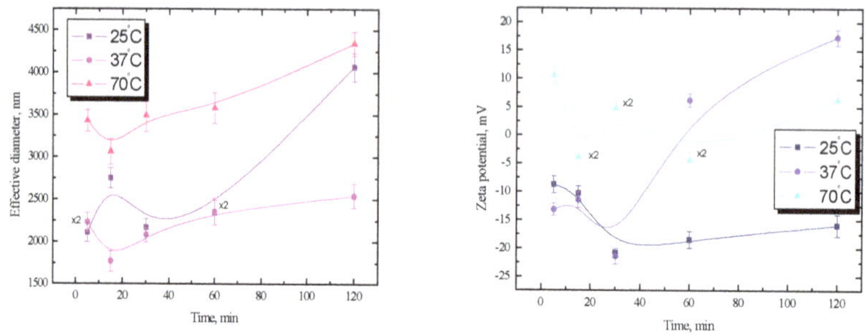

(a) (b)

Figure 4. Dependence of the (**a**) effective diameter and (**b**) zeta potential of *n*-tetradecane/starch dispersion in a 10^{-3} M NaCl solution as a function of time.

The most important difference between proteins and other emulsifiers used is the size. Proteins having a high molecular weight diffuse into the interface much more slowly than a non-protein emulsifier. When the protein reaches the phase boundary, the chains unfold, and fragments are divided between the phases, which takes time. In the case of protein, to achieve the required value of interfacial tension to stabilize the emulsion, such a process may take up to several hours. Protein–polysaccharide combinations show very good stabilizing and emulsifying properties. This makes it possible to replace other low-molecular synthetic emulsifiers, which is advantageous for nutritional and pro-health reasons [30–36]. Therefore, the size will be particularly relevant for samples with lysozymes. In addition to the size and changes in the diameter of the tested systems, for a more complete characterization, the electrokinetic potential dependencies should also be considered. The most stable potential is visible at room temperature in the range from −20.8 mV to −18 mV. Despite quite stable values of the zeta potential, which usually guarantee the stability of the system (Figure 4b), the diameters are not stable (Figure 4a). This may be due to incomplete dispersion of the fat phase at this temperature. The homogeneity of the dispersion of the lipid phase should increase with increasing temperature. At a temperature of 37 °C, after the initial decrease in the zeta potential in 30 min to the value of −21.4 mV, an increase in the potential to the value of about 17.4 mV is visible. The greatest fluctuations of the potential over time were recorded at a temperature of 70 °C in a range from 10.6 mV to −4.5 mV, and the course of the curve resembles a sine wave (Figure 4b).

It is well known that changes to temperature involved in food processing and products can also influence emulsion stability. For example, this type of research has been carried out for phospholipids, and the authors [24] have investigated the effect of the combination of starch and PC on the freeze–thaw stability of an O/W emulsion. The adsorption behavior, rheological properties, and mechanism of the improved stability of the starch–PC complex at the oil–water interface were discussed. The binding energy of the starch–PC complex was significantly lower than that of PC molecules, suggesting that the complex was relatively stable, and their changes were consistent with changes in electrostatic energy, indicating that such energy played a major role in the formation of the starch–PC complex. There were intermolecular hydrogen bonds and hydrophobic interactions in the starch–PC complex, while only hydrophobic interactions existed in the PC molecules [24].

2.2. Starch System without n-Tetradecane Film

One of the aspects of this research, apart from stability, was to investigate how the substrate and the modification of the starch layer will affect its wettability depending on the starch preparation temperature. The changes in bulk properties are usually seen in the surface properties of the starch. The parameters of wettability of the starch surface obtained also suggest topographical differences, as was concluded in our previous papers [26,27]. First, tests were carried out without *n*-tetradecane, and then, based on tests with *n*-tetradecane, the effect of starch hydrophobization was analyzed. Figure 5 shows the average values of the contact angle of starch plates without *n*-tetradecane film at three different temperatures. Liquids differing in polarity were used for wettability measurements, from highly polar water, through polar formamide, to non-polar diiodomethane. Natural starches have no affinity for hydrophobic substances, fats, or oils due to their hydrophilic nature. Specialized preparations consisting of two parts—hydrophilic and hydrophobic—are produced, having an affinity for both oil and water, so they can form stable oil-in-water emulsions. Starch with an additional layer can fulfill an analogous role, e.g., reduce the cholesterol content, and emulsifying substances make it possible to maintain a uniform structure in such products [15–20].

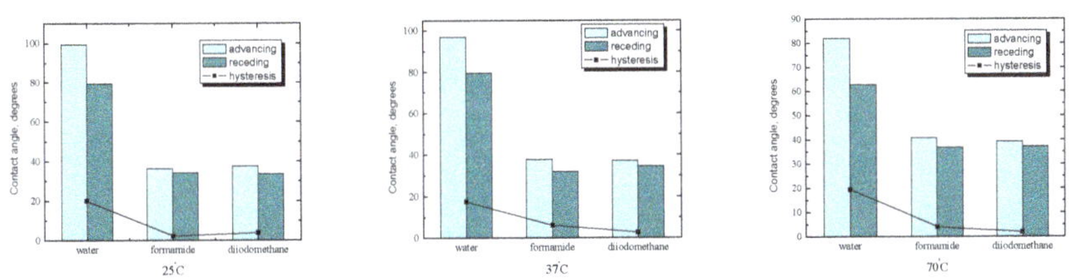

(**a**) (**b**) (**c**)

Figure 5. Comparison of the average value of the contact angle of the test liquid measured for the starch surface without *n*-tetradecane film. Starch preparation temperature (**a**) 25 °C; (**b**) 37 °C; (**c**) 70 °C.

The methodology used for contact angle measurements also indirectly provides information about the changes in the structure of the starch layer before and after modification and the kind (or even strength) of forces between material molecules. The attractive forces between molecules that cannot be explained by ionic or covalent interactions can be caused by polar moments within molecules. During gelatinization, surface flatting can also be observed. Definitely, the highest average values of the contact angle at room temperature (Figure 5a) are noticeable for water. With the reduction in the polarity of the tested liquid, the hysteresis of the angle decreases because the differences between the values of the contact angles are small. For water, the hysteresis value is about 20°, which confirms the great sensitivity of starch to water. Starch absorbs water at any temperature, so even after a few seconds, its surface is completely different after placing a drop of water, while for the diiodomethane angle, hysteresis is only 2°. In the case of formamide and diiodomethane, the values of the contact angles are comparable.

It is well known that the contact angle depends on composition and temperature, as well as on the roughness of the film surface [18–20]. Determined by the rate of water drop absorption, starch films can be divided into two main groups (on the basis of contact angle values below 90° and over 90°). At the beginning of the experiment, starch surfaces prepared at 25 °C and 37 °C showed contact angles near 90°; after hydrophobization, contact angles were usually higher than 90°. Conversely, after PC or lysozyme modification, the contact angles were lower. The classification of starches according to the rate of water drop absorption gives temperature dependence. In our previous papers, the fastest rate of water drop absorption (the highest decrease in contact angle, about 0.1 degree/s) was observed for starch prepared at 70 °C after gelatinization [18–20]. A fast rate of water drop absorption was found for the water-soluble starches. Untreated starch at 70 °C showed a smaller contact angle than that at room and physiological temperatures. A reason might be the different retrogradation rates of the starch–water systems during film formation. Starch at room temperature before gelatinization shows the highest retrogradation rate of these untreated starches; therefore, the film is less hydrophilic and the contact angle increased [26,27].

A comparison of the average values of the contact angles of formamide and diiodomethane for starch-coated plates before hydrophobization is also shown in Figure 5. For formamide and diiodomethane, the angle values are very close to each other. The hysteresis value of the angle decreases with the decrease in the polarity of the tested liquids (the highest is for water and the lowest is for diiodomethane). Analyzing the above graphs, it can be concluded that the starch preparation temperature has practically no effect on the values of the contact angles of formamide and diiodomethane, while the influence is much greater for the contact angles of water. For the highest tested temperature, there was a decrease in the advancing angle of water by about 15 degrees. In addition, at each of the three temperatures, the angle hysteresis value decreases as the polarity of the liquid decreases.

2.3. Starch System with n-Tetradecane Film

For the systems with the addition of *n*-tetradecane, the effect of starch hydrophobization was analyzed. A comparison of the average values of the contact angle of water, formamide, and diiodomethane for starch-coated plates with *n*-tetradecane film is shown in Figure 6. At maximal temperature, both the average value of the contact angle and the hysteresis value depend on the polarity of the liquids used. The more polar the liquid is, the greater the contact angle and hysteresis. In the case of hysteresis, this is caused by differences between the values of the advancing and receding angles. At 70 °C, the hysteresis values of the water and formamide angles are very similar, approx. 10°, and for diiodomethane, they are half as much (about 5°).

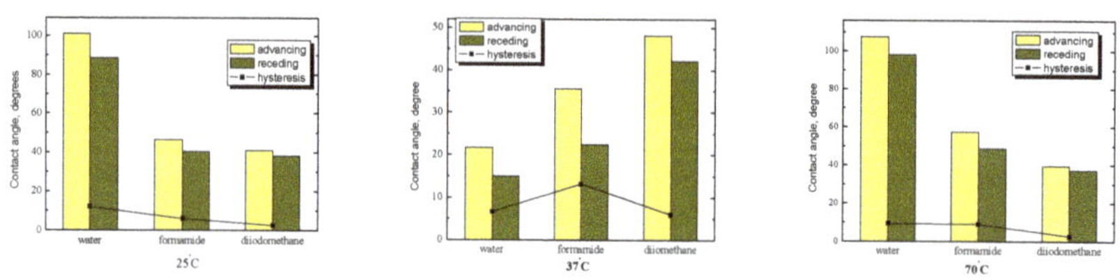

(a) (b) (c)

Figure 6. Comparison of the average value of the contact angle of the test liquid measured for the starch surface with *n*-tetradecane film. Starch preparation temperature (**a**) 25 °C; (**b**) 37 °C; (**c**) 70 °C.

As already mentioned, starch in food products also acts as a fat substitute. In low-fat products, when the degree of fat exchange exceeds 70%, the continuous phase of the product changes into an oil-in-water emulsion, changing the rheological properties and corresponding to condensed emulsions. The replacement of fat comes down to the introduction of water in combination with structure-forming additives in the form of hydrocolloids. The addition of starch has a significant impact on the mechanical and rheological properties of products, thanks to the ability of starch to create stable structures in the presence of water and under the influence of temperature. That is why physicochemical studies of oil/water emulsions with starch, as well as studies of the wettability of starch films, before and after hydrophobization are so important from the application point of view. At 37 °C, the lowest value of the contact angle was obtained for water and the highest for diiodomethane. On the other hand, regarding hysteresis, the highest value is obtained for formamide of approx. 13°, while water and diiodomethane were practically at the same level and amounted to approx. 7°. The chemical nature of the probe liquid, the size of its molecules, and the kind of interactions determine the surface penetration. The differences in the values of the contact angles and hysteresis manifest the strength of the solid/liquid interfacial interactions.

Lately, to prove such a comment, the authors' [47] independent gradient model was created to illustrate non-covalent interactions between starch and PC. For all studies of the the starch–PC complexes, the oil–water interfacial tension decreased quickly, indicating the adsorption of the complex onto the oil–water interface. There may be a formation of an interfacial layer in the starch–PC complex due to interactions between starch and PC, leading to a decrease in interfacial tension greater than that observed in the presence of single emulsifiers. Starch adsorbed slowly to oil surfaces due to its highly branched macromolecular structure and steric contribution and also formed a more extended organization at the interface and, as a result, a relatively thick layer. It is well known that in the case of water and formamide, the interactions are both dispersive and polar in nature, but for apolar diiodomethane, they are practically only dispersive.

Decreasing contact angles of water indicates that during contact between water and the starch surface, its hydrophilicity increases. The largest hysteresis of water contact angles demonstrates that a water droplet penetrates the deepest into the layer structure and depends on surface preparation temperature. Emulsions prepared with the starch–PC complex had a relatively strong electrostatic repulsion that attenuated the oil droplets merging. No creaming and aggregates were observed in the emulsions prepared with starch–PC complexes, indicating that the low and moderate addition of starch helped stabilize PC-based emulsion during the freeze–thaw treatment. The improved stability may be related to the thickness and density of the interfacial layer formed by the complex. A number of studies supported the general idea that emulsifiers, which form a thick interfacial layer, are able to stabilize emulsions [47]. Comparing the average values of the contact angle of the tested liquids on the surface of starch-coated plates with and without the *n*-alkane film, it can be concluded that in this case, the effect of hydrophobization is also visible, mostly at physiological temperatures (Figure 6b).

2.4. *Starch/Phosphatidylcholine Systems without n-Tetradecane Film*

A comparison of the average values of the contact angles of water, formamide, and diiodomethane for starch/PC platelets without *n*-tetradecane film at three different temperatures is shown in Figure 7. As was mentioned earlier, phosphatidylcholine is used, e.g., in pharmaceutical technology due to its unique surface-active properties, and it acts as an emulsifier and a modifier of rheological properties, improving the stability and the consistency of dispersion. Phosphatidylcholine is non-toxic and biodegradable, so it can be used in parenteral nutrition preparations, such as dietary supplements, which lower the concentration of triglycerides and cholesterol in the blood. In addition to drugs, it can be used in the form of ointments, creams, emulsions, or suspensions. It is also important that as a result of combining starch with a phospholipid, the biocompatibility increases because these connections are similar in structure to biological membranes, and they do not cause sensitization or allergic reactions.

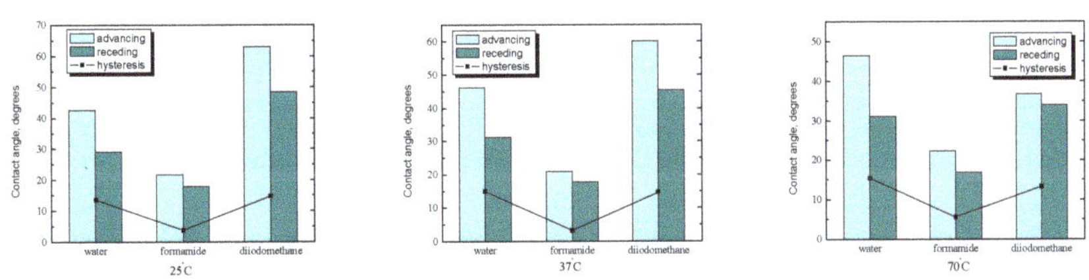

(a) (b) (c)

Figure 7. Comparison of the average value of the contact angle of the test liquid measured for the starch/PC surface without *n*-tetradecane film. Starch preparation temperature (**a**) 25 °C; (**b**) 37 °C; (**c**) 70 °C.

After PC treatment, the changes of advancing contact angles are visible, and the effect grows with the temperature increase. Both polar PC groups and patches of native starch are accessible for water during contact with liquid. Hence, the water contact angle can decrease. Due to the polar nature of water, its molecules exhibit a strong tendency to form hydrogen bonding. Therefore, the structural rearrangement of the PC molecules on the starch surface can occur, and the highest increase in the polarity of the starch surface after PC modification is observed. Wang and co-workers proved that the starch–PC complex at weight ratios of 3:7 and 5:5 remarkably improved the freeze–thaw stability of emulsions (no creaming process and creation of aggregates) and stated that it may be related to a thick and dense interfacial layer [24].

At room temperature (Figure 7a), the highest value of the contact angle was obtained for the non-polar liquid—diiodomethane and the lowest value was for formamide. In the case of non-polar diiodomethane, the value of the contact angle increased by about 20° compared to starch surfaces. The opposite situation is visible in the case of water, i.e., the average value of the angle is lower by more than half. The smallest value of the angle hysteresis was obtained for formamide, while for water and diiodomethane, the values were practically the same. The plates are definitely more hydrophilic compared to the basic ones. This confirms that a small modification can lead to obtaining systems with completely different properties, and the measurements of contact angles are very sensitive and, therefore, extremely useful parameters in determining such changes in a relatively simple and economical way. The hydrophilization of the starch surface after PC modification is caused by the rearrangement of the hydrophobic monolayer into layers with the polar groups of phospholipid heads exposed toward water drop.

For starch/PC plates without *n*-tetradecane film (Figure 7b), at a starch preparation temperature of 37 °C, the graph looks analogous to that for room temperature (Figure 7a). A similar situation also occurs in the case of the highest temperature (Figure 7c), except that the highest value of the contact angle was recorded for water, and not, as in the case of two lower temperatures, for diiodomethane. After exceeding gelatinization, the temperature is completely different, and thus the phospholipid film on the starch surface shows a dissimilar character, which translates to the measured contact angles. The influence of PC is visible, especially for the most polar water and apolar diiodomethane, whose average values of the contact angle changed radically compared to the values of angles measured on plates covered with a layer of starch without modification. Other less sensitive methods would not show these changes, which, once again, confirm the high usefulness of wettability tests. Attractive interactions between starch and PC can cause strong complex formation and, as a result, film condensation and smoothing. Indeed, the low values of angle hysteresis may indicate surface smoothing. During temperature increases the homogeneity of the starch layer increases because the degree of gelatinization grows. An analogous conclusion was also drawn from the decreased values of the roughness parameters in our previous experiment for the same type of starch [26,27].

2.5. Starch/Phosphatidylcholine Systems with n-Tetradecane Film

In Figure 8a, a comparison of the average values of the contact angle of water, formamide, and diiodomethane for starch/PC surfaces with additional *n*-tetradecane film is presented. At room temperature (Figure 8a), the highest value of the contact angle was obtained for water, and the lowest for formamide. As the polarity of the liquid used increases, the hysteresis of the angle also increases. In addition, as a result of hydrophobization, the value of the water contact angle increased by 40°, while for diiodomethane, the value of the contact angle decreased by approx. 20° (Figures 7 and 8). The highest values of the contact angle and hysteresis were obtained for water, while the values of the formamide and diiodomethane angles and the hysteresis values for the contact angle of these two liquids were practically on the same level. The influence of PC on the structure of the starch layer on *n*-tetradecane film and its properties are very visible at a physiological temperature (Figure 8b), especially in the case of water, where the value of the contact angle increased almost four times (see Figure 6b).

At a temperature of 70 °C (Figure 8c), the trend of the measured contact angles and hysteresis changed significantly. This is the reverse of what is obtained at a physiological temperature. The value of the contact angle of formamide is the highest (at lower temperatures, the angle value was much lower), while in the case of water, the average value of the angle is lower, whereas at lower temperatures, it was much higher.

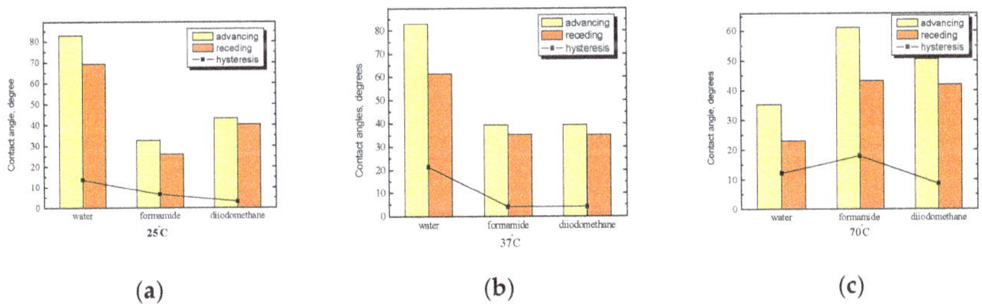

Figure 8. Comparison of the average value of the contact angle of the test liquid measured for the starch/PC surface with *n*-tetradecane film. Starch preparation temperature (**a**) 25 °C; (**b**) 37 °C; (**c**) 70 °C.

2.6. Comparison of the Hysteresis Angle of Test Liquids for the Starch/Phosphatidylcholine System before and after Hydrophobization vs. Temperature

In the next stage of this research, the hysteresis of water angles on the surface of starch with phosphatidylcholine (S/PC) with and without *n*-tetradecane film was compared, depending on the starch preparation temperature. At room temperature (Figure 9), the hysteresis values of the water angles for the S/PC plates without and with *n*-tetradecane film are practically the same and amount to about 13.5°. At a physiological temperature, the hysteresis value for the S/PC plate after hydrophobization is definitely higher (by several degrees) than before hydrophobization.

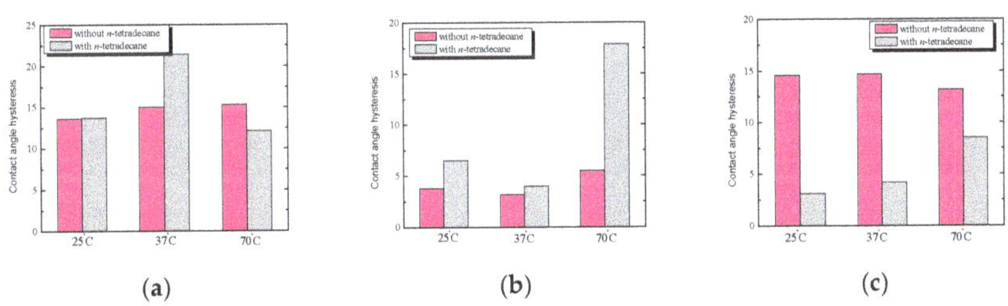

Figure 9. Comparison of the average value of the contact angle hysteresis for the starch/PC surface with and without *n*-tetradecane film for (**a**) water; (**b**) formamide; (**c**) diiodomethane.

The increase in attractive interactions between components causes the formation of more and more tightly packed monolayers characterized by lower values of the roughness parameters. Therefore, such condensed films are less permeable for the probe liquids used for the contact angle measurements. These suggestions have been observed in our previous research based on FTIR analyses [26,27]. Under the influence of modification with bioactive substances or by temperature increase, a change in the spectra of starch is possible. On the other hand, the authors [23] described that the inclusion complexes were formed by hydrophobic interactions between the alkyl chain of phosphatidylcholine and the debranched starch helix cavity. The complexes possessed a mass fractal structure, and a semicrystalline structure formed. After complexation, the stability of phosphatidylcholine was significantly improved. Additionally, this study revealed that debranched starch can be used as an effective carrier of phosphatidylcholine for the purpose of improving its stability [23].

At the highest temperature, the hysteresis value of the water angles for the tested plates without *n*-tetradecane film is higher than for the plates with the film. The obtained hysteresis values of the formamide angles for the starch/PC plates without and with *n*-

tetradecane film, depending on the starch preparation temperature, are shown in Figure 9b. For PC, modified starch OH stretching and skeletal mode were slightly shifted compared to the native starch. Changes in the intensity of specific bands are found to be the major spectral ones during gelation. Gelatinization is primarily a hydration process, and chain conformation and helicity change when crystallinity and molecular orientation are lost. The signal around 1000 cm^{-1} is related to the intermolecular hydrogen bonding of hydroxyl groups or the plasticizing effect of water [26,27].

At each of the tested temperatures, the hysteresis values of contact angle for S/PC surfaces after hydrophobization are higher than before. The smallest difference in the angle values was obtained at the physiological temperature, while at 70 °C, the hysteresis value of the formamide angle for such surface with *n*-tetradecane film is as much as three times higher than for the same plate before hydrophobization.

Completely different relations were obtained in the case of an apolar liquid (Figure 9c). The hysteresis values of the diiodomethane angles for S/PC plates without *n*-tetradecane film are much higher than for the same plates with the film; for example, at room temperature, the hysteresis values after hydrophobization (approx. 3°) are almost five times lower than before (approx. 15°). On this basis, it can be concluded that the change in starch temperature and the presence of *n*-tetradecane significantly affect the nature of the tested surface, and furthermore, a lower hysteresis value usually indicates smaller surface heterogeneity and, as a result better system reproducibility. Some other factors influence the surface parameters, such as the presence of unsaturated bonds in molecules and the phase state in which investigated films exist; therefore, FTIR spectra can be useful [26,27].

2.7. Starch/Lysozyme Systems without n-Tetradecane Film

As was mentioned earlier, the condensed films are less permeable for the probe liquids used for the contact angle measurements. These suggestions can be observed as an effect of temperature. During gelatinization, the intensity of some bands increases, and new peaks can appear; hence, hydrophilic/hydrophobic changes can be observed from the FTIR spectra differences (Figure 10). During the gelatinization process, the most hydrophobic parts of the starch chain are directed into the air. However, steric limitations cause some polar groups to be present on the gel surface. The starch gel surface tends to reach maximal hydrophobic character with polar domains created by the functional glucose groups. It is most probable that the branched chain of amylopectin is directed into the air in a gel film. These are the primary results in the preparation of starch products of special properties, which can influence their solubility or polydispersity in aqueous media [47–51].

Starch is similar to other polysaccharides that absorb in the region 1200–800 cm^{-1}. In this region, the average penetration depth is about 2 μm [26,27]. Because this value can be smaller than the average size of studied starch granules (1–3 μm), the acquired FTIR spectra are probably representative of only the external part of the starch granules. However, such spectra should be adequate for the starch film and compatible with the profilometer results. FTIR analysis of the untreated starch at three different temperatures is shown in Figure 10a–c. All spectra show that the absorption bands in the 900–1300 cm^{-1} region are sensitive to the gelation of starch, which mainly result from the CO and CC vibrational modes being highly coupled. As the absorptions in this region arise largely from C–O stretchings of the ring, linkages (C–O–C), and COH groups, the positions of these bands are similar in all carbohydrates. The native starch surface shows a typical pattern of Raman bands, i.e., skeletal vibrations below 800 cm^{-1}, and the glycosidic linkage assigned to the band at about 940 cm^{-1}. During the temperature increases, the breaking of hydrogen bonds between the chains in the crystallites and hydrophobic bonds may become a major event. Thus, the intensity of some bands changes (Figure 10b,c). The signal around 1000 cm^{-1} is related to the intermolecular hydrogen bonding of hydroxyl groups or the plasticizing effect of water. Absorbance at 1020 cm^{-1} is assigned to the vibration of C–O–H deformation and is related to changes in the amorphous and crystalline parts of the starch granule. The bands in the 1015–1025 cm^{-1} and 1040–1050 cm^{-1} regions are of particular interest because they are associated with the amorphous and crystalline structures of starch, and their ratio can estimate the extent of gelatinization [26,27].

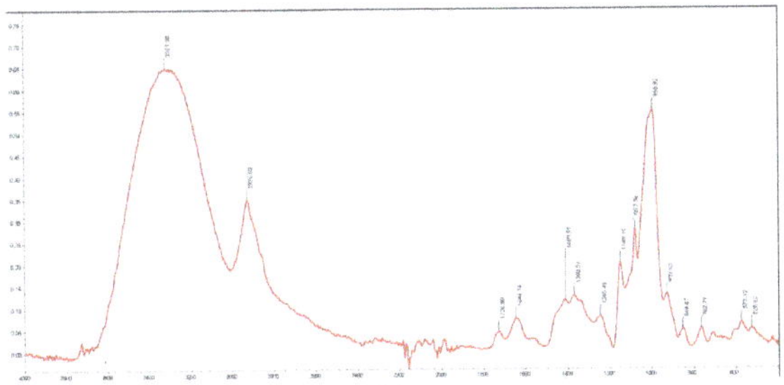

(**a**)

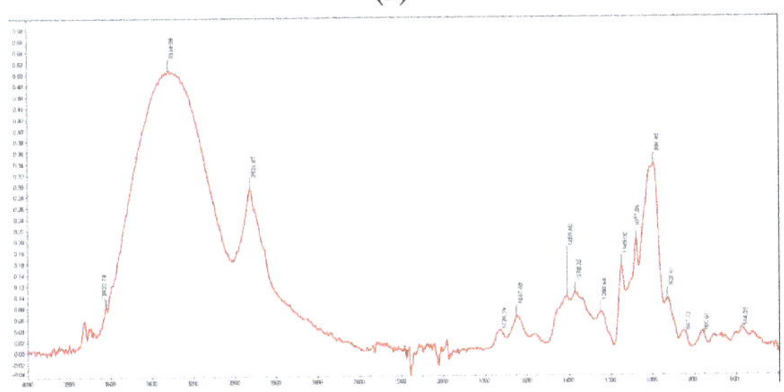

(**b**)

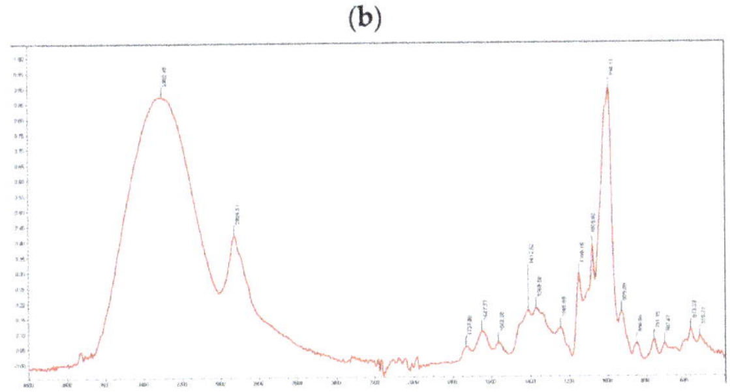

(**c**)

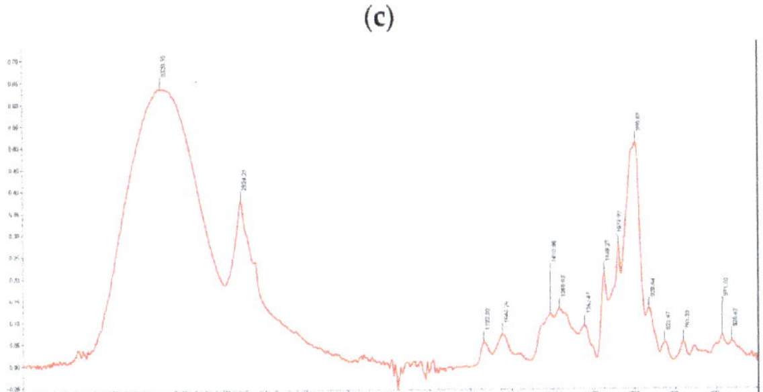

(**d**)

Figure 10. *Cont.*

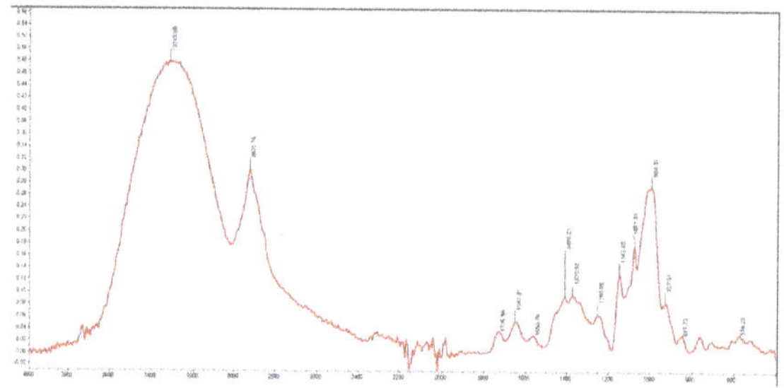

(e)

Figure 10. FTIR analysis for starches at different temperatures of (**a**) 25 °C, (**b**) 37 °C, (**c**) 70 °C, and after lysozyme modification, at (**d**) 25 °C, (**e**) 37 °C.

The absorption bands at 1150, 1080, and 1017 cm^{-1} show a tendency to decrease in the bandwidth during gelation. Their narrowing can be interpreted as an evolution to a more uniform state of the amorphous phase (Figure 10a–c). The bands at 1080 cm^{-1} show the greatest shifts assigned to C–O and C–C stretching and C–C–O bending. During the process of starch gelatinization, the Raman bands are shifted to lower wavenumbers for the CH deformation and the skeletal and CH stretching modes, but for COH deformation, they shift to higher wavenumbers. At room temperature, the Lys-modified starches show differences compared to the native starch (Figure 10d), but at higher temperatures, the lysozyme effect is less visible (Figure 10e) because growing temperature affects the changes occurring during the gelatinization process in the system with and without lysozymes. This process is decisive, and biological additives affect the spectrum to a lesser extent. High temperature causes the quantitative denaturation of a native lysozyme within 20 min. The starch surface with the increasing temperature of preparation becomes more hydrophilic. The contact angles decrease with the temperature of starch preparation, which also increases the hydrophilic character of surfaces. In the case of water, the layer structure changes also result in large contact angle hysteresis. The wettability test together with FTIR analysis and optical profilometry are suitable and powerful methods to characterize the morphological and molecular properties of starch and modified starch.

Accurate calculation of the area of the 1017–1022 cm^{-1} and 1044–1047 cm^{-1} regions and their ratios is commonly used in the literature [26,27] to estimate the extent of gelatinization. A lysozyme is an antibacterial substrate belonging to a large group of proteins and glycoside hydrolases. There are enzymes that damage bacterial cell walls by catalyzing the hydrolysis process of 1,4-beta-linkages between N-acetylmuramic acid and N-acetyl-D-glucosamine residues in a peptidoglycan and also between N-acetyl-D-glucosamine residues in chitodextrins. A lysozyme with these functions is recognized as part of the innate immune system, a natural form of protection against Gram-positive pathogens, like *Bacillus* and *Streptococcus* [16–18]. Therefore, covering starch films with lysozymes or their addition to dispersion systems will result in additional desirable properties both in the food and pharmaceutical industries. On the other hand, the immobilization of lysozymes on solid supports has lately shown positive results, as demonstrated by the increased stability and extended half-life of the enzyme in the research by Anastas and co-workers [5]. Furthermore, the immobilization of the lysozyme improves the ability of the lysozyme to target not only Gram-positive bacteria but also Gram-negative bacteria. The large effect of the lysozyme is noticeable, similar to the case of starch modification with a phospholipid, at 37 °C (Figure 10e). This may be related to conformational changes that occur in the protein structure on the surface of partially gelatinized starch.

A comparison of the average values of the contact angle of water, formamide, and diiodomethane for plates coated with starch and a lysozyme without *n*-tetradecane film is shown in Figure 11. At room temperature, the value of the contact angle increases with increasing polarity of the liquid used; the highest values were obtained for water, and the lowest for diiodomethane. However, for hysteresis, the relationship was not so clear, and the values obtained are as follows. For water, about 10°, for formamide, as much as 17°, and for diiodomethane, only 6°. Applying an additional layer (lysozyme) to the starch-coated plates increased the contact angle values by several degrees in the case of formamide and diiodomethane, while the average value of the water contact angle decreased by several degrees, which may indirectly indicate an increase in surface hydrophilicity.

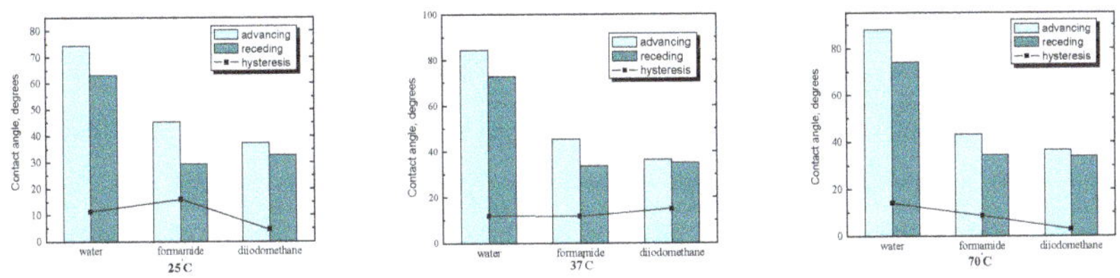

(a) (b) (c)

Figure 11. Comparison of the average value of the contact angle of the test liquid measured for the starch/lysozyme surface without *n*-tetradecane film (starch preparation temperature (**a**) 25 °C; (**b**) 37 °C; (**c**) 70 °C).

At a physiological temperature (Figure 11b), there is an analogy to measurements at room temperature (Figure 11a). It is easy to see that the values of the contact angles change in a similar way, i.e., the more polar the liquid is, the higher the values of the contact angles (Figure 11a). Additionally, the hysteresis value for polar liquids is practically at the same level of about 12°, while for non-polar diiodomethane, it is larger and amounts to about 16°. At a temperature of 70 °C (Figure 11c), the average values of the contact angles did not change compared to the measurements at lower temperatures (Figure 11a,b). Noticeable changes are seen in the hysteresis values, which decrease as the polarity of the liquid used decreases. The greatest difference between the advancing and receding angles is for water, and the smallest is for diiodomethane. The starch preparation temperature did not affect the contact angle values in these systems, while the application of protein on the starch increased the contact angles.

2.8. Starch/Lysozyme Systems with n-Tetradecane Film

The graph (Figure 12) shows the average values of the contact angles together with the hysteresis values of water, formamide, and diiodomethane for S/Lys plates with *n*-tetradecane film. When the starch preparation temperature was 25 °C (Figure 12a), the highest value of the contact angles was obtained for water, and the lowest for formamide. As in most cases, the angle hysteresis varies depending on the polarity of the liquid used. The less polar the liquid, the smaller the hysteresis of the angle. In the case of physiological temperature (Figure 12b), both the value of the angle hysteresis and the average values of the contact angle increase with the increase in the polarity of the liquid used. For diiodomethane, which is non-polar, the lowest values were obtained, while for polar water, the values of the contact angle and hysteresis were the highest.

The values of the contact angles obtained for the S/Lys plates with *n*-tetradecane film (starch preparation temperature of 70 °C) are shown in Figure 12c. Both the value of the advancing and receding angle increases as the polarity of the liquid used decreases. The hysteresis values for water and formamide are very similar. A significantly lower hysteresis value was noted for diiodomethane. The application of an additional layer on a plate with starch-coated *n*-tetradecane film caused an increase in the contact angles of polar water and non-polar diiodomethane.

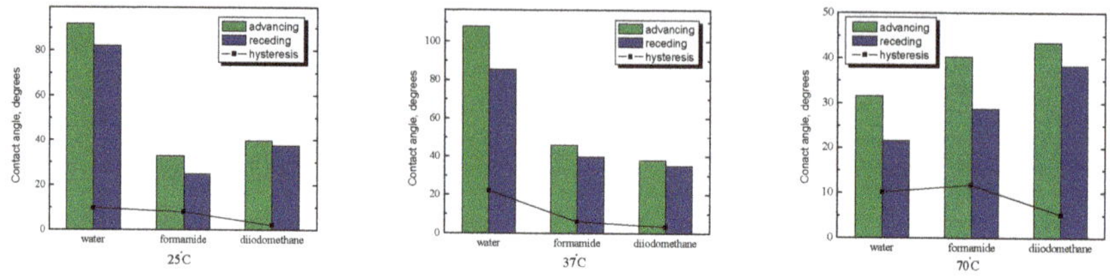

(a) (b) (c)

Figure 12. Comparison of the average value of the contact angle of the test liquid measured for the starch/lysozyme surface with *n*-tetradecane film (starch preparation temperature (**a**) 25 °C; (**b**) 37 °C; (**c**) 70 °C).

2.9. Comparison of the Hysteresis Angle of Test Liquids for the Starch/Lysozyme System before and after Hydrophobization vs. Temperature

In Figure 13, the hysteresis of water, formamide, and diiodomethane angles on the S/Lys surface with and without *n*-tetradecane film was compared depending on the starch preparation temperature. In Figure 13, at 25 °C and 70 °C, the hysteresis values of the water contact angle for S/Lys plates with *n*-tetradecane film are very similar and amount to about 10°. At physiological temperature, the effect of *n*-tetradecane is most noticeable. For room and physiological temperature, the hysteresis values are twice as high for the S/Lys plates without *n*-alkane film compared to the same plates with the film.

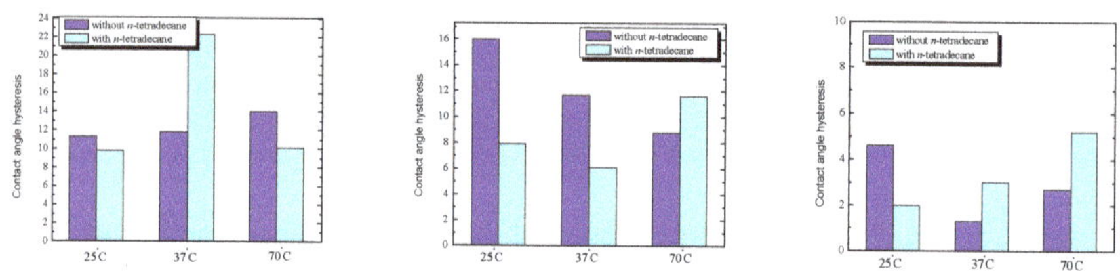

(a) (b) (c)

Figure 13. Comparison of the average value of the contact angle hysteresis for the starch/lysozyme surface with and without *n*-tetradecane film for (**a**) water; (**b**) formamide; (**c**) diiodomethane.

At a temperature of 70 °C, the effect of hydrophobization is more visible; the hysteresis value before hydrophobization is 9°, and after hydrophobization, it is 12°. In addition, it can also be seen that the hysteresis values for the S/Lys plates without *n*-tetradecane film decrease with increasing temperature. This can be explained by the easier spreading of the starch solution at higher temperatures and the concomitant decreasing film thickness. The hysteresis of diiodomethane angles of the starch/lysozyme system before and after hydrophobization is shown in Figure 13c, and these are the smallest hysteresis values that we obtained. As the temperature increases, the hysteresis value for the S/Lys plates with *n*-tetradecane film increases, i.e., the effect of hydrophobization also increases. This can be explained by the fact that large surface changes occur under the influence of *n*-tetradecane and temperature, which are visible on the basis of contact angle measurements. As the temperature increases, the breaking of hydrogen bonds between the poly-(1 $\rightarrow$ 4)-glucan chains in the crystallites and hydrophobic bonds may become a major event. The loss of crystallinity may be considered a manifestation of starch gelatinization.

Another currently explored novel application of starch is the immobilization of enzymes, proteins, or drugs to improve the catalytic activity of enzymes used in food technology or to create more efficient drug delivery systems [19]. The research carried out strictly fits into this trend of potential applications. A detailed description of topographic changes was presented in our previous works [26,27]. The roughness and topography of the starch surface were investigated using an optical profilometer (Contour GT-K1, Veeco) measuring topography with high accuracy from a subnanometer up to 10 mm in size. We have performed such measurements in three different sites for glass plates covered with native starch and its modification. Below, we present some sample surfaces (Figure 14).

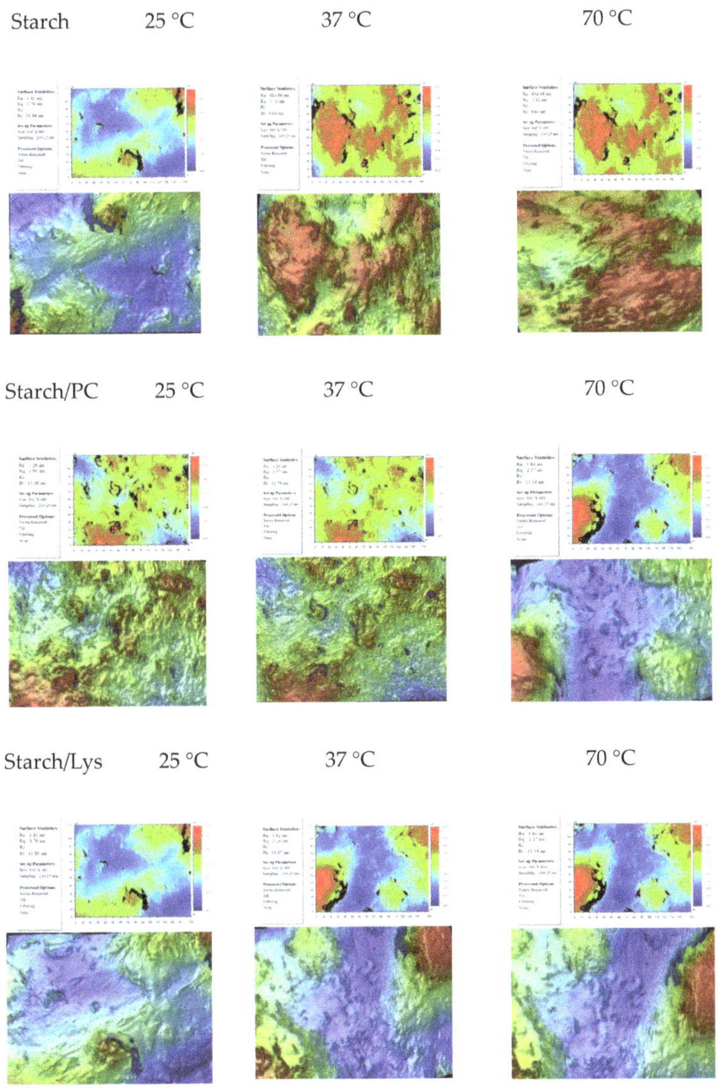

Figure 14. Images obtained from an optical profilometer for starch before and after modification at different preparation temperatures.

The obtained images of the starch surface by profilometer show topographical differences between the native starch surface and that modified by a biological solution. For example, the height distribution shows that the starch surface is covered mostly by patches. The hydrophilization of the starch surface after PC modification is caused by the rearrangement of the hydrophobic monolayer into bilayer patches with the head polar groups exposed toward water drop. Attractive interactions between starch and PC may cause complex formation. This strong association can cause film condensation and smoothing. Indeed, the monolayer of PC on the starch makes the surface smoother. Reduced values of all roughness parameters were obtained compared to the native starch surface; although, in order to explain changes in surface roughness, the type and size of interactions between starch molecules and the additional layer should be taken into account. Due to the polar nature of water, its molecules have a strong tendency to form hydrogen bonds. Therefore, there may be a structural rearrangement of PC or Lys molecules on the starch surface and an increase in the polarity of the starch surface after its modification. This effect decreases as the temperature increases. On the other hand, the lysozyme reduces the hydrophobic nature of the starch surface.

In both cases, during temperature increase, the homogeneity of the starch layer increases because the degree of gelatinization grows. Moreover, one can also find some correlations between the changes in surface roughness and the values of contact angles measured on native and modified starch surfaces. The increase in attractive interactions between both components causes the formation of more and more tightly packed monolayers characterized by lower values of the roughness parameters. Therefore, such condensed films are less permeable for the probe liquids used for the contact angle measurements. The chemical nature of the probe liquid, the size of its molecules, and the kind of interactions determine the surface penetration. It is clear that the differences in the values of the contact angles and hysteresis manifest the strength of the solid/liquid interfacial interactions. It is obviously known that the contact angle measured on such surfaces is of macroscopic quantity, which is, in fact, an averaged (apparent) value of the microscopic contact angles. The reproducibility of the contact angle values and the topography images enable full surface characterization.

2.10. Starch Surface Wettability: Work of Adhesion

The changes in wettability and the hydrophilic/hydrophobic character of the starch surface before and after modification vs. temperature can be also depicted by calculating the work of probe liquids adhesion. Details of this procedure were presented in our previous papers [26,27]. In the case of polar liquids (water or formamide), the changes in the values of work of adhesion indicate changes in polar interactions. However, if the work of adhesion and correlated surface energy is calculated from apolar diiodomethane contact angles, the values characterize the London dispersion interactions on the surface and usually are the most stable. This indirectly informs us what impacts we are dealing with in a specific case. The changes in the work of adhesion calculated from water angles are much greater than in the case of diiodomethane. In all the presented systems (Table 1), both the hydrophobization effect and the influence of temperature are noticeable.

Table 1. Work of adhesion [mJ/m^2] calculated from water (W) or diiodomethane (D) contact angles for the starch surface before and after modification vs. preparation temperature.

Temp./ Surface	Without *n*-Tetradecane			With *n*-Tetradecane		
	S	S/PC	S/Lys	S	S/PC	S/Lys
25 °C W	75.70	43.68	85.3	135.58	129.35	0.41
37 °C W	82.95	41.34	141.94	102.51	141.94	30.79
70 °C W	129.34	0.56	145.55	100.14	11.02	139.39
25 °C D	64.35	70.71	64.35	2.92	30.48	16.92
37 °C D	16.92	2.42	99.32	18.28	16.92	16.92
70 °C D	16.92	89.62	99.32	16.92	88.50	78.99

Also, the temperature effect is more noticeable. For starch, as a function of temperature, an opposite trend of changes in the adhesion work was observed, depending on the substrate, increasing without *n*-tetradecane and decreasing after hydrophobization. For S/PC and S/Lys substrates, such a clear trend of changes is difficult to capture. On the other hand, the effect of the hydrophobic substrate is less visible in the values of the adhesion work calculated on the basis of diiodomethane contact angles. This is understandable due to the type of interactions in this case, and values in terms of 16–18 mJ/m^2 are often repeated

3. Materials and Methods

3.1. Materials

The following reagents were used in the research without further purification: *n*-tetradecane (≥99% GC, Fluka, Steinheim, Germany), rice starch for food purposes, that is naturally sourced, sustainably produced from renewable feedstocks, and GMO free (not genetically modified, Remylgel 663 DR-P, HORTIMEX, Wielsbeke, Belgium), L-α-phosphatidylcholine from egg yolks, p.a. (SIGMA-Aldrich, Steinheim, Germany), lysozymes (SIGMA, St. Louis, MO, USA), NaCl (p.a. POCH S.A. Gliwice, Poland), pentane (≥99% GC, Fluka), diiodomethane (p.a. POCH S.A. Gliwice, Poland), formamide (p.a. POCH S.A. Gliwice, Poland), water from the Millipore Q system, and a resistivity 18.2 MΩ cm. Products marked as GMO free contain, consist of, or have been produced from organisms/raw materials for which there are equivalents entered in the community register of genetically modified foods, and these organisms do not contain, consist of, or have not been produced from GMOs.

3.1.1. Solution Preparation

A total of 1 g of starch (1 mg of lecithin or 0.25 mg of lysozyme) was weighed into a 50 mL volumetric flask and supplemented to the mark with 10^{-3} M NaCl solution. Depending on the purity of the ingredients used, the proportions of the molecular weight of lysozyme to lecithin range from 5 to 15; the lowest proportion of 1:5 was chosen in the research. A solution of *n*-tetradecane in pentane was made as follows: 10 mL of pentane was poured into a 25 mL beaker and 0.1 mL of *n*-tetradecane was pipetted. The solution prepared in this way was poured with a pipette onto a glass plate (optical glass plates, As Polonia, Warsaw, Poland). In the case of hydrophobized plates, after one day, the pentane evaporated, leaving only *n*-tetradecane on the surface.

3.1.2. Preparation of Glass Plates

Optical glass plates were used for the measurements. The solution of *n*-tetradecane prepared for wettability tests, described in Section 3.1.1, was poured with a pipette onto plates; next, the pentane was evaporated. The other solutions were poured onto plates or plates previously coated with a solution of *n*-tetradecane in pentane. Glass plates were used as solid substrata for covering with starch solution at three different temperatures (25 °C, 37 °C, and 70 °C). After reaching the sample preparation temperature, the appropriate volume of the native starch (partially gelatinized or gelatinized) solution was slowly spread on glass plates to obtain a homogeneous layer. Next, the plates were kept at room temperature for drying, and after that, they were placed in desiccators filled with silica gel for about 1 day to perform experiments.

3.1.3. Emulsion Preparation

To prepare the *n*-tetradecane/starch emulsion in NaCl solution, 1 mg of starch was weighed into a 100 mL volumetric flask, and 0.1 cm^3 of *n*-tetradecane was pipetted and filled up to the mark with a 10^{-3} M NaCl solution. It was homogenized for 15 min at 10,000 rpm in a homogenizer (SilentCrusher M, Heidolph, Schwabach, Germany). After homogenization, the emulsion was placed in measuring cuvettes, and measurements were carried out as a function of time. Measurements of analogous systems were made at 25 °C (room temperature), 37 °C (physiological), and 70 °C (after starch gelatinization).

3.2. Methods

During the experimental part, a laboratory scale (Sartorius, Göttingen, Germany), a homogenizer, a zetameter (ZetaPlus, Brookhaven, Holtsville, NY, USA), and a contact angle measurement apparatus (GBX, Rhone, France) were used.

3.2.1. Dynamic Light Scattering

The phenomenon of dynamic light scattering is used to measure the particle size of dispersed systems from a few nm to a maximum of 1 μm. This method is based on measuring changes in a specific light wavelength (in this type of system of 670 nm) and fluctuations in average intensity. The radiation scattered by the particles is registered with a detector placed at a certain angle to the incident beam. The intensity of the scattered radiation fluctuates constantly as the dispersed particles undergo continuous Brownian motion and thermal motion. When the particles are small, they exhibit fast diffusion motions and, therefore, fast fluctuations. When larger particles are present in the sample, they diffuse more slowly and thus cause slower fluctuations.

Two physical processes, i.e., particle size increase due to coalescence or flocculation and particle migration, which leads to creaming or sedimentation, often result in system instability. Detection of these processes at an early stage and kinetic analysis of changes in these destabilizing processes is very important in predicting stability. One of the ways to indirectly determine the stability of the dispersion system is the analysis of changes in particle diameter and zeta potential. The computer coupled with the device (Zeta Plus/Pals) automatically controls the course of the correlation function, including the selection of the sampling time and the measurement time. Using the method of dynamic light scattering, assuming that non-interacting spherical particles are dispersed in a medium with a viscosity η, the diffusion coefficient with the particle size $d(H)$ can be related by the Stokes–Einstein equation (Equation (1)) [33] as follows:

$$D = \frac{kT}{3\pi\eta} d(H) \tag{1}$$

where k is the Boltzmann constant, T is the absolute temperature, η is the medium viscosity, and $d(H)$ is the particle size.

Importantly, for the same portion of the sample, both the particle size and the electrokinetic potential were determined using an apparatus (zetameter) in thermostatic polyacrylic cuvettes as a function of time, i.e., after 5 min, 15 min, 30 min, 1 h, and 2 h from the moment of emulsion (dispersion) preparation, depending on the temperature, i.e., at 25 °C, 37 °C, and 70 °C.

3.2.2. Microelectrophoresis

Electrophoresis is one of the known methods for determining electrokinetic potential. It consists of the movement of charged colloidal particles relative to the solvent under the influence of the applied electric field of constant intensity ε. These particles move with a characteristic speed, which depends on the field strength, the viscosity of the medium, and the zeta potential. According to Henry's theory, the electrophoretic mobility of particles of any size can be expressed by the following formula (Equation (2)) [33–36]:

$$U_e = \frac{\varepsilon\zeta}{\eta} f(\kappa a) \tag{2}$$

where U_e is electrophoretic mobility, ε is permittivity, ζ is electrokinetic potential, η is medium viscosity, and $f(\kappa a)$ is the Henry function (a is particle radius, κ is the Debay—Hückel parameter, depending on electrolyte concentration, $1/\kappa$ is diffusion layer thickness).

For small values of κa ($\kappa a \leq 1$), the Henry function takes the form of the Hückel equation (Equation (3)) as follows:

$$U_e = \frac{2\varepsilon\zeta}{3\eta} \tag{3}$$

For large values of κa, the Smoluchowski equation is used (Equation (4)).

$$U_e = \frac{\varepsilon\zeta}{\eta} \tag{4}$$

The zeta potential can be determined for the same batch of emulsion (dispersion) for which the mean or effective diameter was measured, which is an advantage of the instrument used. After measuring the effective diameter, an electrode is placed in the measuring cuvette and electrophoretic mobility is obtained by microelectrophoresis. The computer then converts the measured mobility into electrokinetic potential according to the Smoluchowski or Hückel equation (Equations (3) and (4)) [34–37].

3.2.3. Contact Angle Measurements

Measurements of contact angles were carried out using the GBX apparatus by the sessile drop method using liquids that differed in polarity, i.e., strongly polar water, polar formamide, and non-polar diiodomethane. The determination of the hydrophilic/hydrophobic character of the starch surface before and after modification can be used in the food and pharmaceutical industries [26–29]. Switchable wettability may be a convenient parameter providing information about even subtle changes of starch surface properties. Additionally, modification, for example, by an antibacterial substrate's film, may give new applications of starch. During these experiments on the surface of glass plates coated with starch, such as *n*-tetradecane (or both) or starch modified by phospholipids or lysozymes, a drop of the test liquid with a volume of 3 μL was placed, and the advancing contact angle was measured. Then, 1/3 of the liquid volume was taken from the drop, and the receding contact angle was measured. Hysteresis was defined as the difference between these two angles. For each system, 10 to 20 contact angle measurements were made, depending on the type of liquid used. For various types of modifications, measurements of both angles and determining the hysteresis provided appropriate results and the possibility of in-depth interpretation.

4. Conclusions and Future Perspective

The studies of dispersion systems with the participation of starch as an emulsifier and additional bioactive substances are current and interesting due to the possibility of their versatile use in the pharmaceutical, medicine, and food/packaging industries. In this aspect, it is also very helpful to determine the hydrophilic/hydrophobic nature of the starch surface. For all systems, changes in zeta potential were usually correlated with changes in the effective diameter. Moreover, the starch dispersions were more stable than analogous *n*-alkane/starch emulsions at room and physiological temperatures. With regard to the combination of starch and lysozyme, it was found that the stability of the *n*-tetradecane/starch Lys system in the electrolyte solution increases with increasing temperature; the most stable one was obtained at 70 °C. Increasing the starch preparation temperature and the use of a lysozyme film contributed to an increase in the average contact angle values.

In the case of the tested *n*-alkane/starch and the PC system, the increase in temperature did not enhance its stability; however, for plates coated with unmodified starch and starch/PC without the *n*-alkane film, the contact angle hysteresis values decreased with a temperature increase, pointing to a more homogeneous surface. The highest wettability was obtained for the starch/PC both with and without the *n*-tetradecane film, where the values of water angles at each of the tested temperatures are the smallest, in the range of 35°–80°, which means moderately or softly hydrophilic character, optimal for biological samples. The effect of hydrophobization was visible for each of the tested liquids and temperatures, decreasing vs. the thickness of the substrate increase. Starch applied to the plate with the *n*-alkane film forms a less uniform layer than that directly applied to glass. The most homogeneous surfaces were acquired for pure starch at 70 °C; the hysteresis angle values decrease with increasing temperature, which can be explained by the fact that starch increases the degree of swelling with increasing temperature and spreads better on glass. However, after starch modification, the highest hysteresis values were recorded for the S/PC and S/Lys plates at physiological temperatures, and the effect of hydrophobization was most visible at this temperature.

To generalize, the obtained results proved that switchable wettability may be a convenient parameter providing information on surface properties. The values of contact angles are apparent but in connection with hysteresis, they provide an opportunity to determine the energetic state of the starch surface via the starch/probe liquid interfacial interactions. The contact angles change with the temperature of starch preparation and different types of modification, which means that the hydrophilic/hydrophobic character of surfaces can be moderated. The wettability test together with stability measurements are suitable and powerful methods to characterize the strength and specificity of interactions that describe many biological processes, whereas a combination of bioactive substances and different treatment parameters led to the improvement of biopolymer wettability and functionality, and the incorporation of additional features into the final product without significantly changing the beneficial properties of native starch. We hope that these results will be helpful for new applications of such starch-based polysaccharide systems.

Author Contributions: Conceptualization A.E.W.; methodology A.E.W. and A.F.; software, A.E.W. and A.F.; validation, A.E.W. and A.F.; formal analysis, A.E.W. and A.F.; investigation, A.E.W. and A.F.; resources, A.E.W. and A.F.; data curation, A.E.W. and A.F.; writing—original draft preparation, A.E.W. and A.F.; writing—review and editing, A.E.W. and A.F.; visualization, A.E.W. and A.F.; supervision, A.E.W. All authors have read and agreed to the published version of the manuscript.

Funding: This research received no external funding.

Institutional Review Board Statement: Not applicable.

Informed Consent Statement: Not applicable.

Data Availability Statement: The data will be shared by the authors.

Acknowledgments: Thanks to Pankiewicz and Sujka for the opportunity to complete the ZUL grant internship (No. SzM/23/222/TŻ) and new inspirations for research on starch systems.

Conflicts of Interest: The authors declare no conflicts of interest.

References

1. Valero-Valdivieso, M.F.; Ortegon, Y.; Uscategui, Y. Biopolymers: Progress and prospects. *Dyna* **2013**, *80*, 171–180.
2. Bertolini, A.C. *Starches Characterization, Properties, and Applications*; CRC Press: Boca Raton, FL, USA, 2010.
3. Pszczola, D.E. Which starch is on first? *Food Technol.* **2006**, *4*, 51–64.
4. Zhang, B.; Pan, Y.; Chen, H.; Liu, T.; Tao, H.; Tian, Y. Stabilization of starch-based microgel-lysozyme complexes using a layer-by-layer assembly technique. *Food Chem.* **2017**, *214*, 213–217. [CrossRef] [PubMed]
5. Anastas, P.T.; Rodriguez, A.; de Winter, T.M.; Coish, P.; Zimmerman, J.B. A review of immobilization techniques to improve the stability and bioactivity of lysozyme. *Green Chem. Lett. Rev.* **2021**, *14*, 302–338. [CrossRef]
6. Ferreira Bonomo, R.C.; Santos, T.A.; Santos, L.S.; da Costa, R.; Fontan, I.; Rodrigues, L.B.; dos Santos Pires, A.C.; Veloso, C.M.; Ramos Gandolfi, O.R.; Bonomo, P. Effect of the Incorporation of Lysozyme on the Properties of Jackfruit Starch Films. *J. Polym. Environ.* **2018**, *26*, 508–517. [CrossRef]
7. Ji, Y. Microgels prepared from corn starch with an improved capacity for uptake and release of lysozyme. *J. Food Eng.* **2020**, *285*, 110088. [CrossRef]
8. Li, S.; Shi, L.; Ye, T.; Huang, B.; Qin, Y.; Xie, Y.; Ren, X.; Zhao, X. Development of Crosslinker-Free Polysaccharide-Lysozyme Microspheres for Treatment Enteric Infection. *Polymers* **2023**, *15*, 1077. [CrossRef]
9. Li, X.; Wang, C.; Lu, F.; Zhang, L.; Yang, Q.; Mu, J.; Li, X. Physicochemical properties of corn starch isolated by acid liquid. *Food Hydrocoll.* **2015**, *44*, 353–359. [CrossRef]
10. Milhaud, J. New insights into water–phospholipid model membrane interactions. *Biochim. Biophys. Acta* **2004**, *1663*, 19–51. [CrossRef]
11. Falade, K.O.; Christopher, A.S. Physical, functional, pasting and thermal properties of flours and starches of six Nigerian rice cultivars. *Food Hydrocoll.* **2015**, *44*, 478–490. [CrossRef]
12. Domian, E.; Brynda-Kopytowska, A.; Oleksza, K. Rheological properties of physical stability of o/w emulsions stabilized by OSA starch with trehalose. *Food Hydrocoll.* **2015**, *44*, 49–58. [CrossRef]
13. Xu, W.; Jin, W.; Wang, Y.; Li, J.; Huang, K.; Shah, B.; Li, B. Effect of physical interactions on structure of lysozyme in presence of three kinds of polysaccharides. *J. Food Sci. Technol.* **2018**, *55*, 3056–3064. [CrossRef] [PubMed]
14. Singh, J.; Kaur, L.; McCarthy, O.J. Factors influencing the physico-chemical, morphological, thermal and rheological properties of some chemically modified starches for food applications—A review. *Food Hydrocoll.* **2007**, *21*, 1–22. [CrossRef]
15. Zhang, Y.; Chi, C.; Huang, X.; Zou, Q.; Li, X.; Chen, L. Starch-based nanocapsules fabricated through layer-by-layer assembly for oral delivery of protein to lower gastrointestinal tract. *Carbohyd. Polym.* **2017**, *171*, 242–251. [CrossRef] [PubMed]

16. Gennari, C.G.; Sperandeo, P.; Polissi, A.; Minghetti, P.; Cilurzo, F. Lysozyme mucoadhesive tablets obtained by freeze-drying. *J. Pharm. Sci.* **2019**, *108*, 3667–3674. [CrossRef] [PubMed]
17. Ragland, S.A.; Criss, A.K. From bacterial killing to immune modulation: Recent insights into the functions of lysozyme. *PLoS Pathog.* **2017**, *13*, e1006512. [CrossRef] [PubMed]
18. Ferraboschi, P.; Ciceri, S.; Grisenti, P. Applications of lysozyme, an innate immune defense factor, as an alternative antibiotic. *Antibiotics* **2021**, *10*, 1534. [CrossRef] [PubMed]
19. Koczoń, P.; Josefsson, H.; Michorowska, S.; Tarnowska, K.; Kowalska, D.; Bartyzel, B.J.; Niemiec, T.; Lipińska, E.; Gruczyńska-Sękowska, E. The Influence of the Structure of Selected Polymers on Their Properties and Food-Related Applications. *Polymers* **2022**, *14*, 1962. [CrossRef]
20. Zhong, J.; Wang, Q.; Qin, X. Improving the stability of phosphatidylcholine-enhanced nanoemulsions using octenyl succinic anhydride-modified starch. *Int. J. Biol. Macromol.* **2018**, *120*, 1500–1507. [CrossRef]
21. Khatun, A.; Waters, D.L.E.; Liu, L. The Impact of Rice Lipid on In Vitro Rice Starch Digestibility. *Foods* **2022**, *11*, 1528. [CrossRef]
22. Zdybel, E.; Wilczak, A.; Kapelko-Żeberska, M.; Tomaszewska-Ciosk, E.; Gryszkin, A.; Gawrońska, A.; Zięba, T. Physicochemical Properties and Digestion Resistance of Acetylated Starch Obtained from Annealed Starch. *Polymers* **2021**, *13*, 4141. [CrossRef]
23. Cheng, W.; Luo, Z.; Li, L.; Fu, X. Preparation and characterization of disbranched-starch/phosphatidylcholine inclusion complexes. *J. Agric. Food Chem.* **2015**, *63*, 634–641. [CrossRef] [PubMed]
24. Zhong, J.; Li, Y.; Qin, X.; Li, J.; Wang, Y. Improving the freeze–thaw stability of emulsions via combining phosphatidylcholine and modified starch: A combined experimental and computational study. *Int. J. Food Sci. Technol.* **2022**, *57*, 1050–1060. [CrossRef]
25. Hao, B.; Narciso, J.O.; Nyström, L. Extraction and identification of phospholipids from whole grain kabog millet flour and predictive effects on starch binding and retrogradation. *LWT* **2021**, *152*, 112406. [CrossRef]
26. Wiącek, A.E. Effect of surface modification on starch biopolymer wettability. *Food Hydrocoll.* **2015**, *48*, 228–237. [CrossRef]
27. Wiącek, A.E.; Dul, K. Effect of surface modification on starch/biosurfactant wettability. *Colloids Surfaces Physicochem. Eng. Aspects* **2015**, *480*, 351–359. [CrossRef]
28. Sołtys, A.; Łazarz, M.; Chibowski, E. Zeta potential and effective diameter of *n*-tetradecane emulsions in n-propanol solutions and in the presence of lysozyme. *Colloids Surf. A Physicochem. Eng. Asp.* **1997**, *127*, 163–173. [CrossRef]
29. Krstonosic, V.; Dokic, L.; Nikolic, I.; Milanovic, M. Influence of xanthan gum on oil-in-water emulsion characteristics stabilized by OSA starch. *Food Hydrocoll.* **2015**, *45*, 9–17. [CrossRef]
30. Fechner, P.M.; Wartewig, S.; Kiesow, A.; Heilmann, A.; Kleinebudde, P.; Neubert, R.H.H. Influence of water on molecular and morphological structure of various starches and starch derivatives. *Starch/Stärke* **2005**, *57*, 605–615. [CrossRef]
31. Dupuy, N.; Laureyns, J. Recognition of starches by Raman spectroscopy. *Carbohydr. Polym.* **2002**, *49*, 83–90. [CrossRef]
32. Schuster, K.C.; Ehmoser, H.; Gapes, J.R.; Lendl, B. FT-Raman spectroscopic monitoring of starch gelatinization and enzyme catalyzed starch hydrolysis. *Vib. Spectrosc.* **2000**, *22*, 181–190. [CrossRef]
33. Adamson, A.W.; Gast, A.P. *Physical Chemistry of Surfaces*, 6th ed.; Wiley-Interscience: Hoboken, NJ, USA, 1997.
34. van Oss, C.J.; Wu, W.; Docoslis, A.; Giese, R.F. The interfacial tensions with water and the Lewis acid–base surface tension parameters of polar organic liquids derived from their aqueous solubility's. *Colloids Surf. B* **2001**, *20*, 87–91. [CrossRef] [PubMed]
35. Horcas, I.; Fernandez, R.; Gomez-Rodriguez, J.M.; Colchero, J.; Gomez-Herrero, J.; Baro, A.M. WS x M: A software for scanning probe microscopy and a tool for nanotechnology. *Rev. Sci. Instrum.* **2007**, *78*, 013705. [CrossRef] [PubMed]
36. Grabowski, E.F.; Morison, J.D.; Dabneke, B. *Particle Size Distributions from Analysis of Quasi-Elastic Light Scattering Data*; Wiley: New York, NY, USA, 1983.
37. Rudnik, E.; Matuschek, G.; Milanov, N.; Kettrup, A. Thermal stability and degradation of starch derivatives. *J. Therm. Anal. Calorim.* **2006**, *85*, 267. [CrossRef]
38. Mali, S.; Grossmann, M.V.E.; Garcia, M.A.; Martino, M.N.; Zaritzky, N.E. Microstructural characterization of yam starch films. *Carbohydr. Polym.* **2002**, *50*, 379. [CrossRef]
39. Zhang, Y.; Han, J.H. Plasticization of Pea Starch Films with Monosaccharides and Polyols. *J. Food Sci.* **2006**, *71*, E253–E261. [CrossRef]
40. Volkert, B.; Lehmann, A.; Greco, T.; Nejad, M.H. Comparison of different synthesis routes for starch acetates and the resulting mechanical properties. *Carbohydr. Polym.* **2010**, *79*, 571. [CrossRef]
41. Xiao, C.M.; Tan, J.; Xue, G.N. Synthesis and properties of starch-g-poly(maleic anhydride-co-vinyl acetate). *Express Polym. Lett.* **2010**, *4*, 9. [CrossRef]
42. Raquez, J.-M.; Nabar, Y.; Narayan, R.; Dubois, P. In situ compatibilization of maleated thermoplastic starch/polyester melt-blends by reactive extrusion. *Polym. Eng. Sci.* **2008**, *48*, 1747. [CrossRef]
43. Bengtsson, M.; Koch, K.; Gatenholm, D. Surface octanoylation of high-amylose potato starch films. *Carbohydr. Polym.* **2003**, *54*, 1. [CrossRef]
44. Zhou, J.; Ren, L.; Tong, J.; Ma, Y. Effect of surface esterification with octenyl succinic anhydride on hydrophilicity of corn starch films. *J. Appl. Polym. Sci.* **2009**, *114*, 940. [CrossRef]
45. Rajan, A.; Abraham, E. Enzymatic modification of cassava starch by bacterial lipase. *Bioprocess Biosyst. Eng.* **2006**, *29*, 65.
46. Rajan, A.; Sudha, J.D.; Abraham, T.E. Enzymatic modification of cassava starch by fungal lipase. *Ind. Crop. Prod.* **2008**, *27*, 50. [CrossRef]

47. McClements, D.J.; Jafari, S.M. Improving emulsion formation, stability and performance using mixed emulsifiers: A review. *Adv. Colloid Interface Sci.* **2018**, *251*, 55–79. [CrossRef] [PubMed]
48. Sevenou, O.; Hill, S.E.; Farhat, I.A.; Mitchell, J.R. Organization of the external region of the starch granule as determined by infrared spectroscopy. *Int. J. Biol. Macro-Mol.* **2002**, *31*, 79–85. [CrossRef] [PubMed]
49. Muljana, H.; Picchioni, F.; Heeres, H.J.; Janssen, L.P.B.M. Supercritical carbon dioxide induced gelatinization of potato starch. *Carbohydr. Polym.* **2009**, *78*, 511–519. [CrossRef]
50. Capron, I.; Robert, P.; Colonna, P.; Brogly, M.; Planchot, V. Starch in rubbery and glassy states by FTIR spectroscopy. *Carbohyd. Polym.* **2007**, *68*, 249–259. [CrossRef]
51. Suzuki, N.; Hanashiro, I.; Fujita, N. Molecular Weight Distribution of Whole Starch in Rice Endosperm by Gel-permeation Chromatography. *J. Appl. Glycosci.* **2023**, *70*, 25–32. [CrossRef]

International Journal of
Molecular Sciences

MDPI

Article

Physicochemical Characteristics of Porous Starch Obtained by Combined Physical and Enzymatic Methods, Part 1: Structure, Adsorption, and Functional Properties

Monika Sujka [1,*] and Agnieszka Ewa Wiącek [2,*]

1 Department of Analysis and Food Quality Assessment, Faculty of Food Sciences and Biotechnology, University of Life Sciences in Lublin, Skromna 8, 20-704 Lublin, Poland
2 Department of Interfacial Phenomena, Faculty of Chemistry, Maria Curie-Skłodowska University, Maria Curie-Skłodowska Sq.3, 20-031 Lublin, Poland
* Correspondence: monika.sujka@up.lublin.pl (M.S.); agnieszka.wiacek@mail.umcs.pl (A.E.W.); Tel.: +48-81-462-33-33 (M.S.); +48-81-537-56-54 (A.E.W.)

Abstract: Porous starch can be applied as an adsorbent and encapsulant for bioactive substances in the food and pharmaceutical industries. By using appropriate modification methods (chemical, physical, enzymatic, or mixed), it is possible to create pores on the surface of the starch granules without disturbing their integrity. This paper aimed to analyze the possibility of obtaining a porous structure for native corn, potato, and pea starches using a combination of ultrasound, enzymatic digestion, and freeze-drying methods. The starch suspensions (30%, *w*/*w*) were treated with ultrasound (20 kHz, 30 min, 20 °C), then dried and hydrolyzed with amyloglucosidase (1000 U/g starch, 50 °C, 24 h, 2% starch suspension). After enzyme digestion, the granules were freeze-dried for 72 h. The structure of the native and modified starches were examined using VIS spectroscopy, SEM, ATR-FTIR, and LTNA (low-temperature nitrogen adsorption). Based on the electrophoretic mobility measurements of the starch granules using a laser Doppler velocimeter, zeta potentials were calculated to determine the surface charge level. Additionally, the selected properties such as the water and oil holding capacities, least gelling concentration (LGC), and paste clarity were determined. The results showed that the corn starch was the most susceptible to the combined modification methods and was therefore best suited for the production of porous starch.

Keywords: porous starch; specific surface area; adsorption; ultrasound; enzymatic treatment; surface charge; zeta potential

1. Introduction

The growing demand for natural, cheap, non-toxic, and biodegradable adsorbents for the food industry has resulted in increased interest among researchers in starch as a raw material for their production. Native starch naturally occurs in the form of semi-crystalline granules of round, spherical, polygonal, elongated, and kidney shapes, depending on the botanical source of starch. Porous starch may be obtained by enzymatic (hydrolysis with e.g., α-amylase (EC 3.2.1.1), amyloglucosidase (EC 3.2.1.3), glycogen branching enzyme (EC 2.4.1.18)), chemical (e.g., solvent exchange, crosslinking, acid hydrolysis) or physical (e.g., extrusion, ultrasonic treatment, pulsed electric field) methods, or a combination of them [1–10].

Parameters characterizing the texture of food materials and influencing the mechanical, diffusion and sensory properties of food include, among others, porosity, pore size and pore size distribution [11]. According to the IUPAC classification, there are three groups of pores in solid materials: (a) macropores (width > 50 nm), (b) mesopores (2 nm < width < 50 nm), (c) micropores (width < 2 nm) [12]. Pores are a natural feature of starch granules, although they can also be formed during mechanical or enzymatic processing of starch [9,13–15].

Low-temperature nitrogen adsorption belongs to the most frequently used methods for determining the starch porosity [2,5,13–16]. This method allows to determine the porosity of the material in the entire range of mesopores and the lower range of macropores (pore diameter 1.7–300 nm) and it involves determining the nitrogen adsorption isotherm at the temperature of liquid nitrogen and calculating the capacity of the monolayer based on the BET adsorption isotherm [13,14,17,18]. Porosity of starch granules influences their functional properties and reactivity, because the presence of pores, channels and cavities increases the surface area that is potentially available for molecules of chemical compounds and amylolytic enzymes (i.e., the greater the specific surface area of the granules, the greater the susceptibility to amylolysis) [15,16,19,20]. Porous starch has many applications in the food industry, mainly as a carrier for ingredients such as polyphenols [21,22], essential oils [23], flavors [24], or probiotics [25].

Starch also has very important biological functions in the human body. It is a source of energy for the central nervous system and red blood cells, it is also necessary in the synthesis of amino acids and the oxidation of fatty acids in the body. Starch contains hydroxyl groups and simple glycosidic bonds that can be easily modified to improve its functional properties or to incorporate specific new features enhancing cooking properties, gel clarity, texture, adhesion, and film formation, increasing freeze–thaw stability or reducing syneresis, retrogradation, and gelling tendencies. Starch amphiphilicity, hydrophobicity, mechanical strength, and thermal stability are among the properties that can be improved as a result of starch modification. Regarding such many aspects wettability tests using probe liquids of various chemical nature may be helpful. Based on the wettability parameters, conclusions can be drawn regarding the potential interactions of starch with biological membranes, physiological fluids, and other substances [26,27].

The purpose of this study was to use three methods of modification: ultrasound, enzymatic hydrolysis, and freeze-drying to prepare porous starch from starches of different botanical origin. Their structural, adsorption, and functional properties, were analysed to select the most effective porous matrix for adsorption of valuable bioactive substances. In assessing the degree of starch modification, changes in the surface charge of granules before and after modification with combined methods, the electrokinetic zeta potential, which indirectly may also inform about changes in the stability of the system and preliminary wettability test, were used.

2. Results and Discussion

2.1. Structure and Morphology of Native and Modified Starch

The first step in assessing the impact of chemical, physical or enzymatic modifications on the structure of starch may be the analysis of its complex with iodine. For example, blue value (BV) and λ_{max} provide information on the degree of polymerization of amylose and the average chain length (CL) of amylose and amylopectin [28]. Results of the following determinations: degree of hydrolysis, BV, A_{680}/A_{545} ratio, and λ_{max} are presented in Table 1. Among the starches tested, the most susceptible to hydrolysis by amyloglucosidase was corn starch, followed by pea starch, and the most resistant was potato starch. These results are consistent with literature data [29]. Iodine forms a coloured complex with linear chains which gradually shifts from red to blue based on the length of the available chains. The value of λ_{max} for long chains (amylose-like chains) is >610 nm, whereas for short chains (amylopectin-like chains) it is 530–575 nm [30]. Pea starch, unlike corn and potato starches, had the A_{680}/A_{545} ratio over 1 which means that it contained a higher portion of amylose-like chains than amylopectin-like chains. This is correlated also with the highest λ_{max} and BV for this starch. For corn and potato starches, the content of both types of chains was almost equal.

As seen in Table 1, the values of all parameters related to the starch-iodine complex decreased after the treatment of granules with three modification methods: ultrasound, enzymatic hydrolysis, and freeze-drying. This means that as a result of physical forces and the enzyme, partial depolymerization of starch occurred [31,32].

Table 1. Degree of starch hydrolysis by amyloglucosidase and results of spectroscopic analysis of starch-iodine complexes obtained for native and modified starches.

Starch		Degree of Hydrolysis (%)	λ_{max}	A_{680}/A_{545}	BV
Corn	native	-	592 ± 4 ^{cd}	0.91 ± 0.06 ^{ab}	0.79 ± 0.04 ^{b}
	modified	48.28 ± 3.83 *^{c}	576 ± 4 ^{a}	0.80 ± 0.12 ^{a}	0.65 ± 0.01 ^{a}
Potato	native	-	586 ± 2 ^{bc}	0.96 ± 0.16 ^{ab}	0.85 ± 0.02 ^{b}
	modified	5.41 ± 0.07 ^{a}	581 ± 3 ^{ab}	0.93 ± 0.22 ^{ab}	0.82 ± 0.02 ^{b}
Pea	native	-	608 ± 7 ^{e}	1.17 ± 0.15 ^{b}	1.07 ± 0.04 ^{c}
	modified	24.00 ± 2.00 ^{b}	603 ± 8 ^{de}	1.02 ± 0.05 ^{b}	1.05 ± 0.02 ^{c}

* values are expressed as mean ± SD. Results with the same letter within a column are not significantly different ($p < 0.05$).

Figure 1 shows SEM micrographs of starch granules before and after the three-step modification. Native starch granules have a shape typical of botanical origin and are rather smooth, except for corn starch, which has a rough surface (Figure 1a,c,d). In the case of modified starches, holes on the surface, giving the granules a spongy structure, are present in corn starch (Figure 1b). Distinct depressions are also visible on the surface of potato starch (Figure 1d), while the pea starch granules are mainly cracked and fragmented (Figure 1e). Surface roughness, pores, and cracks were induced by the force of ultrasonic cavitation, enzymatic digestion and local explosive release of water vapour from the accumulated pressure inside the rigid granules under vacuum conditions during freeze-drying [14,29,33]. All these irregularities increase the surface area potentially available for adsorption.

Figure 2 represents ATR-FTIR spectra of starches obtained in the range of 400–4000 cm^{-1}. The absorption peaks for native starches were similar to those of modified starches indicating that no new chemical bonds were formed during the modification. In the FTIR spectrum, the absorption band of starch at 1047 cm^{-1} is associated with the structural features of the crystalline region, representing the short-range ordered structure of the starch molecule, whereas the absorption band at 1022 cm^{-1} is associated with the structural features of the non-crystalline region of starch. The next absorption band at 995 cm^{-1} reflects the double helix structure in the crystalline regions [34].

The absorbance ratios of 1047 cm^{-1}/1022 cm^{-1} and 1022 cm^{-1}/995 cm^{-1} obtained for native and modified are gathered in Table 2. The results revealed that the ratio of 1047/1022 cm^{-1} was slightly higher and the ratio of 1022/995 cm^{-1} was slightly lower for modified starches than for their native counterparts.

Table 2. IR ratio of the absorbances 1047/1022 (cm^{-1}) and 1022/995 (cm^{-1}) for native and modified starches.

Starch		IR Ratio 1047/1022 (cm^{-1})	IR Ratio 1022/995 (cm^{-1})
Corn	native	0.99 ± 0.01 *	1.01 ± 0.01
	modified	1.02 ± 0.02	0.98 ± 0.01
Potato	native	1.01 ± 0.02	0.99 ± 0.02
	modified	1.05 ± 0.01	0.95 ± 0.01
Pea	native	1.01 ± 0.02	0.99 ± 0.01
	modified	1.02 ± 0.02	0.97 ± 0.02

* values are expressed as mean ± SD.

This means that the amorphous area of starch was reduced as a consequence of the use of combined modification methods. This is in line with the observations of Lacerda et al. [35].

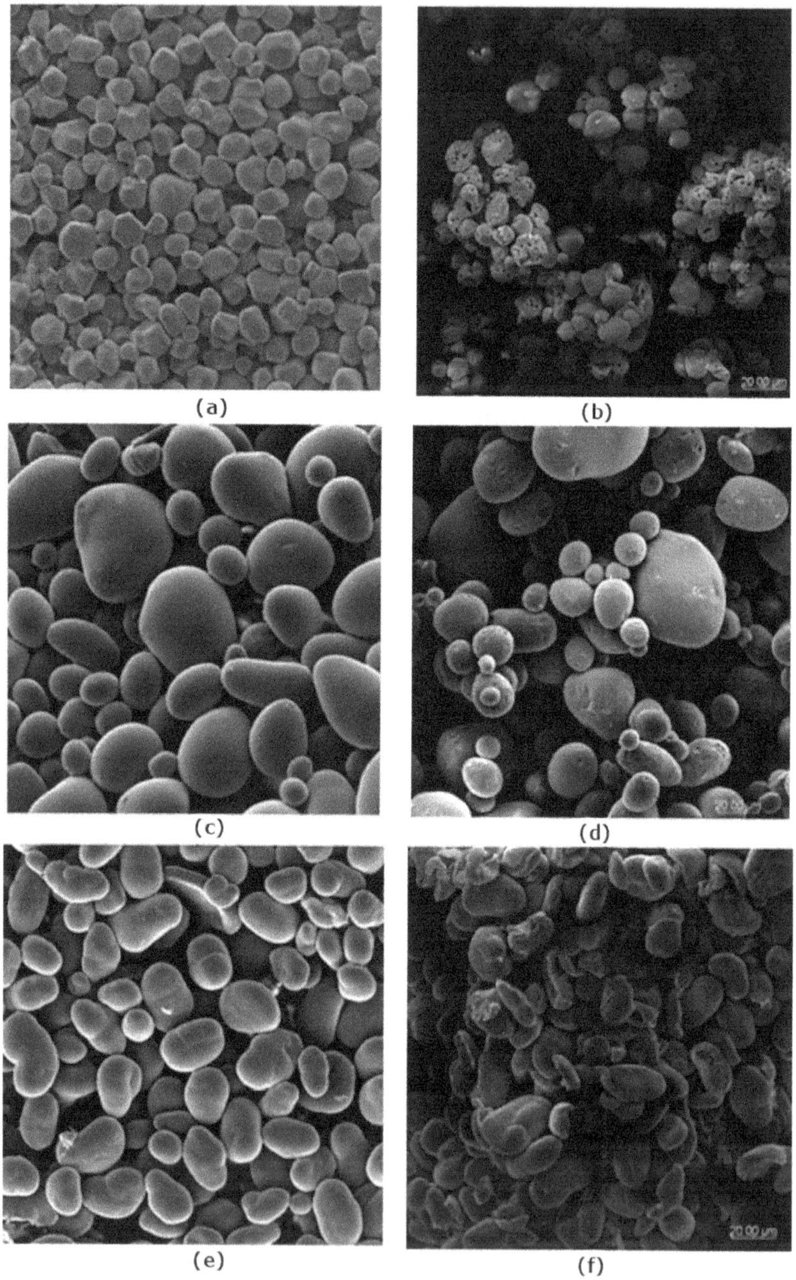

Figure 1. SEM micrographs of native (**a**,**c**,**e**) and modified (**b**,**d**,**f**) starch granules: (**a**,**b**)-corn starch, (**c**,**d**)-potato starch, (**e**,**f**)-pea starch (magnification 1200×).

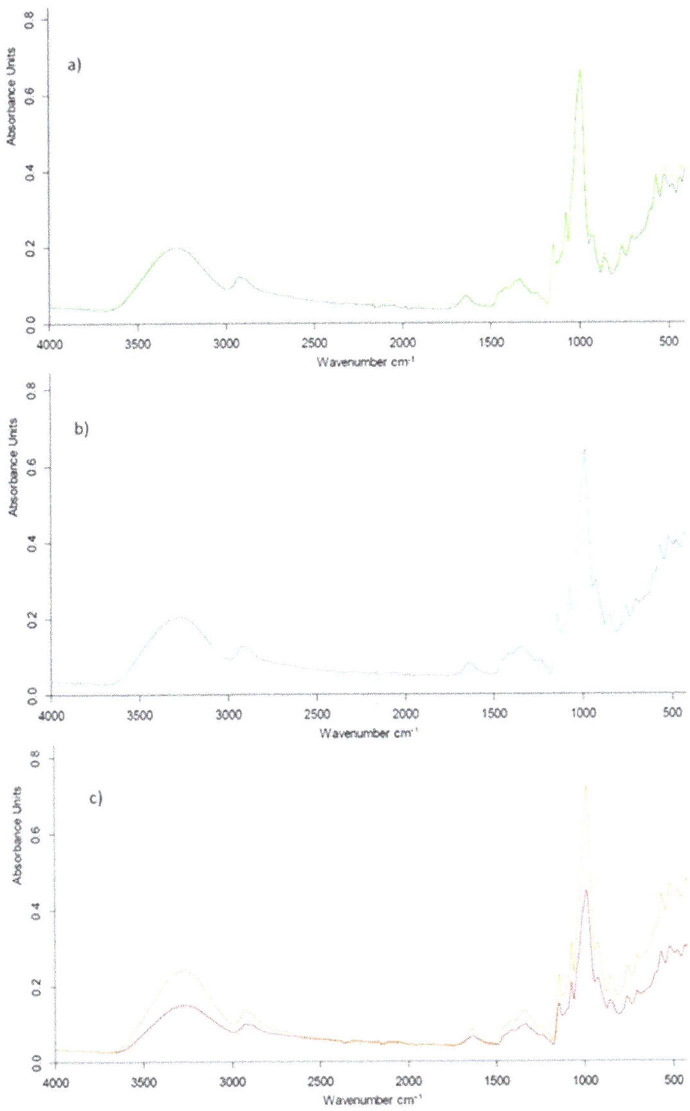

Figure 2. ATR-FTIR spectra of starches: (**a**) native corn (light green) and modified corn (dark green), (**b**) native potato (light blue) and modified potato (dark blue), (**c**) native pea (orange) and modified pea (red).

2.2. Specific Surface Area and Pore Characteristics of Native and Modified Starch

Table 3 shows the results of the determination of specific surface area (S_{BET}), volume and average diameter of pores present in the starch granules. The values of S_{BET} for native starches are in the range of 0.02–0.56 m^2/g and for modified starch between 0.8 and 4.24 m^2/g. These values are consistent with those obtained in previous studies [13,14,36]. The highest S_{BET} among native starches and the rough surface observed in SEM micrographs for corn starch (Figure 1b) correlate well with its greatest susceptibility to hydrolysis with amyloglucosidase. Natural pores on the surface of native starch granules (such as corn starch) can facilitate the formation of pores by enzymes [37]. Using combined methods of modification led to an increase in the volume of mesopores for all examined starches and the average pore diameter of mesopores for corn starch. For pea and potato starches the value of the latter parameter is lower, especially in the case of potato starch, which means that as a result of the modification, the number of pores with smaller diameters increased.

Table 3. Specific surface area (S_{BET}), volume, and average pore diameter for native and modified starches.

Starch		Specific Surface Area (m^2/g)	Volume of Mesopores (cm^3/g) $\times 10^{-3}$	Average Pore Diameter (nm)
Corn	native	0.56 ± 0.04 *	1.23 ± 0.06	8.72 ± 0.06
	modified	1.66 ± 0.08	4.24 ± 0.12	10.21 ± 0.04
Potato	native	0.02 ± 0.01	0.01 ± 0.01	33.87 ± 1.20
	modified	0.22 ± 0.02	0.80 ± 0.06	14.47 ± 0.19
Pea	native	0.16 ± 0.02	0.55 ± 0.04	14.01 ± 0.08
	modified	0.40 ± 0.06	1.24 ± 0.05	12.54 ± 0.10

* values are expressed as mean ± SD.

2.3. Zeta Potential Calculation of the Starch Suspensions and Preliminary Wettability Test

The electrophoretic mobility of suspensions of various types of native and modified starch was measured using the laser Doppler velocimeter method, and zeta potentials were calculated on this basis to determine the level of surface charge. As shown in Table 4 both native and modified potato starch and pea starch provided negative surface charges, which were visualized by zeta potential values.

Table 4. Zeta potential for different kind of starches suspensions as a function of time.

Starch	Zeta Potential [mV]	15 min	30 min	60 min	120 min	Week	Mean
Corn	native	7.41 ± 3.2 *	10.58 ± 4.1	9.86 ± 2.6	7.66 ± 4.0	16.76 ± 3.8	10.45 ± 3.5
	modified	−13.55 ± 4.8	−8.85 ± 0.1	−20.14 ± 3.4	−28.06 ± 7.2	−17.8 ± 1.2	−17.7 ± 3.3
Potato	native	−3.60 ± 1.8	−3.36 ± 1.6	−1.96 ± 0.9	−1.56 ± 0.5	−1.76 ± 0.7	−2.45 ± 1.1
	modified	−1.16 ± 0.9	−1.58 ± 0.7	−1.11 ± 0.2	−2.41 ±0.5	−1.76 ± 0.4	−1.60 ± 0.5
Pea	native	−0.74 ± 0.3	−2.73 ± 1.0	−1.89 ± 0.8	−1.15 ± 0.2	−1.52 ± 0.1	−1.61 ± 0.5
	modified	−1.36 ± 0.5	−1.38 ± 1.8	−4.01 ± 1.5	−3.14 ± 2.1	−3.58 ± 1.8	−2.69 ± 1.5

* values are expressed as mean ± SD.

These values are relatively small and mean values oscillate within the range from −1.6 to 2.7 mV. The change after modification is noticeable, but within the limits of statistical error. However, the values obtained for the dispersion from the third type of starch (corn starch) are completely different. Native corn starch is positively charged, and its surface had a zeta potential of about 10.5 mV. For modified starch, the absolute value of zeta potential increased when compared with native corn starch. Average values of 17.7 mV were obtained. Only in this one case such a large change was achieved after modification. This leads to conclusion that corn starch can be modified relatively easily and as was early mentioned it correlates well with its susceptibility to hydrolysis with amyloglucosidase.

It is well known that starch granules are held together by hydrogen bonds and interchain hydrophobic bonds. During hydrolysis possible process may be the breaking of hydrogen bonds between poly-(1 $\rightarrow$ 4)-glucan chains in crystallites and subsequent the hydrophobic bonds. After breaking hydrogen bonds, the polar groups probably hide inside and the hydrophobic parts remain outside starch granules, so a decrease in zeta potential is noticeable (Table 4). In every starch glucose unit there are three–OH groups, however, not all of them can be present on the granule surface. The hydroxyl groups are the only polar groups available in the biopolymer chain, which are capable of interactions. During hydrolysis the most hydrophobic parts of the starch chain are directed into air. However, steric limitations cause that some polar groups are also present on the surface. The starch granule surface tends to reach maximal hydrophobic character with polar domains created by the functional glucose groups. It is most probable that the branched chain of amylopectin is directed into air and calculated zeta potential changes after modification is probably caused by the increased amount of –OH groups on the starch surface. Taking into account only the zeta potential results, it can be concluded that the modification was most effective in the case of corn starch. This correlates with the porosity changes summarized in Table 3. The specific surface area of corn starch increased threefold and the volume of mesopores increased almost fourfold. In the case of potato and pea starches, these changes were not so noticeable.

The next idea of our research will be, in addition to increasing porosity, adding antioxidant properties through the use of gallic acid. Research that is already promising at this stage will be discussed in Part 2 of the manuscript. Improvement of antioxidant activity of starch film by phenolic acid was described early by Hao et al. [38]. It is important that phenolic acids exert also antimicrobial activity, which is considered beneficial to human health. Lately authors stated that in the treatment with a higher amount of oxidant, most polysaccharides in the starch were degraded to oligosaccharides. Similar process is possible during hydrolysis. It would be reasonable to expect that oxidized starch with a higher oxidation would have a greater negative charge, while a higher amount of oxidant as described in recent papers [38–42] resulted in the higher solubility of starch. To check these properties for our three types of starch, the wettability measurements were performed for their dispersions deposited on glass plates analogous as in our previous papers [26,27,43]. These studies should be considered as preliminary. Milli-Q water was used during wettability measurements as a typical polar tested liquid. Water wettability of starch films before and after modification is also important due to another aspect, e.g., process of microorganism growth. After three-step modification, water wetting times for plates covered with modified starch suspensions increased noticeably, sometimes even several times. This is probably the result of the presence of additional polar groups on starch film surface. For a complete picture, similar tests should be carried out with test liquids of a different chemical nature, i.e., polar formamide and non-polar diiodomethane. These studies are being carried out using Washburn procedure, however due to their time-consuming nature will be described in detail in Part 2 of the manuscript. On the other hand, modified starch-based hydrogel can be also used as an adsorbent for food and non-food application. Therefore, detailed porosity tests were performed on all tested starches (discussed in the Section 2.2).

2.4. Functional Properties of Native and Modified Starches

Modification of starches by combined methods affected their functional properties. Results of WA, OA, LGC and paste clarity determinations are summarized in Table 5. Among native starches, corn starch has the highest water and oil absorption capacity, and LGC, while potato starch is characterized by the highest paste clarity. Modification generally caused an increase in the water absorption (1.3–1.5-fold) and oil absorption (1.1–1.4-fold), and a decrease in the paste clarity (approximately 3-fold) for all the examined starches. LGC of modified corn and potato starches are higher than before modification or stayed unchanged in the case of pea starch.

The increase in water and oil absorption results from the disruption of the hydrogen bonds between the amorphous and crystalline area. Moreover, this increase could also be caused by changes in the structure of the starch granule surface. It became more porous, less "packed", making it easier to absorb water and oil [44]. So, despite the hydrophilic nature of starch, modified starches showed good capacity for oil absorption. Modification of corn and potato starch weakened the gelling properties of amylose to some extent, which is reflected in higher LGC values. In the case of paste clarity, scientists more often report that the value of this parameter increases after the action of e.g., acid hydrolysis or ultrasonics, as a result of partial depolymerization of starch [32,45]. However, there are also contrary reports e.g., lower paste clarity was found by Li et al. after ultrasonic treatment of pea starch [46]. The authors explained this by long-term processing, which produced more short amylose chains that were easier to accumulate, resulting in starch condensation. This study additionally used two other modification methods that could enhance this effect.

Table 5. Functional properties of native and modified starches.

Starch		WA (%)	OA (%)	LGC (%)	Paste Clarity (%)
Corn	native	194.65 ± 0.64 [*c]	173.05 ± 3.46 [c]	6 [b]	21.50 ± 0.50 [e]
	modified	261.15 ± 1.34 [e]	237.55 ± 4.03 [e]	8 [c]	8.30 ± 0.42 [b]
Potato	native	174.85 ± 3.47 [a]	152.10 ± 1.70 [a]	4 [a]	74.85 ± 2.62 [f]
	modified	229.75 ± 2.61 [d]	168.95 ± 2.05 [bc]	6 [b]	25.25 ± 0.92 [d]
Pea	native	186.20 ± 0.71 [b]	163.15 ± 0.50 [b]	4 [a]	15.25 ± 0.21 [c]
	modified	286.95 ± 0.64 [f]	206.45 ± 6.01 [d]	4 [a]	5.20 ± 0.00 [a]

[*] values are expressed as mean ± SD. Results with the same letter within a column are not significantly different ($p < 0.05$).

3. Materials and Methods

3.1. Materials

Starches (corn, potato, and pea) were purchased from Sigma-Aldrich Inc. (St. Louis, MO, USA), WPPZ S.A. (Luboń, Poland), and Roquette (Lestrem, France). Amyloglucosidase was obtained from Merck (Darmstadt, Germany). Distilled water from the Milli-Q system (resistivity ~18.2 MΩ × cm) for electrophoretic mobility and wettability measurements was used.

3.2. Methods

Modification of Starch with Ultrasounds, Enzymatic Hydrolysis, and Freeze-Drying

A 30% starch suspension in distilled water was prepared in a 100 mL beaker. The sample was exposed to ultrasound with a frequency of 20 kHz and a power of 170 W for 15 min at a temperature of 20 °C. During sonication, the sample was cooled by immersing the beaker in ice water. Then it was centrifuged (5000 rpm for 5 min), and the sediment was washed with 96% ethanol and left to dry. After drying, starch samples were used for enzymatic hydrolysis.

A portion of 15 mL of amyloglucosidase solution with an activity of 1000 U/g starch was added to a 2% starch suspension in citrate buffer at pH 4.2. The total sample volume was 300 mL. Hydrolysis was carried out at 50 °C for 24 h in a shaking incubator (150 rpm). After this time, the suspension was centrifuged (5000 rpm for 5 min), the supernatant was collected, and the sediment was washed with 96% ethanol. Then the sample was freeze-dried for 3 days at −45 °C in a vacuum (Labconco, Kansas City, MO, USA). The concentration of released glucose in the supernatant was determined by the enzymatic method (GOPOD) using the Megazyme kit (Bray, Ireland). The assay was performed in duplicate. Based on the results obtained, the degree of hydrolysis (%) was calculated.

3.3. Characteristics of Native and Modified Starch

3.3.1. Analysis of Starch-Iodine Complexes

Blue value (BV) of the starch-iodine complex was determined based on the absorbance measured at 680 nm and calculated according to the procedure of Gilbert and Spragg [47]. Iodine absorption spectra were measured in the range of λ from 400 to 800 nm with a spectrophotometer V-630 (Jasco Analytical Instruments, Tokyo, Japan). Additionally, A_{680}/A_{545} relations were calculated. Samples were analyzed in triplicate.

3.3.2. Scanning Electron Microscopy (SEM)

Dried starch samples were applied to double-sided adhesive tape attached to an aluminum plate. They were then coated with a thin layer of gold using an Emitech K550X sputter coater (Quorum Technologies Ltd., Laughton, UK) and observed in a Vega2 microscope (Tescan, Brno, The Czech Republic) at an accelerating voltage of 10 kV. Images were recorded at 1200× magnification.

3.3.3. ATR-FTIR Analysis

Mid-IR absorption spectra were obtained using attenuated total reflectance-Fourier transform infrared spectroscopy (ATR-FTIR) (Alpha II, Bruker Optics Inc., Billerica, MA, USA). Measurements were performed in damped total reflection mode using a Platinum ATR diamond crystal accessory. A pinch of the dried sample was placed directly on the crystal (with a contact surface diameter of 1.8 mm) and pressed against its surface. Spectra were collected by performing 36 spectral scans at a resolution of 4 cm^{-1} in the wavenumber range between 4000 and 400 cm^{-1}. All spectral analyses were performed using OPUS 8.5 SP1 software (Bruker Optics Inc., Billerica, MA, USA).

3.3.4. Low-Temperature Nitrogen Adsorption

Characteristics of pores of equivalent radius in the range of 1.7–300 nm was investigated based on nitrogen adsorption isotherms. The isotherms were measured using an apparatus ASAP 2420 (Micromeritics Inc., Norcross, GA, USA) at the temperature of liquid nitrogen (−195.85 °C) in the range of relative pressures p/p_0 from 0.004 to 0.997. The measurements were performed on samples of 1 to 3 g, depending on the specific surface area. Before the measurements, starch samples were dried for 24 h in a vacuum at 100 °C, automatically desorbed, and flushed with pure helium. The volume of the pores was calculated with the use of computer software attached to the instrument.

The surface area of the investigated starches was evaluated from sorption isotherms of nitrogen using the BET method [17]. In the first step, the monolayer value (a_m) was calculated within the range of p/p_0 = 0.06–0.20 (from five measurement points) using a linear polynomial form of the BET equation. Equation (1) describes the relationship between relative pressure and the amount of adsorbed/desorbed liquid nitrogen. The slope of this relationship and the intercept were determined.

$$a_m = 1/(tg\alpha + a) \tag{1}$$

where: a—adsorption, a_m—monolayer value [cm^3 N_2/g dry mass].

The specific surface area was calculated from Equation (2):

$$S_{BET} = (a_m \times \omega \times L)/M \tag{2}$$

where: a_m is a monolayer value (cm^3/g dry mass), L is the Avogadro number (6.02 × 10^{23} molecules per mole), M—the molecular weight of nitrogen (28,013 g/mol), and ω is the molecular cross-section area (16.2 × 10^{-20} m^2 for nitrogen molecule).

Pore characteristic was determined according to BJH method [48]. Samples were analyzed in triplicate.

3.3.5. Determination of the Zeta Potential

To perform electrophoretic mobility measurements, it was necessary to prepare a starch and modified starch dispersion of appropriate concentrations. Starch dispersions were measured for their electrophoretic mobility by laser Doppler velocimetry using a zeta potential analyzer (ZetaPlus-Bimass, Brookhaven Instrument Corporation, Holtsville, NY, USA). In each experiment, the same amount of different kinds of starch or starch after modification was dispersed in 100 mL of distilled water from the Milli-Q system (resistivity ~18.2 MΩ × cm) and then sonicated at 20 °C for 15 min (10,000 rpm, Silent Crush homogenizer, Heidolph). The samples were sonicated for 15 min to attain finely dispersed gel particles. The temperature of all samples was controlled at 20 °C. Initially, two starch weights, 0.3 g or 0.5 g, were tested, but ultimately, the higher weight was chosen because it turned out to be more suitable for further stages of the experiment, i.e., wettability measurements. The tests were carried out at time intervals of up to 2 h (15 min, 30 min, 60 min, 120 min) after the end of homogenization and after a week. The electrophoretic mobility of each sample was measured five times in each measurement.

3.3.6. Preliminary Wettability Tests

Glass plates (primary glass, polished edges, optically smooth and clear, ca 76 × 26 × 1 mm, AS Polonia Ltd., Nottingham, UK) [49] were used as solid substrata for covering with native or modified starch dispersion at room temperature. About 4 mL of dispersion was slowly spread on the plate to obtain a homogeneous layer. Next, the plates were kept at room temperature for drying, and after that, were placed in desiccators filled with a silica gel until the measurements were performed, which was a maximum of a week. In the case of the powdered porous solids used in food or pharmaceutical industry, the degree of wetting should be described not by contact angle measurements, but rather by the kinetics of the wetting process, because it is difficult to achieve equilibrium. Therefore, often, the thin-layer wicking method is used regarding the penetration of a probe liquid into a thin, porous layer of the solid powder deposited, for example, on a metal, plastic, or glass support. Such a process may be described by Washburn's equation, based on Poiseuille's law [26]:

$$x^2 = \Delta G\, R\, t/2\, \eta \tag{3}$$

where: x^2 is the square of penetration distance; t is the penetration time, η is the liquid viscosity, R is the effective radius of the intergranular capillaries that are formed in a porous layer or column of a powdered solid body, ΔG is the change in free energy (enthalpy) accompanying the replacement of the unit interface surface, i.e., the solid–gas and solid–liquid phases, during liquid movement (wetting) in a porous film.

Details of the methodology have been described in our previous papers [26,27,43].

The measurement of the time spreading of the probe liquids of different chemical characters provides important information about the starch wettability before and after modification of the relatively uncomplicated experimental procedure. Applying different kinds of probe liquids allows us to estimate liquid interfacial interactions. More advanced methods also allow us to determine the values of apparent surface free energy [26,27]. On the other hand, their changes reveal interesting information about the energetic properties of the investigated starch surface. The selection of liquids of different characters (strongly polar, less polar, or apolar) gives an overview of the various types of interactions, and enables a full description of the phenomenon of the wetting process. On this basis, the hydrophilic/hydrophobic nature of the tested surface for different kinds of starch can be determined, which consequently helps to select specific applications.

3.4. *Functional Properties of Native and Modified Starch*

3.4.1. Oil Absorption (OA)

The starch sample (0.5 g) was weighed into a centrifuge tube and 10 mL of oil was added. The mixture was mixed for 30 s using a homogenizer at 10,000 rpm, and the mixer was rinsed with distilled water. The tubes were left to stand for 5 min and then mixed again. The tubes with their contents were centrifuged (3000 rpm) for 10 min. The unbound oil was gently poured off and the tubes were placed upside down on filter paper for 10 min, then weighed, and calculations were made.

$$OA = (a - d)/W \times 100\% \tag{4}$$

where: OA—oil absorption (%), a—mass of tube with sample after centrifugation (g), d—mass of dry tube (g), W—sample mass (g).

The assay was performed in duplicate.

3.4.2. Water Absorption (WA)

The starch sample (0.5 g) was weighed into a centrifuge tube and 12.5 mL of distilled water was added. The mixture was mixed for 30 s using a homogenizer at 10,000 rpm, and the mixer was rinsed with distilled water. The tubes with their contents were centrifuged (3000 rpm) for 10 min. The unbound water was gently poured off and the tubes were placed upside down on filter paper for 10 min, then weighed and calculations were made.

$$WA = (c - b)/W \times 100\% \quad (5)$$

where: WA—water absorption (%), a—mass of tube with sample after centrifugation (g), d—mass of dry tube (g), W—sample mass (g).

3.4.3. Clarity of Starch Pastes and the Minimum Concentration of Gel-Forming Starch Suspension (LGC)

A 1% starch suspension was prepared, then it was heated in a boiling water bath for 15 min, stirring occasionally, and cooled. The transmittance of the paste was measured at a wavelength of 650 nm against distilled water as a blank sample. The assay was performed in duplicate.

For LGC measurement, starch suspensions with concentrations of 2–10% (every 2%) were prepared in distilled water. The samples were heated in a boiling water bath for 15 min, stirring occasionally, then cooled and placed in a refrigerator (4 °C) for 2 h. After this time, it was assessed at which concentration the sample did not fall out of the inverted test tube. The assay was performed in duplicate.

3.5. Statistical Analysis

The obtained results of degree of hydrolysis, VIS spectroscopic determinations, and functional properties of the starch were subjected to statistical analysis in the astatsa.com online program. One-way analysis of variance (ANOVA) was used to compare the results, and then the Tukey post hoc test was used to determine the significance of differences between the group mean values. Statistical hypotheses were verified at the significance level of $p < 0.05$.

4. Conclusions

The modification of starch by three methods: ultrasound, hydrolysis with amyloglucosidase, and freeze-drying, improved the starch properties in terms of the porosity, oil and water absorption, paste clarity, surface charge, and wettability. The granules of the modified starches had a spongy structure (corn starch), were cracked and fragmented (pea starch), or showed only fissures and depressions on the surface (potato starch). The specific surface area and volume of the mesopores increased after modification for all the examined starches. The three-stage modification resulted in an increase in the average pore diameter only for the corn starch, while in the case of the potato and pea starch, more pores with a smaller average diameter were created. Changes in the electrokinetic potential and wettability parameters correlate quite well with changes in the porosity and are a good tool for assessing the processes occurring on the starch surface. The functional properties of the starches changed as a consequence of the modifications: the absorption of water and oil increased for all the examined starches, whereas the paste clarity decreased. In the case of the LGC, the modification caused an increase in the case of the corn and potato starch, while in the case of the pea starch, no effect was observed.

Author Contributions: Conceptualization, M.S.; methodology, M.S. and A.E.W.; validation, M.S. and A.E.W.; formal analysis, M.S. and A.E.W.; investigation, M.S. and A.E.W.; data curation, M.S. and A.E.W.; writing—original draft preparation M.S. and A.E.W.; writing—review and editing, M.S. and A.E.W.; visualization, M.S. and A.E.W.; supervision, M.S. All authors have read and agreed to the published version of the manuscript.

Funding: This research was partially funded by the Union of Lublin Universities in the framework of the program "Staż za miedzą" (No. SzM/23/222/TŻ).

Institutional Review Board Statement: Not applicable.

Informed Consent Statement: Not applicable.

Data Availability Statement: The data presented in this study are available on request from the corresponding authors.

Acknowledgments: We thank Urszula Pankiewicz for supervision over the research internship that resulted in this publication. We also thank Michał Klepacki, and Melike Damla Yüncü for technical support.

Conflicts of Interest: The authors declare no conflict of interest.

References

1. Xie, Y.; Zhang, B.; Li, M.-N.; Chen, H.-Q. Effects of cross-linking with sodium trimetaphosphate on structural and adsorptive properties of porous wheat starches. *Food Chem.* **2019**, *289*, 187–194. [CrossRef]
2. Gao, F.; Li, D.; Bi, C.; Mao, Z.; Adhikari, B. Application of Various Drying Methods to Produce Enzymatically Hydrolyzed Porous Starch Granules. *Dry. Technol.* **2013**, *31*, 1627–1634. [CrossRef]
3. Jiang, T.; Wu, C.; Gao, Y.; Zhu, W.; Wan, L.; Wang, Z.; Wang, S. Preparation of novel porous starch microsphere foam for loading and release of poorly water soluble drug. *Drug Dev. Ind. Pharm.* **2014**, *40*, 252–259. [CrossRef] [PubMed]
4. Benavent-Gil, Y.; Rosell, C.M. Comparison of porous starches obtained from different enzyme types and levels. *Carbohydr. Polym.* **2017**, *157*, 533–540. [CrossRef] [PubMed]
5. Chen, G.; Zhang, B. Hydrolysis of granular corn starch with controlled pore size. *J. Cereal Sci.* **2012**, *56*, 316–320. [CrossRef]
6. Han, X.; Wen, H.; Luo, Y.; Yang, J.; Xiao, W.; Xie, J. Effects of chitosan modification, cross-linking, and oxidation on the structure, thermal stability, and adsorption properties of porous maize starch. *Food Hydrocoll.* **2022**, *124*, 107288. [CrossRef]
7. Han, Z.; Han, Y.; Wang, J.; Liu, Z.; Buckow, R.; Cheng, J. Effects of pulsed electric field treatment on the preparation and physicochemical properties of porous corn starch derived from enzymolysis. *J. Food Process. Preserv.* **2020**, *44*. [CrossRef]
8. Ji, Y. Synthesis of porous starch microgels for the encapsulation, delivery and stabilization of anthocyanins. *J. Food Eng.* **2021**, *302*, 110552. [CrossRef]
9. Fang, C.; Huang, J.; Yang, Q.; Pu, H.; Liu, S.; Zhu, Z. Adsorption capacity and cold-water solubility of honeycomb-like potato starch granule. *Int. J. Biol. Macromol.* **2020**, *147*, 741–749. [CrossRef]
10. Qian, D.; Chang, P.R.; Ma, X. Preparation of controllable porous starch with different starch concentrations by the single or dual freezing process. *Carbohydr. Polym.* **2011**, *86*, 1181–1186. [CrossRef]
11. Karathanos, V.T.; Saravacos, G.D. Porosity and pore size distribution of starch materials. *J. Food Eng.* **1993**, *18*, 259–280. [CrossRef]
12. Sing, K.S.W. Reporting physisorption data for gas/solid systems with special reference to the determination of surface area and porosity (Recommendations 1984). *Pure Appl. Chem.* **1985**, *57*, 603–619. [CrossRef]
13. Sujka, M.; Jamroz, J. Characteristics of pores in native and hydrolyzed starch granules. *Starch Stärke* **2010**, *62*, 229–235. [CrossRef]
14. Sujka, M. Ultrasonic modification of starch—Impact on granules porosity. *Ultrason. Sonochem.* **2017**, *37*, 424–429. [CrossRef] [PubMed]
15. Keeratiburana, T.; Hansen, A.R.; Soontaranon, S.; Tongta, S.; Blennow, A. Porous rice starch produced by combined ultrasound-assisted ice recrystallization and enzymatic hydrolysis. *Int. J. Biol. Macromol.* **2020**, *145*, 100–107. [CrossRef]
16. Jorge, F.-F.; Edith, C.-C.; Eduardo, R.-S.; Jairo, S.-M.; Héctor, C.-V. Hydrothermal processes and simultaneous enzymatic hydrolysis in the production of modified cassava starches with porous-surfaces. *Heliyon* **2023**, *9*, e17742. [CrossRef] [PubMed]
17. Brunauer, S.; Emmett, P.H.; Teller, E. Adsorption of Gases in Multimolecular Layers. *J. Am. Chem. Soc.* **1938**, *60*, 309–319. [CrossRef]
18. Zhang, L.; Zhong, L.; Wang, P.; Zhan, L.; Yangzong, Y.; He, T.; Liu, Y.; Mao, D.; Ye, X.; Cui, Z.; et al. Structural and Functional Properties of Porous Corn Starch Obtained by Treating Raw Starch with AmyM. *Foods* **2023**, *12*, 3157. [CrossRef]
19. Tahir, R.; Ellis, P.R.; Butterworth, P.J. The relation of physical properties of native starch granules to the kinetics of amylolysis catalysed by porcine pancreatic α-amylase. *Carbohydr. Polym.* **2010**, *81*, 57–62. [CrossRef]
20. Jung, Y.-S.; Lee, B.-H.; Yoo, S.-H. Physical structure and absorption properties of tailor-made porous starch granules produced by selected amylolytic enzymes. *PLoS ONE* **2017**, *12*, e0181372. [CrossRef]
21. Wu, A.; Fang, Z.; Qin, J.; Huang, Z.; Wu, Z. Characterization and adsorption-release property of fermented porous starch as well as its bioactivity protection for guava leaf polyphenols. *Food Biosci.* **2023**, *53*, 102535. [CrossRef]
22. Hu, X.; Du, X. Adsorption of Tea Polyphenols using Microporous Starch: A Study of Kinetics, Equilibrium and Thermodynamics. *Molecules* **2019**, *24*, 1449. [CrossRef] [PubMed]
23. Ju, J.; Chen, X.; Xie, Y.; Yu, H.; Cheng, Y.; Qian, H.; Yao, W. Simple microencapsulation of plant essential oil in porous starch granules: Adsorption kinetics and antibacterial activity evaluation. *J. Food Process. Preserv.* **2019**, *43*, e14156. [CrossRef]

24. Belingheri, C.; Ferrillo, A.; Vittadini, E. Porous starch for flavor delivery in a tomato-based food application. *LWT Food Sci. Technol.* **2015**, *60*, 593–597. [CrossRef]
25. Li, H.; van Thuy Ho, T.; Turner, M.S.; Dhital, S. Encapsulation of Lactobacillus plantarum in porous maize starch. *LWT* **2016**, *74*, 542–549. [CrossRef]
26. Wiącek, A.E. Effect of surface modification on starch biopolymer wettability. *Food Hydrocoll.* **2015**, *48*, 228–237. [CrossRef]
27. Wiącek, A.E.; Dul, K. Effect of surface modification on starch/phospholipid wettability. *Colloids Surf. A Physicochem. Eng. Asp.* **2015**, *480*, 351–359. [CrossRef]
28. Hj Latip, D.N.; Samsudin, H.; Utra, U.; Alias, A.K. Modification methods toward the production of porous starch: A review. *Crit. Rev. Food Sci. Nutr.* **2021**, *61*, 2841–2862. [CrossRef]
29. Li, C.; Gong, B.; Hu, Y.; Liu, X.; Guan, X.; Zhang, B. Combined crystalline, lamellar and granular structural insights into in vitro digestion rate of native starches. *Food Hydrocoll.* **2020**, *105*, 105823. [CrossRef]
30. Rundle, R.E.; French, D. The Configuration of Starch in the Starch—Iodine Complex. III. X-Ray Diffraction Studies of the Starch—Iodine Complex 1. *J. Am. Chem. Soc.* **1943**, *65*, 1707–1710. [CrossRef]
31. Guo, L.; Li, J.; Gui, Y.; Zhu, Y.; Yu, B.; Tan, C.; Fang, Y.; Cui, B. Porous starches modified with double enzymes: Structure and adsorption properties. *Int. J. Biol. Macromol.* **2020**, *164*, 1758–1765. [CrossRef] [PubMed]
32. Sujka, M.; Jamroz, J. Ultrasound-treated starch: SEM and TEM imaging, and functional behaviour. *Food Hydrocoll.* **2013**, *31*, 413–419. [CrossRef]
33. Zhang, B.; Wang, K.; Hasjim, J.; Li, E.; Flanagan, B.M.; Gidley, M.J.; Dhital, S. Freeze-Drying Changes the Structure and Digestibility of B-Polymorphic Starches. *J. Agric. Food Chem.* **2014**, *62*, 1482–1491. [CrossRef] [PubMed]
34. Hong, T.; Yin, J.-Y.; Nie, S.-P.; Xie, M.-Y. Applications of infrared spectroscopy in polysaccharide structural analysis: Progress, challenge and perspective. *Food Chem. X* **2021**, *12*, 100168. [CrossRef] [PubMed]
35. Lacerda, L.D.; Leite, D.C.; Da Silveira, N.P. Relationships between enzymatic hydrolysis conditions and properties of rice porous starches. *J. Cereal Sci.* **2019**, *89*, 102819. [CrossRef]
36. Sujka, M.; Cieśla, K.; Jamroz, J. Structure and selected functional properties of gamma-irradiated potato starch. *Starch Stärke* **2015**, *67*, 1002–1010. [CrossRef]
37. Zhao, Y.; Sun, Y.; Meng, Z.; Guo, J.; Peng, G.; Sun, W.; Zhu, X.; Gu, R.; Wu, Z.; Gan, H.; et al. Effects of enzymatic treatments on different crystal types of starch. *Food Chem. Adv.* **2023**, *3*, 100334. [CrossRef]
38. Hao, J.; Lu, J.; Xu, N.; Linhardt, R.J.; Zhang, Z. Specific oxidation pattern of soluble starch with TEMPO-NaBr-NaClO system. *Carbohydr. Polym.* **2016**, *146*, 238–244. [CrossRef]
39. Zhao, Y.; Huerta, R.R.; Saldaña, M.D. Use of subcritical water technology to develop cassava starch/chitosan/gallic acid bioactive films reinforced with cellulose nanofibers from canola straw. *J. Supercrit. Fluids* **2019**, *148*, 55–65. [CrossRef]
40. Zhao, Y.; Saldaña, M.D. Use of potato by-products and gallic acid for development of bioactive film packaging by subcritical water technology. *J. Supercrit. Fluids* **2019**, *143*, 97–106. [CrossRef]
41. Zhang, L.; Zhao, L.; Bian, X.; Guo, K.; Zhou, L.; Wei, C. Characterization and comparative study of starches from seven purple sweet potatoes. *Food Hydrocoll.* **2018**, *80*, 168–176. [CrossRef]
42. Ji, Y.; Lin, X.; Yu, J. Preparation and characterization of oxidized starch-chitosan complexes for adsorption of procyanidins. *LWT* **2020**, *117*, 108610. [CrossRef]
43. Wiącek, A.E.; Przykaza, K. Wettability and Stability of Naproxen, Ibuprofen and/or Cyclosporine A/Silica Delivery Systems. *Colloids Interfaces* **2022**, *6*, 11. [CrossRef]
44. Han, X.; Wen, H.; Luo, Y.; Yang, J.; Xiao, W.; Ji, X.; Xie, J. Effects of α-amylase and glucoamylase on the characterization and function of maize porous starches. *Food Hydrocoll.* **2021**, *116*, 106661. [CrossRef]
45. Jafari, M.; Koocheki, A. Impact of ultrasound treatment on the physicochemical and rheological properties of acid hydrolyzed sorghum starch. *Int. J. Biol. Macromol.* **2023**, *256*, 128521. [CrossRef] [PubMed]
46. Li, G.; Ge, X.; Guo, C.; Liu, B. Effect of Ultrasonic Treatment on Structure and Physicochemical Properties of Pea Starch. *Foods* **2023**, *12*, 2620. [CrossRef] [PubMed]
47. Gilbert, G.A.; Spragg, S.P. *Methods in Carbohydrate Chemistry 4: Iodimetric Determination of Amylose*; Academic Press: Orlando, FL, USA, 1964.
48. Barrett, E.P.; Joyner, L.G.; Halenda, P.P. The Determination of Pore Volume and Area Distributions in Porous Substances. I. Computations from Nitrogen Isotherms. *J. Am. Chem. Soc.* **1951**, *73*, 373–380. [CrossRef]
49. *ISO 8037-1:1986*; Optics and Optical Instruments—Microscopes—Slides—Part 1: Dimensions, Optical Properties and Marking. ISO: Geneva, Switzerland, 1986.

International Journal of
Molecular Sciences

MDPI

Article

A Comparative and Critical Analysis for In Vitro Cytotoxic Evaluation of Magneto-Crystalline Zinc Ferrite Nanoparticles Using MTT, Crystal Violet, LDH, and Apoptosis Assay

Juan Luis de la Fuente-Jiménez [1], César Iván Rodríguez-Rivas [2], Irma Beatriz Mitre-Aguilar [3], Andrea Torres-Copado [1], Eric Alejandro García-López [3], José Herrera-Celis [4], María Goretti Arvizu-Espinosa [1], Marco Antonio Garza-Navarro [5], Luis Gerardo Arriaga [4], Janet Ledesma García [2], Domingo Ixcóatl García-Gutiérrez [5], Alejandro Zentella Dehesa [3,6,*], Ashutosh Sharma [1,*] and Goldie Oza [4]

1 Centre of Bioengineering, School of Engineering and Sciences, Tecnologico de Monterrey, Campus Queretaro, Av. Epigmenio González, No. 500, Fracc. San Pablo, Querétaro 76130, Mexico; juan.luis.delafuente@tec.mx (J.L.d.l.F.-J.); a01700238@tec.mx (A.T.-C.); goretti@tec.mx (M.G.A.-E.)
2 División de Investigación y Posgrado, Facultad de Ingeniería, Universidad Autónoma de Querétaro, Cerro de Las Campanas S/N, Querétaro 76010, Mexico; cesarmediuaq@gmail.com (C.I.R.-R.); janet.ledesma@uaq.mx (J.L.G.)
3 Unidad de Bioquímica, Instituto Nacional de Ciencias Médicas y Nutrición Salvador Zubirán, Vasco de Quiroga 15, Belisario Domínguez secc 16, Tlalpan, Mexico City 14080, Mexico; irma.mitrea@incmnsz.mx (I.B.M.-A.); garcialp007@gmail.com (E.A.G.-L.)
4 Centro de Investigación y Desarrollo Tecnológico en Electroquímica, Parque Tecnológico, Sanfandila, Pedro Escobedo, Querétaro 76703, Mexico; jlherrera@cideteq.mx (J.H.-C.); larriaga@cideteq.mx (L.G.A.); goza@cideteq.mx (G.O.)
5 Facultad de Ingeniería Mecánica y Eléctrica, FIME, Universidad Autónoma de Nuevo León, UANL, Av. Universidad S/N, Cd. Universitaria, San Nicolás de los Garza 66450, Mexico; marco.garzanr@uanl.edu.mx (M.A.G.-N.); domingo.garciagt@uanl.edu.mx (D.I.G.-G.)
6 Departamento de Medicina Genómica y Toxicología Ambiental, Programa Institucional de Cáncer de Mama, Instituto de Investigaciones Biomédicas IIB & Red de Apoyo a la Investigación, Universidad Nacional Autónoma de México, Ciudad Universitaria, Ave. Universidad 3000, Col. Copilco Universidad, Del. Coyoacán Ciudad de México, Mexico City 04510, Mexico
* Correspondence: azentell@iibiomedicas.unam.mx (A.Z.D.); asharma@tec.mx (A.S.)

Abstract: Zinc ferrite nanoparticles (ZFO NPs) are a promising magneto-crystalline platform for nanomedicine-based cancer theranostics. ZFO NPs synthesized using co-precipitation method are characterized using different techniques. UV-visible spectroscopy exhibits absorption peaks specific for ZFO. Raman spectroscopy identifies Raman active, infrared active, and silent vibrational modes while Fourier transforms infrared spectroscopic (FTIR) spectra display IR active modes that confirm the presence of ZFO. X-ray diffraction pattern (XRD) exhibits the crystalline planes of single-phase ZFO with a face-centered cubic structure that coincides with the selected area electron diffraction pattern (SAED). The average particle size according to high-resolution transmission electron microscopy (HR-TEM) is 5.6 nm. X-ray photoelectron spectroscopy (XPS) signals confirm the chemical states of Fe, Zn, and O. A superconducting quantum interference device (SQUID) displays the magnetic response of ZFO NPs, showing a magnetic moment of 45.5 emu/gm at 70 kOe. These ZFO NPs were then employed for comparative cytotoxicity evaluation using MTT, crystal violet, and LDH assays on breast adenocarcinoma epithelial cell (MCF-7), triple-negative breast cancer lines (MDA-MB 231), and human embryonic kidney cell lines (HEK-293). Flow cytometric analysis of all the three cell lines were performed in various concentrations of ZFO NPs for automated cell counting and sorting based on live cells, cells entering in early or late apoptotic phase, as well as in the necrotic phase. This analysis confirmed that ZFO NPs are more cytotoxic towards triple-negative breast cancer cells (MDA-MB-231) as compared to breast adenocarcinoma cells (MCF-7) and normal cell lines (HEK-293), thus corroborating that ZFO can be exploited for cancer therapeutics.

Keywords: zinc ferrite; co-precipitation; magneto-crystalline; triple-negative breast cancer; cytotoxicity; MTT; LDH; flow cytometry; apoptosis

1. Introduction

Spinel zinc ferrites ($ZnFe_2O_4$ or ZFOs) are magneto-crystalline nanoparticles (MCNs) with several properties and applications in biomedicine and nanotechnology. They have a great demand in drug delivery, tissue engineering, and hyperthermia, and act as a contrast agent in magnetic resonance imaging due to their stability in colloidal suspensions, non-immunogenicity, non-toxicity, and biocompatibility [1,2]. Hence, ZFOs have vastly contributed to biomedical functions mostly due to their biocompatible nature [3]. Moreover, spinel ferrites have a structural formula (A) $[B]_2O_4$ (also known as MFe_2O_4) in which (A) is a divalent metal cation from 3D transition metals such as zinc, copper, cobalt, and nickel, among others, and possesses availability for cations at the tetrahedral site, while [B] occupies the octahedral sites. These spinel ferrites are in a unit cell of 32 oxygen atoms of a cubic-closed packed arrangement. Zinc ferrite NPs have a normal spinel ferrite structure in which all the Zn^{2+} ions are in the A site and Fe^{3+} ions enter in the B sites [4,5]. The most common structure of spinel ferrites is the inverse structure, in which half of the Fe^{3+} ions stay in the octahedral site and the other half replaces the (A) cations in the tetrahedral site [4]. This imparts a strong paramagnetic behavior, since Zn^{2+} ions do not have a magnetic moment in the tetrahedral site. However, nanocrystalline ZFO NPs have an inverse spinel structure in which the Zn^{2+} and Fe^{3+} ions are partially dispersed in both the sites having antiferromagnetic behavior [4–6]. In addition, MNPs produced from iron oxide are composed of a core either from maghemite (γ-Fe_2O_3) or magnetite (Fe_3O_4) and coated with biocompatible layers. Magnetite is the most common crystalline phase of iron oxide for biomedical purposes [2]. Furthermore, MNP synthesis can be performed with several procedures such as thermal decomposition, sol-gel, hydrothermal synthesis, microemulsion, and co-precipitation [2]. From all the above-mentioned synthesis procedures, co-precipitation has several advantages, such as water dispersibility, facile route of functionalization, is the most employed synthesis in biomedical applications due to its hydrophilicity and is an accessible and low-cost large-scale manufacturing practice. NPs obtained from this procedure have high purity suitable for magnetic characterization [2,4]. During co-precipitation synthesis, aqueous solutions of salts from Fe^{2+} and Fe^{3+} are co-precipitated after the addition of a base. The control of ZFO composition, shape, and size changes according to the Fe^{2+} and Fe^{3+} ratio used, ionic strength of the medium, and pH [2].

In this present study, ZFO NPs were synthesized using a co-precipitation method. The characterization of the structural composition, size, and morphology by UV-visible spectroscopy, X-ray diffraction (XRD), Fourier transform infrared spectroscopy (FTIR), high-resolution transmission electron microscopy (HR-TEM), superconducting quantum interference device (SQUID), X-ray photoelectron spectroscopy (XPS), and RAMAN spectroscopy was performed. Additionally, the cytotoxicity of the ZFO NPs was assessed against triple negative human breast cancer cell line MDA-MB-231 (defined as tumors that lack expression of estrogen, progesterone receptor, and human epidermal growth factor [7]), human breast cancer cell line (MCF-7), and human embryonic kidney 293 (HEK-293) cell lines with different cytotoxicity assays such as 3-(4,5-dimethylthiazol-2-yl)-2,5 diphenyltetrazolium bromide (MTT), crystal violet, and lactate dehydrogenase (LDH). In addition, a count of cells entering early and late apoptosis was measured with flow cytometry. The main goal of this article was to compare the assays for studying cytotoxicity or viability to avoid any interference, leading to under- or over-estimation of results for the proper comprehension of the ZFO effect.

2. Results and Discussion

The mechanism of ZFO NP synthesis based on co-precipitation reaction is as follows [5]: In the first step, as in Reaction 1 below, there is an interaction between Fe^{3+} and OH^-, which moves in the forward direction to form an increasing concentration of FeOOH that possesses low solubility. Reaction 2 is a slow-and-rate determining step in which the intermediate FeOOH reacts with zinc ions at an alkaline pH to form zinc ferrite, spontaneously.

$$Fe^{3+} + 3OH^- \rightarrow FeOOH + H_2O \quad (1)$$

$$FeOOH + Zn^{2+} + 2OH^- \rightarrow ZnFe_2O_4 + 2H_2O \quad (2)$$

2.1. Optical and Molecular Vibrational Characterization of ZFO NPs

The UV-visible absorption spectroscopy is a characterization technique employed to determine the electronic structure and optical properties of the NPs. The obtained absorption peaks are normally related to the diameter, shape, and size of the NPs [8]. Figure 1A shows the UV-visible spectra of ZFO solid nanopowder in the range of 200–800 nm at room temperature. These ZFO NPs are photoresponsive in both UV and visible light ranges, especially in the wavelength ranging from 300 to 600 nm [9]. In Figure 1, the predominant absorption features of ZFO NPs are noticed at 430, 500, and 550 nm. The evidence suggests optical transitions around 430 and 500 nm that are assigned to ($^6A_{1g} \rightarrow {}^4T_{2g}(D)$) and ($^6A_{1g} \rightarrow {}^4A_{1g}(G)$) transitions, respectively. The 550 nm band corresponds to $O^{2-} + Fe^{3+} \rightarrow O^- + Fe^{2+}$ indirect transition [10]. Likewise, the obtained dark brown color of ZFO proved the visible light absorption capacity of these NPs [8,10].

Raman spectra in Figure 1B corroborates the structure and arrangement of the cubic phase of ZFO NPs. Figure S1 shows the microscopic image of ZFO NPs. The space group of the cubic structure of ZFO NPs belongs to the $O^7{}_h$ (*Fd-3m*) space group with eight molecules per unit cells ending in 56 atoms; however, the lowest Bravais cell includes two formula units for a total of 14 atoms. According to the factor group theory it is possible to derive the next 42 modes of ZFO NPs spinel with 3 acoustic and 39 optical modes at the Γ-point of the Brillouin zone [6,11]:

$$A_{1g}\ (R) + E_g\ (R) + F_{1g} + 3F_{2g}\ (R) + 2A_{2u} + 2E_u + 4F_{1u}\ (IR) + 2F_{2u}$$

where the (IR) and (R) identify infrared active and Raman active vibrational species, respectively, while the rest of the modes are silent modes [6,12]. Thus, the five subsequent vibrational modes of ZFO NPs at room temperature anticipated for spinel with a cubic structure are A_{1g}, E_g, and $3F_{2g}$ Raman active while the rest are silent modes, except F_{1u} which is infrared-active [3,6]. The three Raman-active modes of $3F_{2g}$ are identified as F_{2g} (1), F_{2g} (2), and F_{2g} (3), F_{2g} (1) being the mode with the lowest frequency and F_{2g} (3) the mode with the highest frequency of the vibrational species [6,12]. A visual review of the spectra shows several Raman bands in Figure 2 which can be categorized according to the mode region. The A_{1g} mode region is between 600–800 cm^{-1}, in which a band at 670 cm^{-1} can be observed; in the F_{2g} (3) mode region between 410–550 cm^{-1}, two bands at 459 cm^{-1} and 501 cm^{-1} can be identified; and in the F_{2g} (2) mode region between 260–380 cm^{-1}, bands at 316 cm^{-1} and 373 cm^{-1} were detected. At lower frequencies, the E_g and F_{2g} (1) modes are occasionally detected (Figure 2) [6,12]. The remaining Raman bands at higher energies, between 1000 and 1800 cm^{-1}, can reveal second-order spectral characteristics [6]. In Figure 2, several second-order features can be noticed, such as 1061 cm^{-1}, 1137 cm^{-1}, 1173 cm^{-1}, 1327 cm^{-1}, and 1538 cm^{-1}. Nevertheless, the 1337 cm^{-1} could also be linked to hematite structures [13].

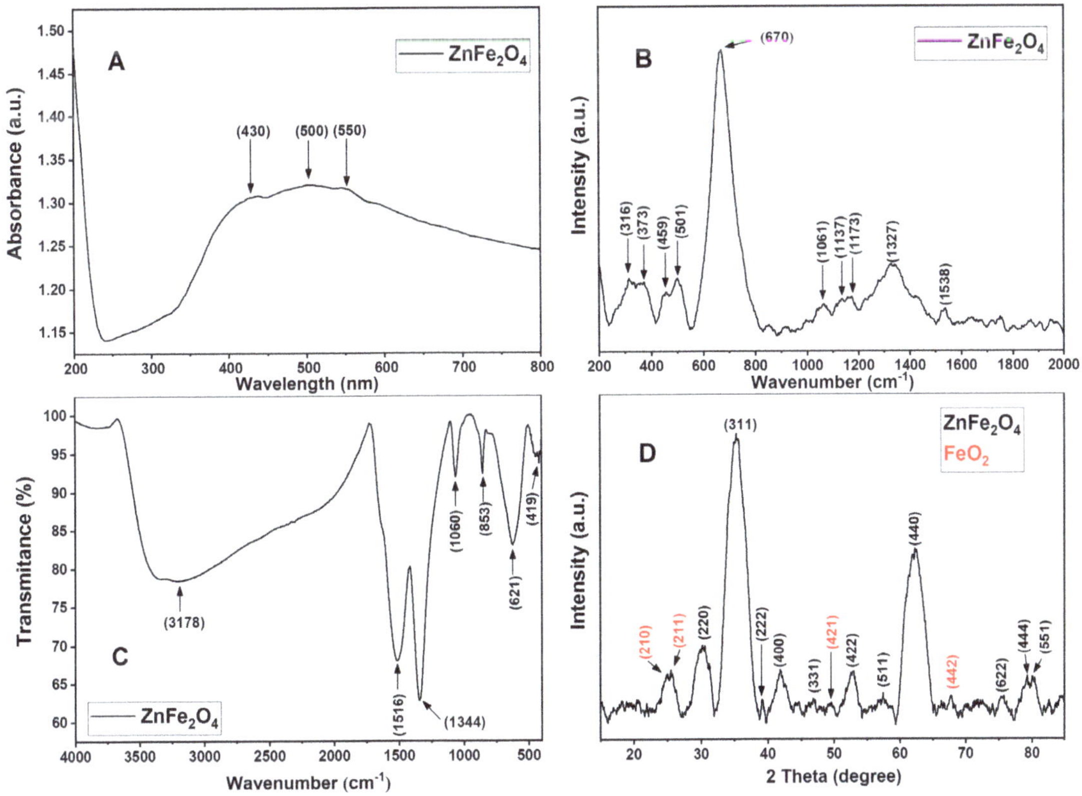

Figure 1. (**A**) UV-visible spectra of zinc ferrite nanoparticles exhibiting three absorption peaks at 430, 500, and 550 nm that correspond to three different transitions. (**B**) Raman spectra of ZFO NPs at room temperature from 200 to 2000 cm^{-1}. (**C**) FTIR spectrum of zinc ferrite nanoparticles. (**D**) X-ray diffractogram of zinc ferrite nanoparticles exhibiting crystalline planes of (220), (311), (222), (400), (331), (422), (511), (440), (622), (444), and (551) from the cubic system of zinc ferrite JCPDS pattern no. 01-089-4926 and 00-022-1012 as well as crystalline planes of (210), (211), (421), and (442), according to JCPDS pattern no. 00-039-1346 that corresponds to iron oxide (Fe_2O_3) + $Fe^{3+} \rightarrow O^-$ + Fe^{2+} indirect transition [10]. Likewise, the obtained dark brown color of ZFO proved the visible light absorption capacity of these NPs [8,10].

Furthermore, it is known that in cubic spinel such as ferrites, lower frequency modes are attributed to metal–oxygen vibrations in the octahedral BO_6 sites, while frequencies above 600 cm^{-1} correspond to metal–oxygen vibration in tetrahedral AO_4 sites. Therefore, it is implied that zinc cations are favorably located in tetrahedral A sites, as there is a shift in the A_{1g} mode as zinc is integrated in the structure [3,12]. Hence, the mode at 670 cm^{-1} is from the Zn–O vibration which is a predominant peak approximately in 650 cm^{-1} at the tetrahedral site and the rest is from the Fe–O octahedral vibration site (Figure 2) [3]. Finally, the Raman frequency varies on the bond length of Fe (Zn)–O which fluctuates with the particle size and lattice parameter variation [3]. In addition, zinc ferrite NPs have a partially inverted spinel structure which suggests a modification of the allocation of Fe^{3+} and Zn^{2+} ions between the octa- and tetrahedral sites. Thus, at the same time, two categories of cations can vibrate guiding two distinctive Raman modes with a very near frequency, and consequently, the appearance of broad peaks in the spectra are generally perceived [11,12]. This could be the reason for the proximity of peaks 316 cm^{-1} and 373 cm^{-1}, 459 cm^{-1} and 501 cm^{-1}, and 1137 cm^{-1} and 1173 cm^{-1} (Figure 2).

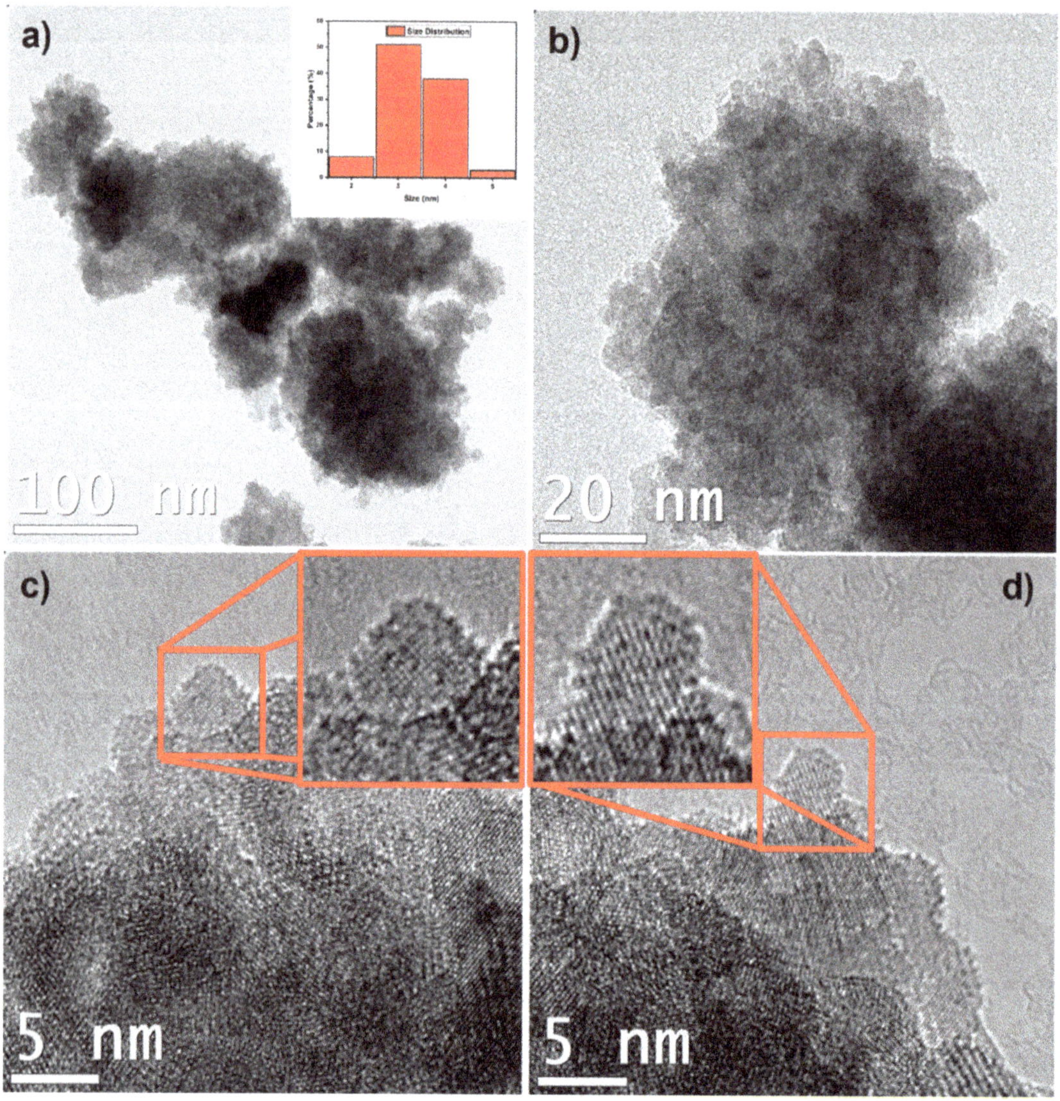

Figure 2. Transmission electron microscopic images of the ZFO NPs nanoparticles: (**a**,**b**) images showing the agglomerates formed by the nanoparticles. Figure (**a**) shows an inset of the histogram of the size distribution of the nanoparticles. (**c**,**d**) High-resolution transmission electron microscopic images (HR-TEM) of the ZFO NPs. Insets: HR-TEM images of individual nanoparticles.

FTIR spectra aids in confirming the spinel structure of ZFO NPs, which is shown in Figure 1C. The normal spinel ferrites are known to display four IR active modes (v_1, v_2, v_3, v_4,) in the vibration spectra. The high frequency bands (v_1 and v_2) are very sensitive to interaction between cations and oxygen in the tetrahedral (A-site) with stretching vibration of the metal–oxygen bonds and octahedral (B-site) with the bending vibration of the metal–oxygen bond positions. Normally, A-site appears between 540–600 cm^{-1} and B-site between 400–450 cm^{-1} [8,13]. Additionally, high frequency bands around 667 cm^{-1} are attributed to tetrahedral groups and the position of the characteristic bands robustly depends on the structural changes of the NPs, such as size of the powder grain, atomic mass, and cation radii, making them sensitive to any alterations when there is interaction between cations and oxygen in the tetra- and octahedral positions [8,13]. According to the possible interaction change between the tetra- and octahedral position, in this study the absorption band at 419 cm^{-1} is related to a bending vibration of Fe-O in the octahedral site, while the band at 621 cm^{-1} is related to the Zn-O stretching vibration in the tetrahedral site, both created by the interaction of metal ions with oxygen [13–16]. This 621 cm^{-1} Zn-O vibration band in the tetrahedral site is consistent with the results of the RAMAN spectroscopy (Figure 2). Furthermore, peaks at 831 cm^{-1} are linked to vibration frequencies of the organic functional groups in the NPs corresponding to C-O stretching and bending vibrations, which in this study correspond to the 853 cm^{-1}, and the peaks around 1105 cm^{-1} are related to the Fe-M ferrite system linked to the obtained peak at 1060 cm^{-1} [17–19]. In addition, two types of O-H interactions with zinc ferrite NPs are suggested. Type 1 is linked to the bending vibrations of the absorbed water at 1520 cm^{-1} and Type 2 bending vibration caused by a synchronized hydrogen absorption at both Zn and O sites, yielding a peak at 1327 cm^{-1} [20]. These peaks are exhibited in Figure 3, in which approximate absorption peaks at 1344 cm^{-1} and 1516 cm^{-1} can be seen, implying an interaction with water. These peaks could be related to absorbed water molecules or humidity, since a strong and sharp absorption peak at approximately 1650 cm^{-1} is assigned to stretching and bending vibrations of O-H functional groups on the surface of zinc ferrite NPs, which in this case resembles to the strong and sharp peak obtained at 1516 cm^{-1} [13,21]. Moreover, the presence of these absorption peaks and other strong and broad absorption peaks between 3200–3600 cm^{-1} are attributed to hydrogen bonded stretch vibration of O-H functional groups [14]. These absorption peaks exhibit H-O-H and O-H vibrations, confirming the presence of water molecules adsorbed on the surface of the NPs caused by the specific area of the ZFO nanopowders [14]. Hence, the band of 3178 cm^{-1} may be due to O-H stretching vibration groups of adsorbed water on the surface of the ZFO NPs (Figure 3) [4,18,22].

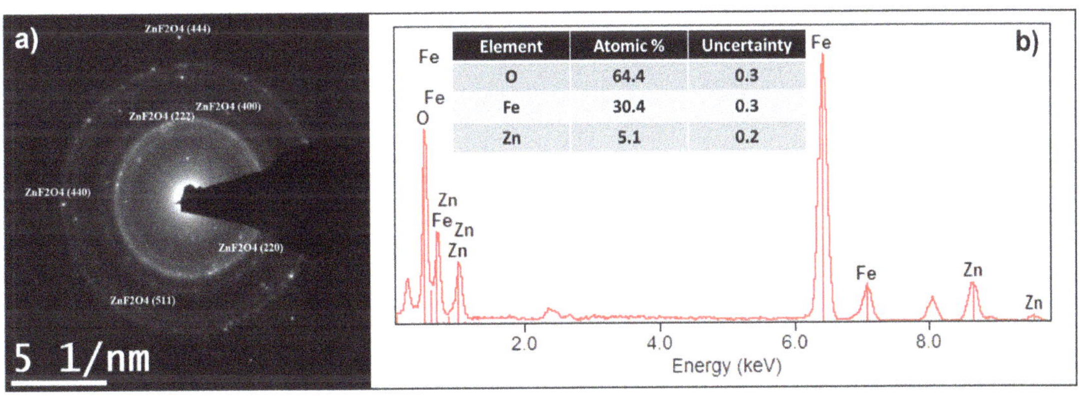

Figure 3. (**a**) SAED pattern from the ZFO NPs nanoparticles. (**b**) EDXS spectra showing the composition of the nanoparticles.

2.2. Structural Characterization of ZFO NPs

Figure 1D shows the result for the crystalline phase and structure of the ZFO NPs. The observed peaks in the XRD diffraction patterns and their intensity and position are very similar to the standard powder diffraction for the cubic spinel structure of zinc ferrite [23]. The diffraction peaks at 2θ values of ZFO NPs shown in Figure 4 are 30°, 35.36°, 39.2°, 41.88°, 46.8°, 52.84°, 57.44°, 62.68°, 75.52°, 79.24°, and 80.12°, which are in strong agreement with the crystalline planes of (220), (311), (222), (400), (331), (422), (511), (440), (622), (444), and (551), respectively, from the cubic system of zinc ferrite JCPDS pattern no. 01-089-4926 and 00-022-1012, confirming the presence of a single ZFO NPs phase with a face-centered cubic structure with space group *Fd3m*. The most intense peak of ZFO nanopowder is

(311) [2,21,23,24]. Moreover, sodium hydroxide of 1.5 M or above, as well as 80 °C used during the single-phase ZFO NPs synthesis method, produces stronger, narrower, and sharper diffraction peaks. This is also caused by the rise of crystallinity, crystallite size, and small size of the NPs, which can be seen in the peaks of Figure 4 [4,5,21]. On the other hand, some minor peaks related to the impurities of Fe_2O_3 phases (hematite) were detected in the sample; these impurities happen naturally or due to a possible incomplete reaction during the co-precipitation synthesis [23,25]. XRD peaks of iron oxide Fe_2O_3 were also detected at 24.84° (210), 25.45° (211), 49.52° (421), and 67.8° (442), according to JCPDS pattern no. 00-039-1346 (Figure 4). The reasons behind the appearance of these peaks are that ZFO NPs having Fe^{3+} as the main element and its manifestation signifies that there were more than 50 mole percent of the normal Fe_2O_3 content of spinel ferrite [22]. Moreover, one of the reasons for the hematite formation is due to the energy deficiency for attaching the Zn^{2+} ions with iron oxide to determine the crystal structure [22]. Finally, these impurities could be the result of using temperatures below the calcination range of 800 °C, since the co-precipitation synthesis method involves synthesis of NPs at low temperatures without the necessity of calcination at higher temperatures [21,26]. Furthermore, since the crystalline planes of ZFO NPs are properly indexed to a cubic spinel structure of the zinc ferrite, the crystallite size of the ZFO was estimated. An average crystallite size was calculated by employing the Scherrer's equation over all the diffraction peaks:

$$D = (0.9\lambda)/(\beta\cos\theta)$$

where λ is the wavelength of the X-ray radiation (λ = 0.154 nm), β is the full width at half-maxima of the peak in radians corrected for instrumental broadening, θ is the Bragg diffraction angle in radians, and D is the mean crystallite size [5,27]. The obtained average particle size was found to be approximately 5.6 nm for the NPs. This size is consistent with the HR-TEM results that confirm the size of ZFO NPs between 2.5 and 5 nm (Figure 4), as will be discussed in the following section. Additionally, MNPs typically consist of an iron oxide core between 5 to 20 nm, which is accomplished during the co-precipitation synthesis [1]. This size is correlated well since, as the NPs enter the blood circulation, they can be identified by the reticuloendothelial system (RES) and correspondingly eliminated from the vascular system. In the case of a particle size above 200 nm, the ability to enter tumor tissues decreases significantly and they are more vulnerable to suffer phagocytosis, thus leading to RES clearance. Hence, the size of the particle is crucial to improve the endocytic uptake by cells and their accumulation in tumor areas. For an effective cancer treatment, the usual size of the drug delivery system should be below 100 nm [1,28,29].

Figure 2 shows the TEM characterization results for the ZFO NPs sample to elucidate its morphology. In this figure, ZFO NPs were observed to form agglomerates (Figure 2a,b), showing dimensions of hundreds of nm to microns in size. These agglomerates were formed by "smaller" ZFO NPs with sizes between 2.5 nm and 5 nm, approximately, in close agreement with the calculated result by the Scherrer formula. In addition, the size distribution of the ZFO NPs is plotted in Figure 2a inset. The results showed a polydispersed size distribution with a ZFO NP mean size of 3.84 nm. These results agree with the HR-TEM and XRD average sizes, ranging from 2.5 and 5.6 nm. Furthermore, HR-TEM images (Figure 2c,d) of these NPs showed their crystalline nature. In these "small" ZFO NPs, a clear interplanar distance of ~0.24 nm could be observed, possibly related to the (222) planes in the ZFO NP cubic phase.

Moreover, Figure 3a shows the Selected Area Electron Diffraction (SAED) pattern of the ZFO sample with clear diffraction spots corresponding to the interplanar distances related to the crystal planes (220), (222) (400), (511), (440), and (444) in the ZFO NPs cubic phase (pattern no. 00-022-1012). These peaks coincide with the planes identified in the XRD characterization, confirming the interplanar distances observed. Additionally, Energy Dispersive X-Ray Spectroscopy (EDXS) spectra shown in Figure 3b and the table shown in the inset of 3b confirmed the composition of ZFO NPs to be Zn, Fe, and O. Nonetheless, in order to elucidate the chemical state of the identified elements, we proceed to record the high resolution Fe2p, Zn2p, and O1s XPS spectra.

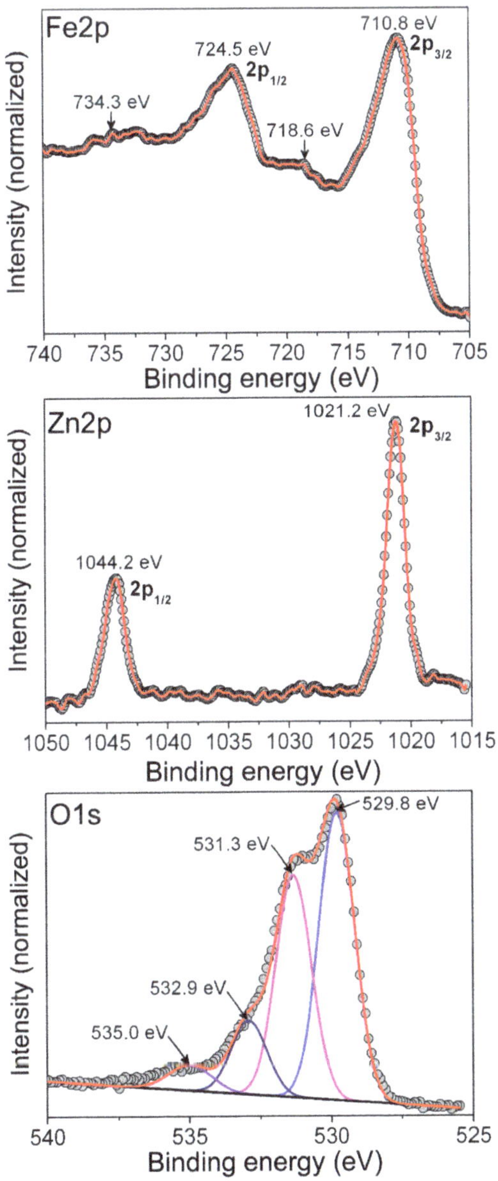

Figure 4. X-ray photoelectron spectroscopy showing high resolution Fe2p, Zn2p, and O1s spectra recorded for the powdered ZFO sample.

Figure 4 displays the X-ray photoelectron spectroscopy showing a high resolution Fe2p, Zn2p and O1s spectra recorded for the powdered ZFO NPs and Figure S2 shows complete survey scan of ZFO NPs (Figure S2). Accordingly, the Fe2p spectrum shows peaks at 710.8 and 724.5 eV related with core emissions from $2p_{3/2}$ and $2p_{1/2}$ states, respectively, for Fe(III) cations. In addition, there are satellite peaks observed at 718.6 and 734.3 eV which are attributed to shake-up-like emissions from Fe(III) cations [30]. Moreover, the Zn2p spectrum displays signals related to photoelectrons emitted from the $2p_{3/2}$ and $2p_{1/2}$ states that are located at 1021.2 and 1044.2 eV, respectively, which are congruent with those expected for Zn(II) cations [30]. Finally, the O1s spectrum can be deconvoluted into four peaks at 529.8, 531.3, 532.9 eV, and 535.0 eV associated with the (a) lattice oxygen in ZFO, (b) surface adsorbed organic molecules, (c) humidity, and (d) Auger electrons from Na, respectively. From these signals, we are assured that the chemical state of Fe, Zn, and O agrees with the expected chemical composition of as-synthesized spinel ferrite.

2.3. Magnetic Characterization of ZFO NPs

Figure 5 shows the m(H) curves recorded at 5 and 312 K for the ZFO NPs sample. As it can be noticed, this sample displays a magnetic response that fits with that expected from a soft ferromagnet, such as remanence (m_R = 5.9 emu/g), and coercivity (H_C = 0.33 kOe) at 5 K (see inset in Figure 5a). It is also observed that the magnetic moment reaches m_{max} = 45.5 emu/g at 70 kOe, although the m(H) does not show saturation even at this high applied field (Figure 5a). This lack of saturation can be attributed to the existence of a spin-glass-like surface layer in ZFO NPs not contributing to the magnetic response of the sample even at high magnetic fields, as it has been proposed for other spinel oxide NPs [31]. In addition, it is noticed that there is lack of hysteretic characteristics in the m(H) curve measured at 312 K (Figure 5b). This feature can be understood as the magnetic response of a ferromagnet in the superparamagnetic regime. Accordingly, above certain critical temperatures, known as blocking temperature (T_B), the thermal energy provokes the relaxation time of spins to become smaller than the relaxation time of the measuring apparatus, hence, their magnetic response resembles that of a Langevin's paramagnet. Figure 6 shows the m(T) curves recorded under zero field-cooled (ZFC) and field-cooled (FC) conditions. The ZFC curve displays a typical deblocking response of the spins as temperature increases, showing a maximum at T_M = 59.2 K. Above this temperature, the measured magnetic moment starts to decrease as the temperature moves towards 312 K. This diminishing magnetic moment can be attributed to the entry of the sample into the superparamagnetic regime; hence, T_M can be a rough estimation of its correspondent T_B. Moreover, it is also observed that the FC curve diverges from the ZFC one at T_{IRR} = 154.7 K and depicts an increase in the magnetic moment as the temperature decreases. In fact, the magnetic moment reaches a magnitude higher than that recorded at T_B. This result implies that additional blocked spins contribute to the magnetic response of sample during the FC process. These spins could be those in spin-glass-like surface layer in NPs that were identified from the measuring of the m(H) curves, but also a cooperative blockage of interacting nanoparticle from the observed clusters in the TEM images.

2.4. Comparative Cytotoxicity Evaluation Using Different Viability Assays (MTT, Crystal Violet and Lactate Dehydrogenase)

After a cautious characterization and estimation of the stability of nanoparticles, a biocompatibility and safety assessment is desirable for the NPs, beginning with in vitro assays for biomedical purposes. However, NPs can interfere with the cytotoxicity assays, leading to under- or overestimation of the NPs effect. Some examples of interferences are interaction of NPs with one of the components of the assay altering the functions of NPs or cellular metabolism, enhancing the optical activity of the reagents, or adsorbing the reagent on the surface of the NPs [32]. Therefore, there is a precise need for analyzing the cytotoxicity of nanoparticles with different assays and their combinations will give a more reliable result. To evaluate the cytotoxicity of the ZFO NPs, there were three viability assays that were tested, which are as follows:

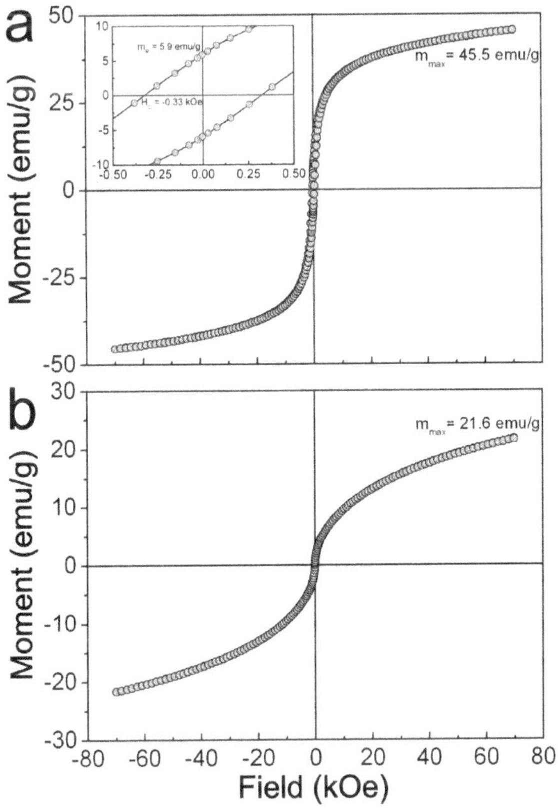

Figure 5. Isothermal field-dependent magnetic moment curves recorded for ZFO NPs sample at: (**a**) 5 K; (**b**) 312 K. The axes of the plot in the inset have the same units as seen in the plots (**a**,**b**).

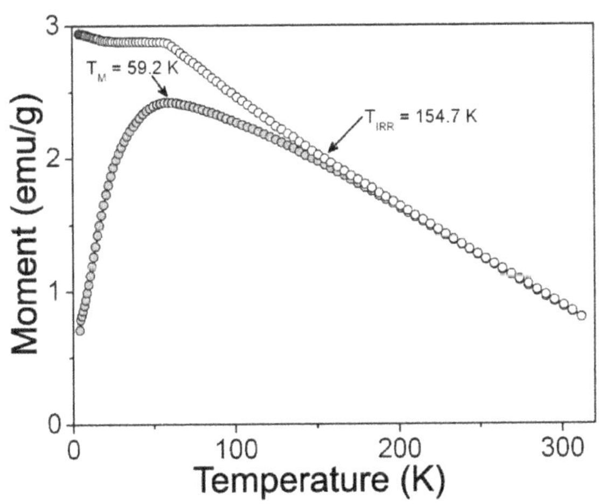

Figure 6. Zero field-cooled (ZFC) (gray circles) and field-cooled (FC) (white circles) temperature-dependent magnetic moment curves measured for ZFO NPs sample at 100 Oe.

2.4.1. MTT Assay

The standard MTT assay was performed on the MDA-MB-231, MCF-7, and HEK-293 cell lines after 48 h incubation with the ZFO at several concentrations (4, 12, 20, 50, 100, 200, 300, 400, 500, and 600 μg/mL). The viability profiles of the treated cells are shown in Figure 7. Likewise, MTT evaluates the mitochondrial activity exhibited by the viable cells after the treatment. The major contribution of NPs towards cytotoxicity is the formation of reactive oxygen species (ROS) causing oxidative stress. ROS produces several radicals such as superoxide anions, hydrogen peroxide, as well as hydroxyl and alkoxy radicals which damage lipids, DNA, and proteins, among others [15]. In addition, the interaction of ZFO NPs with the cell membrane causes a rupture and deformation of the membrane, generating ROS and particle dissolution, following the release of free intracellular Zn^{2+} and subsequently leading to failure of the cellular redox system and thus causing ROS production inside the cell [15]. In Figure 7A, it can be observed that the most damaged cell line at lower concentrations of ZFO NPs is MCF-7, between 4 and 50 μg/mL while MDA-MB-231 has higher cell viability but is lower than untreated cells. Moreover, in some cases at higher concentration such as 100, 200, 400, and 500 μg/mL, MCF-7 exhibits more cell viability than the MDA-MB-231 cell line, but at 300 and 600 μg/mL MDA-MB-231 has higher cell viability. Hence, a variation of cell viability exposed to different ZFO NPs concentration among the two different breast cancer cell lines creates doubts about which cell line is more sensitive to the exposure of zinc ferrite NPs. Nevertheless, the HEK-293 cell line exhibits statistically similar or higher cell viability in comparison to the untreated cells in almost all the ZFO NP concentrations, suggesting that ZFO NPs have a good safety profile, showing more than 75% of viability in almost all the cases.

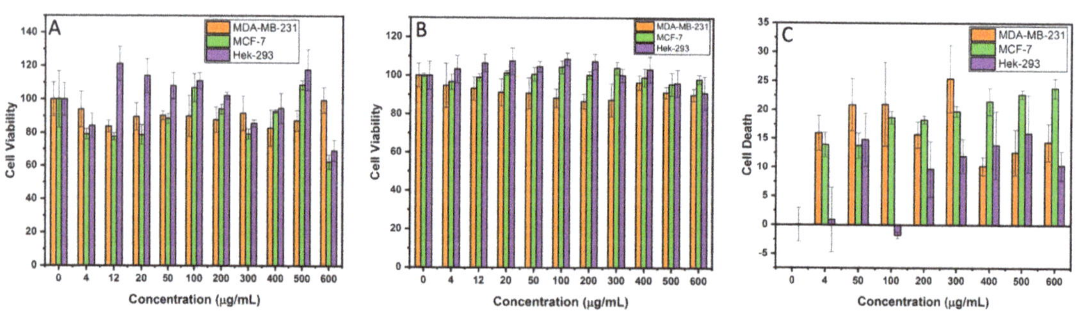

Figure 7. (**A**) Cytotoxicity profile of ZFO NPs using a MTT assay of MDA-MB-231, MCF-7, and HEK-293 cell lines. (**B**) Cytotoxic profile of Crystal violet assay of MDA-MB-231, MCF-7, and HEK-293 cell lines using ZFO NPs. (**C**) Dose-response data of LDH release from MDA-MB-231, MCF-7, and HEK-293 cells after the ZFO NP exposure.

Nonetheless, based on the TEM, XRD, and HR-TEM results, the size of the NPs are very small, in the range of 2.5 to 5.6 nm unravels certain exclusive properties of NPs, as size-dependent surface reactivity rendering higher ambiguities in toxicological NP evaluation. There is a critical need to evaluate different assays and confirm the effect of the size of NPs, considering different assays may react differently to small-sized particles [33]. Hence, the reduced cell viability of HEK-293 at 4, 300, and 600 μg/mL could be inferred from the NP size, since the calculated size is an approximation around 5 nm and an agglomeration could have formed [34]. The higher cell viability at 12 and 500 μg/mL in the HEK-293 cell line could be either due to the absorption of the substrate, interference of Zn^{2+} with the reduction reaction of MTT, interaction with the substrate depleting free MTT and creating false negative results, or oxidation of MTT to formazan due to the high NP concentration [34,35]. Therefore, other cytotoxic assays need to be done for reducing this existing variance be-tween breast cancer cell lines and cell viability.

2.4.2. Crystal Violet Assay

The crystal violet assay was employed with the same ZFO NPs concentrations as used in the MTT assay. These results confirm that the ZFO NPs damage more cancerous MDA-MB-231 and MCF-7 cell lines than the normal HEK-293 cell lines, as can be seen in the standard deviation presented in the graphs (Figure 7B). In this case, in all exposed concentration, the most damaged breast cancer cell line was MDA-MB-231 compared to MCF-7. This result creates a discrepancy with the obtained MTT results between which breast cancerous cell line is more sensitive to the several ZFO NP concentrations. However, HEK-293 cell line still exhibits a higher cell viability for nanoparticle treated cells as compared to the untreated ones, especially in some concentrations such as 4, 12, 20, 50, 100, 200, 300, and 400 μg/mL, which could be caused by the size, shape, aggregation, optical and magnetic properties, adsorption capacity, surface charge, and solubility, among others [31, 35], as can be noted from the characterizations performed in this work. Nevertheless, an interesting result is that the HEK-293 cell line exhibited a reduced cell viability in both MTT as well as crystal violet assay at the same concentrations of ZFO NPs. Moreover, NPs may aggregate at the bottom of the culture plate and sediment affecting the outcome of cell viability due to the mass composition and concentration and spectrum absorbance, leading to a false interpretation [31,36,37]. In addition, MNPs are known to enhance the signal due to electrical interactions, or absorb UV-visible light, inducing a false positive viability result [31,34]. However, the results of this assay reinforce the suggestion that ZFO NPs showed a good safety profile in non-cancerous cell lines exhibiting more than 90% of viability in nearly all the cases. Therefore, another cytotoxic assay was performed to obtain a clear picture of the ZFO NP cytotoxicity and determine as to which breast cancer cell line was more sensitive and strengthen the fact that ZFO NP has a good safety profile against benign cells.

2.4.3. Lactate Dehydrogenase Assay

Thus, the Lactate dehydrogenase assay (LDH) realizes the cytotoxicity of NPs by measuring the extracellular content of released lactate dehydrogenase from damaged cells [33]. In this study, the ZFO NP exposure in all cell lines demonstrate LDH release in nearly the same ZFO NP concentration as observed in the MTT and crystal violet assay, except for 12 and 20 μg/mL concentrations. In this dose-response curve, the LDH concentration in the MDA-MB-231 cell line increases at lower concentrations such as 4, 50, 100, and 300 μg/mL as compared to MCF-7. This indicates a rise in dead cells, while at higher concentrations 200, 400, 500, and 600 μg/mL MDA-MB-231 showed a decrement in the LDH release and MCF-7 exhibited a higher LDH concentration resulting in more dead cells, thus switching positions with MDA-MB-231 for more dead cells (Figure 7C). This reduced quantity of LDH release could be due to the possibility of ZFO NP interaction with the LDH enzyme, thus causing its inactivation. One more possibility is that MNPs create a metal-catalyzed oxidation of LDH producing ROS and causing a site-specific damage to proteins [33]. Another possibility could be that the NPs could adsorb the LDH molecules on their surface, inactivate them, and then either release them and adsorb new LDH molecules (dynamic adsorption) or attach them strongly to the surface without subsequent release (static adsorption) of the enzyme. In dynamic adsorption, if the rate of LDH degradation by ZFO NPs is higher, then the LDH levels will decrease over time in a rapid manner, while in the static adsorption, the LDH molecules are unavailable for measurement due to the surface attachment [33]. Hence, after obtaining the results, it can be implied that due to the rise of ZFO NP concentrations there was a higher adsorption of LDH molecules onto their surface, thus decreasing the dose-response curve measurements by either dynamic or static adsorption. Furthermore, the toxicity of ZFO NPs previously shown in the earlier cytotoxicity assays suggests that at higher NP concentrations there was an increase in cell death, thus a higher LDH release occurred by the limited number of available cells. Therefore, the high NP concentration range during the exposure time could offer a bigger surface for dynamic or static adsorption for LDH molecules, thus reducing the viability of the cells. This can be interpreted from the release of LDH by the cells and then the LDH is adsorbed on the surface of the NPs, which could show false positive results. On the other hand, in the HEK-293 cell line, the dose-response curve exhibits an increase in the LDH release as the concentrations of ZFO NPs increases (Figure 7C). The plausible explanation as to why, in the 400 and 500 µg/mL, the amount of HEK-293 cell deaths is higher than in MDA-MB-231 cell line, which could be due to the NP absorption in the MDA-MB-231 cell line. So, these results suggest that LDH results must be read with caution, and they confirm the importance of linking several different cytotoxic assays for the safety and cytotoxic evaluation of NPs. Finally, these results confirm ZFO NPs are more toxic to the MDA-MB-231 and MCF-7 cancer cell line than HEK-293 normal cell lines.

Based on the results of all the three assays, it was corroborated that the there is no inimical effect of ZFO NPs on HEK-293 non-cancerous cell lines but there were many discrepancies with respect to both MDA-MB-231 and MCF-7 cancer cell lines sensitivity. It was quite difficult to determine the degree of cytotoxicity of ZFO NPs on each cancer cell line, i.e. Triple-negative breast cancer (MDA-MB-231) and breast adenocarcinoma cells (MCF-7). Since there were irregularities in dose-dependent cytotoxicity of cancer cell lines, there was a need of automated cell counting and sorting system such as flow cytometric analysis to determine which cancer cell line was more sensitive to ZFO NPs.

2.5. Apoptosis Assay Using Flow Cytometry

In order to corroborate the good safety profile of benign cells (HEK-293) towards ZFO NP exposure, and to distinguish as to which breast cancer cell line (MDA-MB-231 and MCF-7) is more sensitive, flow cytometry studies were performed. Different concentrations of ZFO NPs (0, 8, 30, 50, and 250 µg/mL) were used for the flow cytometry-based assays. The two modes of cell death, apoptosis or necrosis, were measured using Annexin V-FLUOS (FITC) and propidium iodide (PI) dyes. The cellular uptake can be detected soon after 5–6 h, however, toxicity is generally seen at 24 h [38]. A two-dimensional contour density plot of early apoptosis (FITC) against late apoptosis (PI) was made. Each plot is separated into four quadrants and changes are measured based on no uptake or if there is uptake of either one or both the dyes. Every single cell will occupy one of the four spaces of the plot depending upon the process the cell is going through.

Cells not stained with either of the dyes, FITC and PI, are living cells. These cells are in the live cell quadrant of the contour plot (down left or Q3). Cells stained only with FITC dye are early apoptotic cells and are in the apoptotic cell quadrant of the plot (downright or Q4). Cells stained with both the dyes, FITC and PI, are late apoptotic cells (upper right or Q2) and necrotic cells (upper left or Q1). A total number of 20,000 cells were measured per ZFO NP concentration. Each cell was located in each quadrant depending on the status of the cell; live cells, early apoptosis, late apoptosis and necrosis (Table 1, Figure 8).

Table 1. Obtained percentages from integration of cells (MDA-MB-231, MCF-7, and HEK-293) in each quadrant (live, early apoptosis, and late apoptosis) normalized to the total cell number 20,000 when exposed to different concentrations of zinc ferrite nanoparticles (0, 8, 30, 50, and 250 µg/mL).

	Live (%)			Early Apoptosis (%)			Q1 and Late Apoptosis (%)		
Concentration	**MDA-MB-231**	**MCF-7**	**HEK-293**	**MDA-MB-231**	**MCF-7**	**HEK-293**	**MDA-MB-231**	**MCF-7**	**HEK-293**
0 µg/mL	33	53	89	24	1	7	43	47	5
8 µg/mL	4	62	34	33	0	26	63	38	40
30 µg/mL	6	61	33	28	0	31	66	39	36
50 µg/mL	3	65	24	28	0	27	70	35	49
250 µg/mL	4	61	24	22	0	23	74	39	54

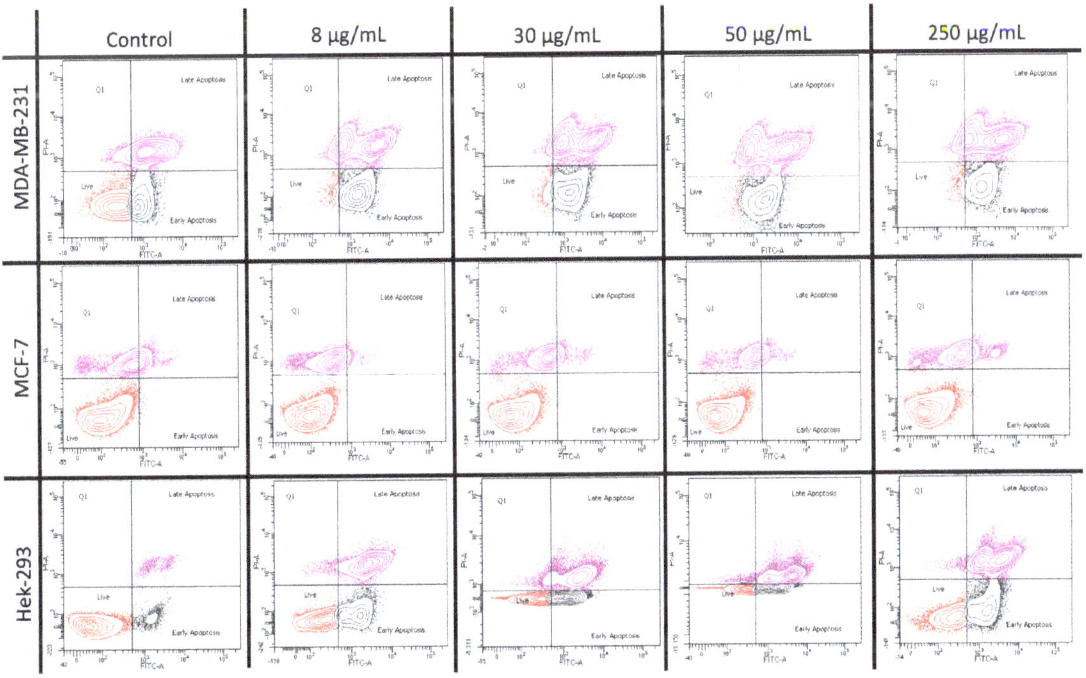

Figure 8. Representative two-dimensional contour density plots to determine fraction of live, early apoptosis, and late apoptosis of cells (MDA-MB-231, MCF-7, and HEK-293) when exposed to different concentrations (0, 8, 30, 50, and 250 µg/mL) of zinc ferrite nanoparticles for 24 h. Early and late apoptosis were measured by using Annexin V-FLUO and propidium iodide dyes.

Moreover, the key point of this research was to identify if the non-cancerous cell line was susceptible to the treatment of ZFO NPs and as to which breast cancer cell line was more sensitive. All the different ZFO NP concentrations induced a higher percentage of the cell line, MDA-MB-231, in the quadrants of early apoptosis, late apoptosis and necrosis as compared to MCF-7 and HEK-293 (Figure 9). Late apoptosis and Q1 quadrant for the MDA-MB-231 cell line confirmed that this cell line is more sensitive towards the exposition of NPs with a higher percentage of dead cells as compared to the MCF-7 cell line after 24 h exposure (Figure 9).

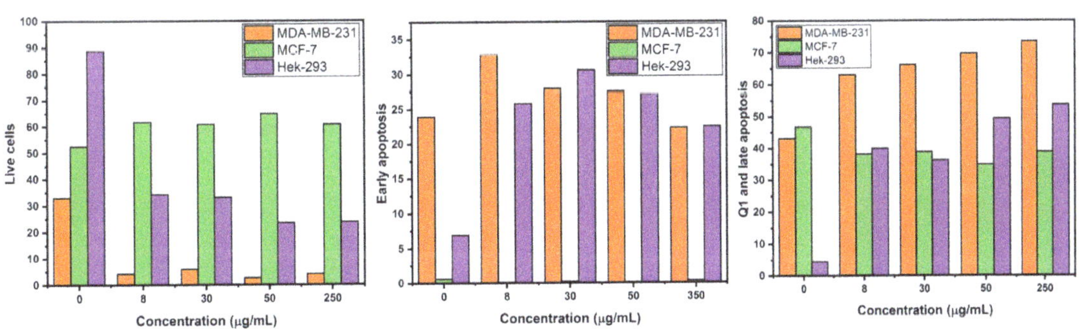

Figure 9. Percentages of live cells and cells entering in early and late apoptosis of MDA-MB-231, MCF-7, and HEK-293 in a flow cytometry study.

Additionally, the non-cancerous cell line HEK-293 presented a lower percentage of cells in the quadrants of early apoptosis, necrosis, and late apoptosis in comparison to MDA-MB-231 (Figure 9). Hence, the flow cytometry results confirmed and correlated well with the scores from the MTT, CV and LDH assay, confirming triple-negative breast cancer cell line (MDA-MB-231) as more sensitive to several ZFO NP concentrations exposure. Also, HEK-293 exhibits a lower sensitivity to enter an early and late apoptotic process exhibiting a good-safety profile against non-cancerous cells. Finally, flow cytometry results also showed the resistance of the breast cancer cell lines (MCF-7) wherein very lower percentage of cells enter in late apoptotic phase as compared to MDA-MB-231 (Figure 9).

An early signature marker of apoptosis is the damage in the asymmetry of the phospholipid layer from the cell membrane resulting in the translocation of phosphatidylserine from inner face to the outer surface of the cell. The employed FITC binds specifically to the exterior ligands of the phosphatidylserine in the cell membrane forecasting an early apoptotic stage of the cell [39,40]. This grade of apoptosis from nanoparticle-exposed cells was quantified using the FITC dye in flow cytometry. Also, PI intercalates with the double strand of the nucleic acid during the excitation of blue light producing red fluorescence. However, PI cannot pass into the living cell when the cell membrane is intact. When the cells enter an apoptotic stage, the cell membrane becomes debilitated, and PI can enter the cell and stain the nucleic acids [39]. Therefore, both dyes can be used to discriminate between living and dead cells. Furthermore, several methods are employed for NP cytotoxicity and to determine the cell viability of cells exposed to different NP concentrations. The most common is the MTT assay which measures the metabolic activity of the cell. This method along with crystal violet and LDH assay generates a colored product, which can be easily quantified with spectrophotometric techniques [40]. Nevertheless, accurate quantification of NPs is not always precisely done since their optical properties can create an absorbance in an equivalent manner thus leading to a false estimation of the cell viability [40]. In addition, nanoparticles can have some interference in the colorimetric and absorbance cytotoxic assays, so the flow cytometry assay which is an automated cell counting and sorting system can give us an in-depth information and confirm the results of all the three cytotoxic assays such as MTT, crystal violet, and LDH. Additionally, with these experiments, the cell viability can be precisely quantified using flow cytometry-based assays since it is the cumulative output from thousands of individually measured cells. [40]. Therefore, flow cytometry has several advantages such as a low possibility of interference from NPs during the measurements, two different cell death mechanisms that can be differentiated as early and late apoptosis, and an understanding of the biochemical and molecular mechanisms of cell death after the exposure to different concentrations of magnetic nanoparticles [40]. This is a breakthrough since triple-negative breast cancer expresses a poor prognosis, is a more aggressive disease, has a low overall survival, has a mesenchymal phenotype, is multidrug resistance disease, is insensitive to treatments with antiestrogens, is more immunogenic, and has fewer options for treatments due to the lack of receptors or enough protein for hormone therapy. All of these restrict the therapies with radiation, immunotherapy, chemotherapy, and lastly with surgery [41,42]. Consequently, future treatments with ZFO NPs against triple-negative breast cancer could be an option since the results showed a good safety profile, a low benign cell death, and more sensitivity for this type of breast cancer.

2.6. Optical Microscopy Studies for Cell Morphological Analysis

Furthermore, the rigorous use of NPs in diagnosis and therapies in the medical field has led us to exploit such NPs to enter inside the cells based on their size. Once they enter, the NPs start accumulating within the cells, thus leading to cellular dysfunctions such as cytoskeletal damages as well as altered cellular metabolism by ROS mediation [43]. These cellular alterations are involved in protein and gene regulation, controlling cell migration, differentiation, proliferation, apoptosis, and tissue development [43]. Additionally, once the cytoskeleton is damaged, the focal adhesion of proteins and their following adhesion are affected, which begins beneath the filipodia (finger-like protrusions) and lamellipodia (branched actin filaments) [43]. In this study, the cytotoxic results of the different assays were confirmed by observing the morphology of the cells using digital photographs obtained by Nikon ECLIPSE TS100 microscope attached to the ProgRes® CT3 at 20×. In the ProgRes® CapturePro 2.10.0.1—JENOPTIK Optical Systems GmbH software, the images exhibit a reduction in the cell viability of HEK-293, MDA-MB-231, and MCF-7 as per the Figure 10A–C, respectively. Furthermore, it is clearly seen that the untreated cells were highly confluent, and they are spreading efficiently with the aid of lamellipodia and filipodia (as indicated by white circles in the figure). At 300 µg/mL, all cell lines exhibited stunted growth of lamellipodia and filipodia, and further, as the concentrations of ZFO NPs start increasing, especially at 600 µg/mL, all cell lines were demonstrated to possess much smaller cell area inhibiting the formation of lamellipodia and filipodia (as indicated by white circles in the figure). At that higher concentration, the cell morphology was completely damaged in comparison to the untreated ones and were completely round, causing cell shrinkage. There was higher dead cell debris observed in cancer cell lines in comparison to normal cell lines (as indicated by white circles in the figure). These results confirm that the lamellipodia and filipodia are vastly affected by the NPs, thus blocking their formation, leading to the decrement in the tumor dissemination and cancer invasion [43]. Ketebo and colleagues (2020) had the same result exhibiting an inhibition of filipodia and lamellipodia at 0.1 and 1 µg/L of silica-coated magnetic NPs with rhodamine B isothiocyanate [43]. Hence ZFOs provide the opportunity to generate a new and functional platform for Cancer Therapeutics.

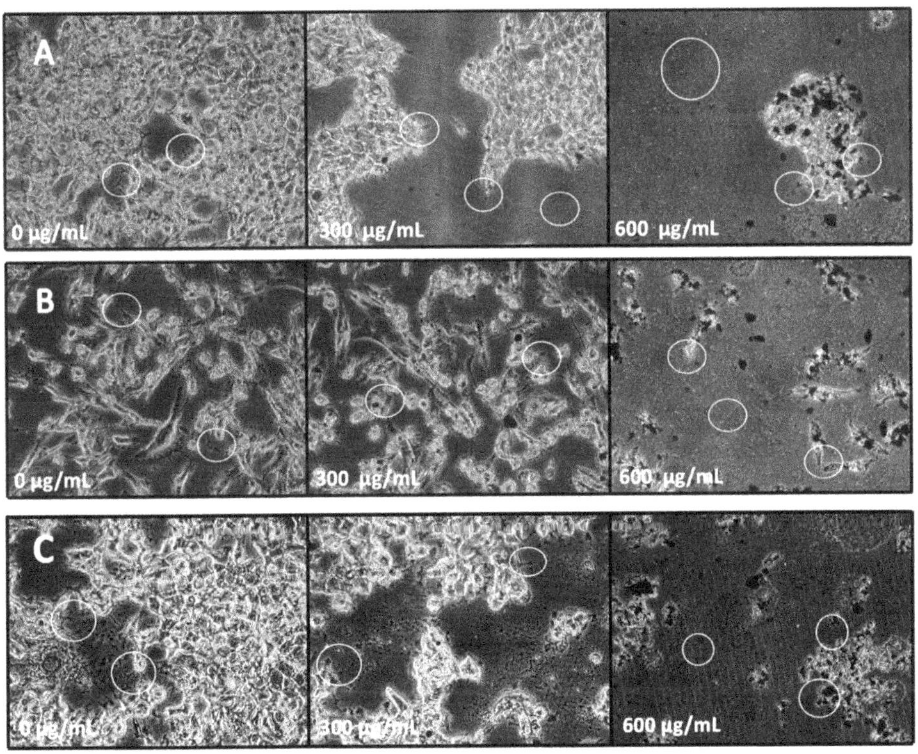

Figure 10. (**A**) Digital microscopic images of HEK-293 cell line, (**B**) Digital microscopic images of MDA-MB-231 cell line, and (**C**) Digital microscopic images of MCF-7 cell line after the exposure of 0 µg/mL, 300 µg/mL, and 600 µg/mL of ZFO NPs.

3. Materials and Methods

3.1. Synthesis of ZFO NPs

5.2 gm of (0.5 M) of iron (III) chloride hexahydrate ($FeCl_3$. $6H_2O$; Sigma-Aldrich 97%, Toluca, Mexico) and 2 gm (0.25 M) of zinc (II) chloride, granular ($ZnCl_2$; J. T. Baker Analyzed ACS reagent, Radnor, PA, USA) were mixed in 25 mL of deoxygenated water under a nitrogen atmosphere. A total of 0.85 mL of 12.1 M hydrochloric acid (HCl; Sigma Aldrich) was added to the solution for 30 min under nitrogen atmosphere and stirred for 2 h at 85–90 °C. In addition, 15 gm of aqueous solution of 1.5 M sodium hydroxide pellets (NaOH; Meyer >97.0%) were added to 250 mL of deoxygenated water and mixed. Solutions of Fe (III), Zn (II), and HCl were added dropwise into the NaOH solution and mixed under magnetic stirring at 70 °C for 1 h under N_2 atmosphere. The synthesis is based on the precipitation and the obtained precipitates were recollected and centrifuged for 20 min at 4000 rpm. The supernatant was thrown, and the precipitate was washed with 50 mL of deionized water. This procedure of washing was repeated three times. At the end of the last wash, the precipitated ZFO NPs were dried at 80–100 °C for 4 h. [2].

3.2. Characterization

The as-synthesized ZFO NPs were characterized using a BK-UV 1800 pc (BIOBASE, Jinan, Shandong, China) UV-vis spectrometer between 200 and 800 nm. Raman spectrum from 100 to 2000 cm^{-1} was acquired with a Xplora (Horiba, Palaiseau, France) spectrometer with an exciting radiation of 638 nm, power of 20–25 mW, objective ×10, and a CCD camera used as a detector. FTIR spectra were recorded with an IRA-affinity-1S spectrophotometer (SHIMADZU, Kyoto, Japan) and the LabSolution IR software. 5 µL of the NP ZFO NPs samples were added on a module ATR Specac Quest with a diamond prism and measured between the wavenumber interval of 400 cm^{-1} and 4000 cm^{-1}, with a resolution of 4 cm^{-1}. X-ray diffraction (XRD) analysis of the crystalline phase of ZFO were performed with a X-ray diffractometer D8 Advance (Bruker, Karlsruhe, Germany) with a Cu Kα lamp (Measurements parameters: radiation wavelength, 1.54 A; acceleration voltage, 40 KV; current 40 mA; using a coupled method). The obtained results were compared with the Joint Committee Powder Diffraction Standards (JCPDS) database for phase affinity. Transmission electron microscopy (TEM) analyses were performed with a FEI, Titan G2 80–300 instrument operated at 300 kV as acceleration voltage, equipped with an EDAX energy-dispersive X-ray spectroscopy (EDXS) detector. The ZFO NPs sample was also characterized by X-ray photoelectron spectroscopy in a Thermo-Scientific, K-Alpha spectrometer using AlKα radiation (E = 1.5 keV). With this technique we could record high resolution Fe2p, Zn2p, and O1s spectra to revise the chemical state of Fe, Zn, and O in the synthesized sample. The magnetic properties of the powdered NP ZFO sample were measured in a Quantum Design, MPMS3 magnetometer from the recording of isothermal field-dependent magnetic moment curves [m(H)] at 5 and 312 K, using a magnetic field interval between −70 and 70 kOe. In addition, zero field-cooled (ZFC) and field-cooled (FC) temperature-dependent magnetic moment curves [m(T)] were obtained for a temperature interval between 4 and 312 K at 100 Oe.

3.3. Cell Culture and MTT, Crystal Violet, and LDH Cytotoxic Assay

Human epithelial breast cancer cell line (MDA-MB-231) and breast cancer cell line (MCF-7) were obtained from Instituto Nacional de Ciencias Médicas y Nutrición Salvador Zubirán (INCMNSZ), Mexico, and human embryonic kidney 293 cell line (HEK-293) was obtained from Universidad Nacional Autónoma de México (UNAM). Both cell lines were cultured with high glucose Dulbecco's modified Eagle's medium (DMEM) (Sigma Aldrich, St. Louis, MO, USA) with 10% fetal bovine serum (Sigma Aldrich, USA) and incubated at 37 °C, 5% CO_2. The 17,500 MDA-MB-231, 20,000 MCF-7 and 6000 HEK-293 cells were plated in wells of 48 well-tissue culture plates. The cells were kept for recovery for one day after the addition of the cells, and on the third day, they were stimulated. ZFO NPs were added in different concentrations in triplicates, as well as doxorubicin hydrochloride

(Doxylid 50, CelonLabs, Nagpur, India) was added as a positive control, and untreated cells served as negative controls. ZFO concentrations were taken as follows: 4, 12, 20, 50, 100, 200, 300, 400, 500, and 600 μg/mL in MTT and crystal violet and 4, 50, 100, 200, 300, 400, 500, and 600 μg/mL in LDH assay. The treated cells were incubated at 37 °C, 5% CO_2, for 48 h.

MTT assay: After the incubation period with the ZFO NPs and doxorubicin, the media was aspirated, then the wells washed with 1× phosphate-buffered saline (PBS) solution, and 300 μL of MTT (Sigma-Aldrich 98%) was added to every well, mixed gently and later incubated for 4 h at 37 °C, 5% CO_2. The MTT assay in each well was carefully removed and replaced with 700 μL ethanol and mixed gently to dissolve the formazan crystals. 100 μL of the resolubilized formazan of each well was placed in a 96-well tissue culture plate and read in a plate reader at 570 nm (Thermo Scientific MULTISKAN Sky, Marsiling Industrial Estate Road 3, Singapore). The OD at 570 nm for the blue-purple formazan was used to evaluate the anti-proliferative effect of compounds. The OD at 570 nm for control cells were taken as 100% viability.

Crystal violet assay was performed directly in the 48-well tissue culture plate by adding 250 μL of 5% crystal violet stain solution and then incubated in an orbital shaker at 100 rpm for 5 min, at room temperature. The stain was removed, and the wells were washed four times with tap water and then the plates were left to dry. A total of 500 μL of a 10% *v*/*v* acetic acid solution was then added to each well and incubated in an orbital shaker at 100 rpm for 5 min, at room temperature. A total of 200 μL was removed and transferred to a 96-well tissue culture plate to read the optical density in a plate reader at 590 nm.

The Lactate dehydrogenase (LDH) cytotoxicity assay was performed as is described in In Vitro Toxicology Assay Kit, Lactic Dehydrogenase based, reference TOX7-1KT (Sigma-Aldrich), which can be explained as follows: ss mentioned above in the MTT and Crystal violet assay, after the incubation of the nanoparticles and Doxorubicin with the cell cultures and the medium, aliquots of the medium were removed. Later Lactate Dehydrogenase (LDH) assay mixture was added to the medium that was removed for the analysis. The medium volume was twice the LDH mixture. The plates were covered by an aluminum foil to avoid any interference of the light and the incubation is performed for 20–30 min. After the termination of the reaction using 1N HCl, the absorbance of the solution can be measured spectrophotometrically at 490 nm.

Flow cytometry-based assay was performed with a BD LSRFortessa™ (BD Biosciences, San José, CA) flow cytometer containing a blue laser 488 nm laser, forward scatter (FSC) diode detector, and a side scatter (SSC) detector was used. MDA-MB-231, MCF7, and HEK-293 cell lines were seeded in 12 plate-wells for 24 h and treated with ZFO at different concentrations (0, 8, 30, 50, and 250 μg/mL) for 24 h at 37 °C. Cells without nanoparticles were considered as the control group. All plates were removed from DMEM and washed with 1 mL of PBS. The 1 mL of versene was added into each well of the plate and incubated for 10 min, then removed from each well and added into an independent Falcon tube with 1.5 mL of previously added PBS. Each Falcon tube was centrifuged for 3 min at 1500 rpm, resuspended with 1 mL of binding buffer, and centrifuged for 5 min at 4000 rpm. Lastly, each tube was resuspended with binding buffer with propidium iodide (Sigma-Aldrich) and Annexin V-FLUOS (Roche; 11828681001) and incubated for 30 min. All data analyses from 20,000 cells per sample were carried out in FACSDiva v. 8.1 flow cytometer (BD Bioscience) unless otherwise stated. Samples were gated as exhibited in representative example Figure S3.

4. Conclusions

In the present study, ZFO NPs are synthesized using a co-precipitation method. The results of the different techniques used for the characterization show that ZFO NPs were single phase with normal spinel structure. The crystallite size was determined by XRD and HR-TEM analysis, which was approximately 5 nm and showed patterns of successful synthesis with peaks related to the cubic phase. The UV-vis absorbance peaks exhibited three characteristic electronic transitions peaks of ZFO, thus confirming the presence of ZFO. Additionally, FTIR and RAMAN spectra presented bands related to Zn-O and Fe-O bonds. XPS analysis confirmed the presence of Zn, Fe, and O. Moreover, the observed maximum magnetization of ZFO powders was at 45.5 emu/g at 70 kOe. Finally, during the cytotoxic assays, MTT, crystal violet, LDH, and flow cytometry, some interactions could be noticed but all the assays showed zinc ferrite NPs damages more the cancerous cell lines MDA-MB-231 and MCF-7 than human embryonic kidney cell line HEK-293. These results were confirmed with the digital photographs showing the lamellipodia and filipodia inhibition as the ZFO NPs concentration raised. These studies will help in optimizing cytotoxicity assays for future nanoparticle-based drug-delivery, contrast agent, biolabeling, and Theranostics.

Supplementary Materials: The following supporting information can be downloaded at: https://www.mdpi.com/article/10.3390/ijms241612860/s1.

Author Contributions: Conceptualization, J.L.d.l.F.-J., A.S. and G.O.; methodology, J.L.d.l.F.-J., A.Z.D., A.S., G.O., C.I.R.-R., I.B.M.-A., A.T.-C., E.A.G.-L., J.H.-C., M.G.A.-E., M.A.G.-N. and D.I.G.-G.; validation, J.L.d.l.F.-J., C.I.R.-R., I.B.M.-A., A.T.-C., E.A.G.-L., J.H.-C., M.G.A.-E., M.A.G.-N., D.I.G.-G., L.G.A. and J.L.G.; formal analysis, J.L.d.l.F.-J., A.Z.D., A.S., G.O., M.A.G.-N. and D.I.G.-G.; investigation, J.L.d.l.F.-J., A.Z.D., A.S., G.O., C.I.R.-R., I.B.M.-A., A.T.-C., E.A.G.-L., J.H.-C., M.G.A.-E., M.A.G.-N. and D.I.G.-G.; data curation, J.L.d.l.F.-J., C.I.R.-R., A.Z.D., A.S. and G.O.; writing—original draft preparation, J.L.d.l.F.-J., A.Z.D., A.S., G.O., M.A.G.-N. and D.I.G.-G.; writing—review and editing, J.L.d.l.F.-J., A.Z.D., A.S., G.O., M.G.A.-E., M.A.G.-N., D.I.G.-G., L.G.A. and J.L.G.; supervision, J.L.d.l.F.-J., A.Z.D., A.S., G.O., M.G.A.-E., M.A.G.-N., D.I.G.-G., L.G.A. and J.L.G.; funding acquisition, A.Z.D., A.S. and G.O. All authors have read and agreed to the published version of the manuscript.

Funding: This research was funded by Ciencias de Fronteras (Conacyt) 2096029 as well as Laboratorio Nacional micro y nanofluidica-LABMyN (Project number 321116).

Institutional Review Board Statement: Not applicable.

Informed Consent Statement: Not applicable.

Data Availability Statement: Data shared are in accordance with consent provided by participants.

Acknowledgments: One of the authors Juan Luis de la Fuente acknowledges the financial support by Conacyt. We also acknowledge Francisco Cuevas from Instituto de Neurobiologia, UNAM, Mexico for the provision of HEK-293 cell lines for this work. We also acknowledge Guillermo Juárez Vega from Red de Apoyo a la Investigación, UNAM and INCMNSZ for helping us in the flow cytometry analysis for the manuscript.

Conflicts of Interest: The authors declare no conflict of interest.

References

1. Zaloga, J.; Janko, C.; Nowak, J.; Matuszak, J.; Knaup, S.; Eberbeck, D.; Tietze, R.; Unterweger, H.; Friedrich, R.P.; Heimke-Brinck, R.; et al. Development of a lauric acid/albumin hybrid iron oxide nanoparticle system with improved biocompatibility. *Int. J. Nanomed.* **2014**, *9*, 4847–4866. [CrossRef] [PubMed]
2. Mamani, J.; Costa-Filho, A.; Cornejo, D.; Vieira, E.; Gamarra, L. Synthesis and characterization of magnetite nanoparticles coated with lauric acid. *Mater. Charact.* **2013**, *81*, 28–36. [CrossRef]
3. Paz-Díaz, B.; Vázquez-Olmos, A.R.; Almaguer-Flores, A.; García-Pérez, V.I.; Sato-Berrú, R.Y.; Almanza-Arjona, Y.C.; Garibay-Febles, V. ZnFe2O4 and CuFe2O4 Nanocrystals: Synthesis, Characterization, and Bactericidal Application. *J. Clust. Sci.* **2021**, *34*, 111–119. [CrossRef]
4. Shahraki, R.R.; Ebrahimi, M.; Ebrahimi, S.S.; Masoudpanah, S. Structural characterization and magnetic properties of superparamagnetic zinc ferrite nanoparticles synthesized by the coprecipitation method. *J. Magn. Magn. Mater.* **2012**, *324*, 3762–3765. [CrossRef]
5. Ebrahimi, M.; Shahraki, R.R.; Ebrahimi, S.A.S.; Masoudpanah, S.M. Magnetic Properties of Zinc Ferrite Nanoparticles Synthesized by Coprecipitation Method. *J. Supercond. Nov. Magn.* **2014**, *27*, 1587–1592. [CrossRef]
6. Galinetto, P.; Albini, B.; Bini, M.; Mozzati, M.C. Raman spectroscopy in Zinc Ferrites Nanoparticles. *Raman Spectrosc.* **2018**. [CrossRef]
7. Torres-Pérez, S.A.; Ramos-Godínez, M.D.P.; Ramón-Gallegos, E. Glycosylated one-step PAMAM dendrimers loaded with methotrexate for target therapy in breast cancer cells MDA-MB-231. *J. Drug Deliv. Sci. Technol.* **2020**, *58*, 101769. [CrossRef]

8. Gurunathan, S.; Han, J.; Park, J.H.; Kim, J.-H. A green chemistry approach for synthesizing biocompatible gold nanoparticles. *Nanoscale Res. Lett.* **2014**, *9*, 248. [CrossRef]
9. Das, P.; Dutta, A.; Bhaumik, A.; Mukhopadhyay, C. Heterogeneous Ditopic $ZnFe_2O_4$ Catalyzed Synthesis of 4H-Pyrans: Further Conversion to 1,4-DHPs and Report of Functional Group Interconversion from Amide to Ester. *Green Chem.* **2014**, *16*, 1426–1435. [CrossRef]
10. Shams, S.; Sheibanizadeh, Z.; Khalaj, Z. Ternary nanocomposite of $ZnFe_2O_4/\alpha$-Fe_2O_3/ZnO; synthesis via coprecipitation method and physical properties characterization. *Appl. Phys. A* **2021**, *127*, 459. [CrossRef]
11. Kasparis, G.; Sangnier, A.P.; Wang, L.; Efstathiou, C.; LaGrow, A.P.; Sergides, A.; Wilhelm, C.; Thanh, N.T.K. Zn doped iron oxide nanoparticles with high magnetization and photothermal efficiency for cancer treatment. *J. Mater. Chem. B* **2022**, *11*, 787–801. [CrossRef] [PubMed]
12. Singh, J.P.; Srivastava, R.C.; Agrawal, H.M.; Kumar, R. Micro-Raman investigation of nanosized zinc ferrite: Effect of crystallite size and fluence of irradiation. *J. Raman Spectrosc.* **2011**, *42*, 1510–1517. [CrossRef]
13. Rivero, M.; del Campo, A.; Mayoral, Á.; Mazario, E.; Sánchez-Marcos, J.; Muñoz-Bonilla, A. Synthesis and structural characterization of $Zn_xFe_{3-x}O_4$ ferrite nanoparticles obtained by an electrochemical method. *RSC Adv.* **2016**, *6*, 40067–40076. [CrossRef]
14. Abbasian, A.R.; Afarani, M.S. One-step solution combustion synthesis and characterization of $ZnFe_2O_4$and $ZnFe_{1.6}O_4$ nanoparticles. *Appl. Phys. A* **2019**, *125*, 721. [CrossRef]
15. Sarala, E.; Naik, M.M.; Vinuth, M.; Reddy, Y.V.R.; Sujatha, H.R. Green synthesis of Lawsonia inermis-mediated zinc ferrite nanoparticles for magnetic studies and anticancer activity against breast cancer (MCF-7) cell lines. *J. Mater. Sci. Mater. Electron.* **2020**, *31*, 8589–8596. [CrossRef]
16. Caruntu, G.; Bush, G.G.; O'Connor, C.J. Synthesis and characterization of nanocrystalline zinc ferrite films prepared by liquid phase deposition. *J. Mater. Chem.* **2004**, *14*, 2753–2759. [CrossRef]
17. Singh, C.; Devika, D.; Malik, R.; Kumar, V.; Singhal, S. CdS nanorod–MFe_2O_4 (M = Zn, Co and Ni) nanocomposites: A heterojunction synthesis strategy to mitigate environmental deterioration. *RSC Adv.* **2015**, *5*, 89327–89337. [CrossRef]
18. Maksoud, M.I.A.A.; El-Sayyad, G.S.; El-Bastawisy, H.S.; Fathy, R.M. Antibacterial and antibiofilm activities of silver-decorated zinc ferrite nanoparticles synthesized by a gamma irradiation-coupled sol–gel method against some pathogenic bacteria from medical operating room surfaces. *RSC Adv.* **2021**, *11*, 28361–28374. [CrossRef]
19. Tadjarodi, A.; Imani, M.; Salehi, M. $ZnFe_2O_4$ nanoparticles and a clay encapsulated $ZnFe_2O_4$ nanocomposite: Synthesis strategy, structural characteristics and the adsorption of dye pollutants in water. *RSC Adv.* **2015**, *5*, 56145–56156. [CrossRef]
20. Ali, R.; Aziz, M.H.; Gao, S.; Khan, M.I.; Li, F.; Batool, T.; Shaheen, F.; Qiu, B. Graphene oxide/zinc ferrite nanocomposite loaded with doxorubicin as a potential theranostic mediu in cancer therapy and magnetic resonance imaging. *Ceram. Int.* **2022**, *48*, 10741–10750. [CrossRef]
21. Anooj, K.P.E.S.; Sreelekshmi, S.J.; Gopukumar, S.T.; Praseetha, P.K. Evaluation of the Zinc Ferrite Nano particles for Bio-applications. *Int. J. Pharm. Sci. Rev. Res.* **2017**, *46*, 22–26.
22. Zahi, S. Synthesis, Permeability and Microstructure of the Optimal Nickel-Zinc Ferrites by Sol-Gel Route. *J. Electromagn. Anal. Appl.* **2010**, *02*, 56–62. [CrossRef]
23. Wan, J.; Jiang, X.; Li, H.; Chen, K. Facile synthesis of zinc ferrite nanoparticles as non-lanthanide T1 MRI contrast agents. *J. Mater. Chem.* **2012**, *22*, 13500–13505. [CrossRef]
24. Naseri, M.G.; Saion, E.B.; Hashim, M.; Shaari, A.H.; Ahangar, H.A. Synthesis and characterization of zinc ferrite nanoparticles by a thermal treatment method. *Solid State Commun.* **2011**, *151*, 1031–1035. [CrossRef]
25. Medina, M.A.; Oza, G.; Ángeles-Pascual, A.; González, M.M.; Antaño-López, R.; Vera, A.; Leija, L.; Reguera, E.; Arriaga, L.G.; Hernández Hernández, J.M.; et al. Hyperthermia of Monodispersed Cobalt Ferrite Nanoparticles for Cancer Therapeutics. *Molecules* **2020**, *25*, 4428. [CrossRef] [PubMed]
26. Kahmei, R.D.R.; Seal, P.; Borah, J.P. Tunable heat generation in nickel-substituted zinc ferrite nanoparticles for magnetic hyperthermia. *Nanoscale Adv.* **2021**, *3*, 5339–5347. [CrossRef] [PubMed]
27. Jang, J.S.; Hong, S.J.; Lee, J.S.; Borse, P.H.; Jung, O.-S.; Hong, T.E.; Jeong, E.D.; Won, M.S.; Kim, H.G. Synthesis of Zinc Ferrite and Its Photocatalytic Application under Visible Light. *J. Korean Phys. Soc.* **2009**, *54*, 204–208. [CrossRef]
28. Al-Qubaisi, M.S.; El Zowalaty, M.E.; Rasedee, A.; Flaifel, M.H.; Ahmad, S.H.; Hussein-Al-Ali, S.H.; Hussein, M.Z.; Eid, E.E.; Zulkarnain, Z.; Saeed, M.; et al. Cytotoxicity of nickel zinc ferrite nanoparticles on cancer cells of epithelial origin. *Int. J. Nanomed.* **2013**, *8*, 2497–2508. [CrossRef]
29. Chen, X.; Tang, Y.; Cai, B.; Fan, H. 'One-pot' synthesis of multifunctional GSH–CdTe quantum dots for targeted drug delivery. *Nanotechnology* **2014**, *25*, 235101. [CrossRef]
30. Guo, Y.; Zhang, N.; Wang, X.; Qian, Q.; Zhang, S.; Li, Z.; Zou, Z. A facile spray pyrolysis method to prepare Ti-doped $ZnFe_2O_4$ for boosting photoelectrochemical water splitting. *J. Mater. Chem. A* **2017**, *5*, 7571–7577. [CrossRef]
31. Misra, R.; Gubbala, S.; Kale, A.; Egelhoff, W. A comparison of the magnetic characteristics of nanocrystalline nickel, zinc, and manganese ferrites synthesized by reverse micelle technique. *Mater. Sci. Eng. B* **2004**, *111*, 164–174. [CrossRef]
32. Pem, B.; González-Mancebo, D.; Moros, M.; Ocaña, M.; Becerro, A.I.; Pavičić, I.; Selmani, A.; Babič, M.; Horák, D.; Vrček, I.V. Biocompatibility assessment of up-and down-converting nanoparticles: Implications of interferences with *in vitro* assays. *Methods Appl. Fluoresc.* **2018**, *7*, 014001. [CrossRef] [PubMed]

33. Han, X.; Gelein, R.; Corson, N.; Wade-Mercer, P.; Jiang, J.; Biswas, P.; Finkelstein, J.N.; Elder, A.; Oberdörster, G. Validation of an LDH assay for assessing nanoparticle toxicity. *Toxicology* **2011**, *287*, 99–104. [CrossRef] [PubMed]
34. Costa, C.; Brandão, F.; Bessa, M.J.; Costa, S.; Valdiglesias, V.; Kiliç, G.; Fernández-Bertólez, N.; Quaresma, P.; Pereira, E.; Pásaro, E.; et al. *In vitro* cytotoxicity of superparamagnetic iron oxide nanoparticles on neuronal and glial cells. Evaluation of nanoparticle interference with viability tests. *J. Appl. Toxicol.* **2015**, *36*, 361–372. [CrossRef] [PubMed]
35. Kroll, A.; Pillukat, M.H.; Hahn, D.; Schnekenburger, J. Current in vitro methods in nanoparticle risk assessment: Limitations and challenges. *Eur. J. Pharm. Biopharm.* **2009**, *72*, 370–377. [CrossRef] [PubMed]
36. Vrček, I.V.; Pavičić, I.; Crnković, T.; Jurašin, D.; Babič, M.; Horák, D.; Lovrić, M.; Ferhatović, L.; Ćurlin, M.; Gajović, S. Does surface coating of metallic nanoparticles modulate their interference with in vitro assays? *RSC Adv.* **2015**, *5*, 70787–70807. [CrossRef]
37. Almutary, A.; Sanderson, B.J.S. The MTT and Crystal Violet Assays: Potential Confounders in Nanoparticle Toxicity Testing. *Int. J. Toxicol.* **2016**, *35*, 454–462. [CrossRef]
38. Rosenholm, J.M.; Meinander, A.; Peuhu, E.; Niemi, R.; Eriksson, J.E.; Sahlgren, C.; Lindén, M. Targeting of porous hybrid silica nanoparticles to cancer cells. *ACS Nano* **2009**, *3*, 197–206. [CrossRef]
39. Suman, T.; Rajasree, S.R.; Kirubagaran, R. Evaluation of zinc oxide nanoparticles toxicity on marine algae chlorella vulgaris through flow cytometric, cytotoxicity and oxidative stress analysis. *Ecotoxicol. Environ. Saf.* **2015**, *113*, 23–30. [CrossRef]
40. Kumar, G.; Degheidy, H.; Casey, B.J.; Goering, P.L. Flow cytometry evaluation of in vitro cellular necrosis and apoptosis induced by silver nanoparticles. *Food Chem. Toxicol.* **2015**, *85*, 45–51. [CrossRef]
41. Theodossiou, T.A.; Ali, M.; Grigalavicius, M.; Grallert, B.; Dillard, P.; Schink, K.O.; Olsen, C.E.; Wälchli, S.; Inderberg, E.M.; Kubin, A.; et al. Simultaneous defeat of MCF7 and MDA-MB-231 resistances by a hypericin PDT–tamoxifen hybrid therapy. *NPJ Breast Cancer* **2019**, *5*, 13. [CrossRef]
42. Won, K.; Spruck, C. Triple-negative breast cancer therapy: Current and future perspectives (Review). *Int. J. Oncol.* **2020**, *57*, 1245–1261. [CrossRef]
43. Ketebo, A.A.; Shin, T.H.; Jun, M.; Lee, G.; Park, S. Effect of silica-coated magnetic nanoparticles on rigidity sensing of human embryonic kidney cells. *J. Nanobiotechnol.* **2020**, *18*, 170. [CrossRef]

International Journal of
Molecular Sciences

MDPI

Article

Insight into the Mechanism of Interactions between the LL-37 Peptide and Model Membranes of *Legionella gormanii* Bacteria

Katarzyna Pastuszak [1], Bozena Kowalczyk [2], Jacek Tarasiuk [2], Rafal Luchowski [3], Wieslaw I. Gruszecki [3], Małgorzata Jurak [1,*] and Marta Palusinska-Szysz [2]

1 Department of Interfacial Phenomena, Institute of Chemical Sciences, Faculty of Chemistry, Maria Curie-Skłodowska University, Maria Curie-Skłodowska Sq. 3, 20-031 Lublin, Poland; katarzyna.pastuszak@mail.umcs.pl

2 Department of Genetics and Microbiology, Institute of Biological Sciences, Faculty of Biology and Biotechnology, Maria Curie-Skłodowska University, Akademicka 19, 20-033 Lublin, Poland; b.kowalczyk746@wp.pl (B.K.); jacek.tarasiuk@mail.umcs.pl (J.T.); marta.palusinska-szysz@mail.umcs.pl (M.P.-S.)

3 Department of Biophysics, Institute of Physics, Faculty of Mathematics, Physics and Informatics, Maria Curie-Skłodowska University, Radziszewskiego 10, 20-031 Lublin, Poland; rafal.luchowski@mail.umcs.pl (R.L.); wieslaw.gruszecki@mail.umcs.pl (W.I.G.)

* Correspondence: malgorzata.jurak@mail.umcs.pl

Abstract: *Legionella gormanii* is a fastidious, Gram-negative bacterium known to be the etiological agent of atypical community-acquired pneumonia. The human cathelicidin LL-37 exhibits a dose-dependent bactericidal effect on *L. gormanii*. The LL-37 peptide at the concentration of 10 μM causes the bacteria to become viable but not cultured. The antibacterial activity of the peptide is attributed to its effective binding to the bacterial membrane, as demonstrated by the fluorescence lifetime imaging microscopy. In this study, to mimic the *L. gormanii* membranes and their response to the antimicrobial peptide, Langmuir monolayers were used with the addition of the LL-37 peptide to the subphase of the Langmuir trough to represent the extracellular fluid. The properties of the model membranes (Langmuir monolayers) formed by phospholipids (PL) isolated from the *L. gormanii* bacteria cultured on the non-supplemented (PL−choline) and choline-supplemented (PL+choline) medium were determined, along with the effect of the LL-37 peptide on the intermolecular interactions, packing, and ordering under the monolayer compression. Penetration tests at the constant surface pressure were carried out to investigate the mechanism of the LL-37 peptide action on the model membranes. The peptide binds to the anionic bacterial membranes preferentially, due to its positive charge. Upon binding, the LL-37 peptide can penetrate into the hydrophobic tails of phospholipids, destabilizing membrane integrity. The above process can entail membrane disruption and ultimately cell death. The ability to evoke such a great membrane destabilization is dependent on the share of electrostatic, hydrogen bonding and Lifshitz–van der Waals LL-37−PL interactions. Thus, the LL-37 peptide action depends on the changes in the lipid membrane composition caused by the utilization of exogenous choline by the *L. gormanii*.

Keywords: *Legionella gormanii*; LL-37; Langmuir monolayer technique

1. Introduction

Legionella spp. are ubiquitous Gram-negative bacteria that can cause a range of diseases, including self-limited influenza-like syndrome (Pontiac fever) and pneumonia. Pneumonia is the most common clinical manifestation of Legionnaires' disease. *L. pneumophila* is widely recognized as a primary causative agent of both community-acquired and nosocomial pneumonia. The clinically relevant non-pneumophila species include *L. longbeachae*, *L. micdadei*, *L. bozemanae*, *L. dumoffii*, and *L. gormanii*. *L. gormanii* primarily affects children and immunodeficient patients [1,2]. However, immunocompetent individuals can also be susceptible to infection by this pathogen [3]. The pathogenicity of *Legionella* bacteria, which involves numerous virulence factors, is attributed to their capacity to proliferate within human macrophages, evading their bactericidal mechanisms. Genetic analysis has proven that *L. gormanii* encodes approximately 130 putative virulence factors from 300–400 putative virulence genes present in *L. pneumophila* [4]. Among these virulence factors, *L. gormanii* harbors genes that encode homologs of diverse secretion systems, including the Dot/Icm (defective in organelle trafficking/intracellular multiplication) type IV secretion system [5]. This multiprotein system, involved in establishing an intracellular replication niche (LCV-*Legionella* containing vacuole) by translocating macromolecules across the host cell membranes, consists in the effective functioning of the bacterial cell envelope.

The cell envelope of Gram-negative bacteria consists of two membranes: the outer membrane and the cytoplasmic membrane. The cytoplasmic membrane is composed of a phospholipid bilayer, whereas the outer membrane is made up of an inner leaflet of phospholipids and an outer leaflet of lipopolysaccharide (LPS). The lipidomic analysis of *L. gormanii*'s outer (OM) and inner (IM) membranes using liquid chromatography and mass spectrometry revealed that the bacteria synthesized a wide range of lipids. The most abundant lipid fraction in both layers were phospholipids (phosphatidylcholine, PC; phosphatidylethanolamine, PE; phosphatidylglycerol, PG; cardiolipin, CL), followed by glycerolipids (triglyceride, TG; diglyceride, DG) and sphingolipids (ceramides, Cer; hexosylceramides, Hex1Cer) [6]. The distinctive feature of *L. gormanii* is its ability to synthesize PC and ceramides, the characteristic constituents of eukaryotic membranes. *L. gormanii* can produce PC via two independent pathways, i.e., the phospholipid N-methylation (PMT) pathway and the phosphatidyl-choline synthase (PCS) pathway. In the PMT pathway, PE is subjected to a series of methylation reactions catalyzed by the phospholipid *N*-methyltransferase encoded by the *pmtA* gene. However, *L. gormanii* lacks the *pmtA* gene homologous to other *Legionella*, *Rhodobacter*, and *Rhizobium* bacteria. The second pathway is promoted by the PcsA enzyme, which condenses exogenous choline and CDP-diacylglyceride (cytidine diphosphate) to produce PC in a one-step reaction. The *L. gormanii* PcsA protein is 254 amino acids long and highly hydrophobic, containing up to 8 transmembrane helices with the N- and C-termini inside the bacterial cell [7]. PcsA depends on the microorganism interactions with the eukaryotic hosts or access to the eukaryotic metabolites since choline is not a biosynthetic product of prokaryotes.

L. gormanii cultured on the choline-supplemented medium (PL+choline) demonstrates quantitative differences in the molecular profile of phospholipids compared to the bacteria cultured without the addition of exogenous choline (PL−choline) [6]. The lipids isolated from *L. gormanii* cultured on the medium without the addition of choline contained 26% PC, 50% PE, 21% CL, and 3% PG, while those grown on the medium with choline contained 47% PC, 38% PE, 12% CL, and 3% PG [7]. These differences determine the physicochemical properties of the membranes formed by lipids. The monolayers built of lipids isolated from *L. gormanii* cultured with the addition of exogenous choline were more ordered and densely packed at 20 °C and 37 °C compared to the monolayers formed from the bacteria cultured without the addition of choline, as indicated by the higher values of the compression modulus of these monolayers. The tighter packing of the molecules contributed to the increased ordering of the acyl chains. The higher degree of condensation of molecules in the monolayers formed from the *L. gormanii* lipids cultured on the medium supplemented with choline resulted from the favorable interactions mediated by the hydrogen bonds and Lifshitz–van der Waals forces between the polar groups of phospholipids and the long chains of C19–C21 fatty acids (FA), respectively. The Brewster angle microscope images revealed that the monolayers formed from the lipids isolated from *L. gormanii* cultured in the medium with exogenous choline were more homogeneous than those formed from the lipids of bacteria cultured without the addition of exogenous choline. The monolayers of lipids extracted from bacteria cultured on the medium without the addition of choline displayed increased flexibility attributed to the presence of shorter chains (C14–C18) of unsaturated fatty acids, as well as a larger content of cardiolipin (21%), leading to the formation of heterogeneities (domains) [8].

The utilization of exogenous choline by *L. gormanii* leads not only to alterations in the lipid profile but also affects proteins, enabling these bacteria to modulate the immune response to infection. When *L. gormanii* is cultured on the medium supplemented with choline, it induces the production of pro-inflammatory cytokines IL-6 and TNF-α to a lesser extent compared to the bacteria grown on the medium without exogenous choline [7].

To combat *Legionella* spp. in the lungs, cationic antimicrobial peptides like human β-defensin 3 (hBD-3) and the LL-37 peptide play a crucial role. Pulmonary cells produce hBD-3 during an *L. pneumophila* infection via the TLR-JNK-AP-1 (Toll-Like Receptor-c-Jun N-terminal Kinase-Activator Protein-1) pathway [9]. The LL-37 peptide, known as human cathelicidin, is expressed in various cell types, including epithelial cells, macrophages, neutrophils, and natural killer (NK) cells [10]. The LL-37 peptide exhibits antimicrobial activity by interacting with bacterial cell targets, including membrane lipids, thereby initiating and regulating the innate immune response [11]. Legionnaires' disease caused by *L. gormanii* is a severe infection, particularly among immunocompromised patients. Making a rapid diagnosis and effective antibiotic treatment are crucial for reducing mortality rates. The LL-37 peptide, with its broad-spectrum activity including bactericidal effects and immune modulation potential, is considered a potential therapeutic alternative to conventional antibiotics [12].

The aim of this study is to demonstrate that the LL-37 peptide, depending on the applied concentration, exhibits either bactericidal activity against *L. gormanii* cells or induces their transition into a viable but non-culturing state. The ability of *L. gormanii* cells to interact with the LL-37 peptide is demonstrated by measuring the excitation energy transfer of fluorophore located on the surface of the bacteria and the labeled LL-37 peptide. Membrane penetration tests are carried out using the Langmuir monolayer technique to investigate the bactericidal mechanism of the LL-37 peptide on *L. gormanii* membranes. The lipids used to form the membranes are isolated from the *L. gormanii* cells cultured on the choline-supplemented and non-supplemented media, aiming at the investigation of how membrane composition modulates the interactions with this peptide.

2. Results

2.1. The LL-37 Peptide Exerts an Antibacterial Effect on L. gormanii Cells

The antibacterial activity of the LL-37 peptide on *L. gormanii* cells was assessed using double staining with SYTO9 and propidium iodide, enabling differentiation between live and dead cells based on membrane integrity. This peptide exerted a dose-dependent killing effect on *L. gormanii* cultured with and without choline supplementation (Figure 1A,B). At the concentration of 10 μM, the LL-37 peptide caused an approximate 40% decrease in the viability of *L. gormanii* cultured on the medium without the addition of choline, and an approximate 50% decrease in the viability of the bacteria grown on the medium with exogenous choline. The LL-37 peptide at a concentration of 20 μM further decreased cell viability, with a less pronounced difference observed between the bacteria cultured with and without extracellular choline.

The colony-counting plate method showed that the LL-37 peptide at the concentration of 20 μM resulted in a 68% decrease in cell viability of *L. gormanii* cells cultured on the medium without the addition of choline and a 72% decrease in cell viability grown on the medium supplemented with exogenous choline (Figure 1C). The survival rate of bacteria after treatment with the LL-37 peptide at a concentration of 20 μM was the same as that obtained in the live/dead staining test. However, after applying a 10 μM concentration of the LL-37 peptide, the rate of mortality was 83% and 91% for the bacteria cultured without and with the addition of choline, respectively, and was higher than that obtained in the live/dead staining method. The disparity in the bacterial survival rate between the two methods is likely due to the fact that the LL-37 peptide at the concentration of 10 μM, besides the killing effect, renders the bacteria viable but incapable of forming colonies on the BCYE medium (VBNC, viable but nonculturable).

The bacteria grown on the medium without and with exogenous choline after incubation with 10 µM of the LL-37 peptide and plated on the medium with the addition of 1% monosodium glutamate and 0.5% sodium pyruvate formed 17% and 23% more colonies, respectively, compared to those grown on the medium without these additives. These findings suggest that the viable but non-culturable forms of *L. gormanii* induced by the LL-37 peptide have been revived to the viable and cultivable state.

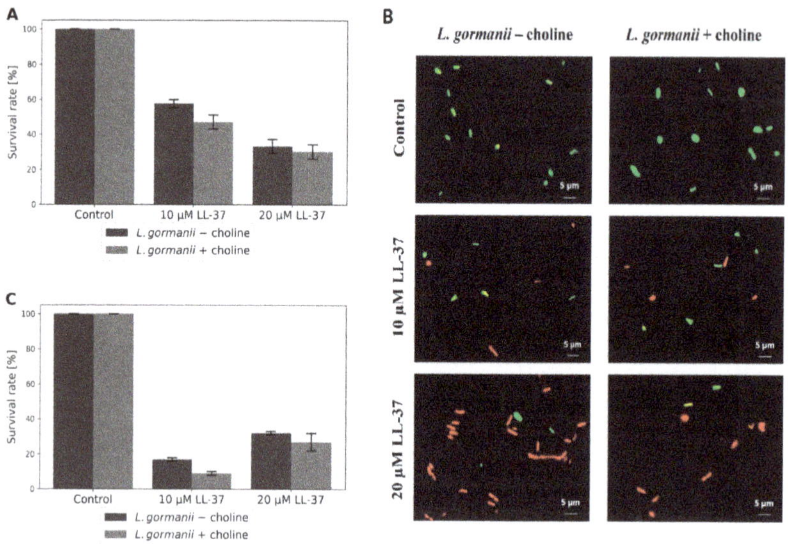

Figure 1. Effect of the LL-37 peptide on the survival rate (%) of *L. gormanii* cultured with and without choline supplementation using (**A**) live/dead staining and (**C**) colony-counting assay. (**B**) Live/dead staining *L. gormanii* cultured with and without choline supplementation after 1-h treatment with 10 µM and 20 µM LL-37 peptide, respectively, and imaged with laser scanning confocal microscopy (LSCM). The green signal is due to the dye SYTO9, indicating live cells, while the red signal is due to PI, which designates dead cells. Magnification 1000×.

2.2. FLIM-FRET Measurements

The phenomenon of nonradiative excitation energy transfer was applied to examine the interactions of the LL-37 peptide with bacterial membranes. Both the protein and membranes were labeled with fluorophores that satisfy the conditions for an efficient excitation energy transfer, providing preferable orientation and close distance between the energy donor and the acceptor molecules: FAM labeled the LL-37 peptide, and Syto 9 labeled the membranes. The fluorescence labels for emission spectrum of the donor and the absorption spectrum of the acceptor are shown in Figure 2.

Among the significant spectroscopic and dynamic effect diagnostics for an efficient energy transfer is shortening of the fluorescence lifetime of the donor in the presence of the acceptor. Such a situation was observed in the case of Syto 9 and FAM fluorescence labels situated on the bacterial membranes and the LL-37 protein, respectively (Figure 2B). The fluorescence lifetime situated in the membranes of bacteria, determined to be 2.25 ± 0.31 ns, was shortened to 0.95 ± 0.19 ns after the bacteria were exposed to the FAM-labeled LL-37 peptide. This effect, interpreted in terms of the strong interactions between the LL-37 peptide and the bacteria, can be visible directly from the fluorescence lifetime imaging microscopy (FLIM) images (Figure 3). Interestingly, the magnitude of such an effect depends on whether the bacteria were grown in the presence of exogenous choline (reported above) or without exogenous choline: the change from 2.78 ± 0.16 ns to 1.22 ± 0.09 ns (Figure 3). This difference can be attributed to the influence of bacterial-synthesized phospholipids and their physicochemical properties on the efficient binding of the LL-37 protein to the *L. gormanii* cells.

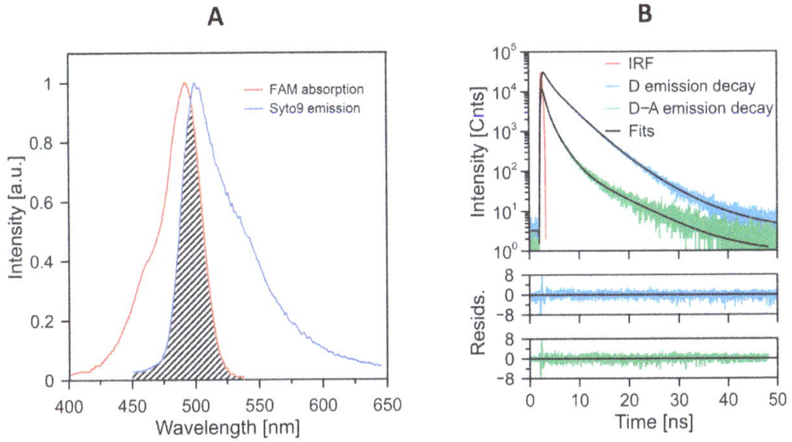

Figure 2. Spectroscopic and dynamic characterization of the pair of fluorophores FAM and Syto 9. (**A**) Fluorescence emission spectrum of excitation energy donor (Syto 9) and absorption spectrum of energy acceptor (FAM). The spectra were normalized at the maximum. The spectral overlap area is designated. (**B**) The original fluorescence decay traces and multiexponential fits of Syto 9 incorporated into the bacterial membrane (D) and the same system after the exposition to the LL-37 peptide labeled with FAM (D−A). The IRF trace is also shown. The residuals confirming the accuracy of the fits are shown in the lower panels.

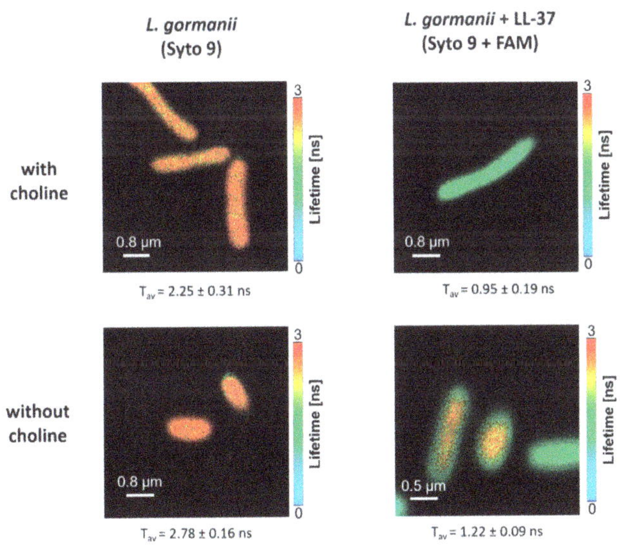

Figure 3. Original FLIM images of the bacteria labeled with Syto 9. The bacteria were either not exposed (**left hand panels**) or exposed (**right hand panels**) to the LL-37 peptide labeled with FAM and grown with (**upper panels**) or without (**lower panels**) exogenous choline. The average fluorescence lifetime values ± S.D. determined from six images for each kind of sample are also reported below the images.

2.3. Surface Pressure—Mean Molecular Area ($\pi - A$) Isotherms and Compression Modulus (C_S^{-1}) Changes

In order to determine the mechanism of the bactericidal action of the LL-37 peptide on the *L. gormanii* model membranes at the molecular level, the Langmuir monolayer technique was used. The Langmuir monolayers composed of phospholipids isolated from the bacteria cultured on the medium without and with exogenous choline were treated with this peptide. During the compression of the PL−choline and PL+choline monolayers, changes in the surface pressure as a function of the mean molecular area ($\pi - A$ isotherms) resulting from the formation of more condensed and ordered structures were registered. The $\pi - A$ isotherms obtained for the phospholipid monolayers in the absence or presence of the LL-37 peptide in the subphase at 20 °C or 37 °C are shown in Figure 4A,B. In contrast to PL, the $\pi - A$ compression isotherm for the acetic acid subphase with LL-37 was not acquired due to the high solubility of the peptide in the 0.01% acetic acid solution. When the PL molecules were present on the subphase, the injected peptide gathered at the interface and interacted with them, affecting the course of the PL compression isotherms. They made it possible to determine the behavior of the PL molecules in the model bacterial membranes affected by the antimicrobial agent.

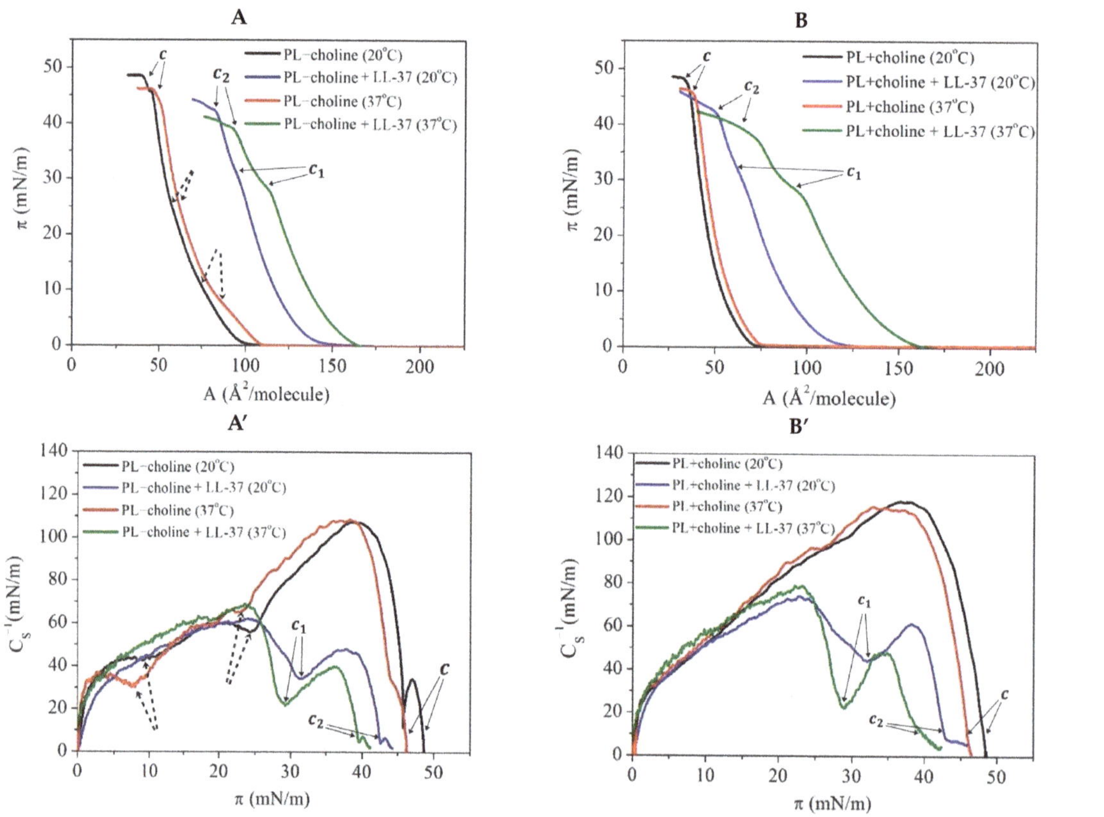

Figure 4. *Cont.*

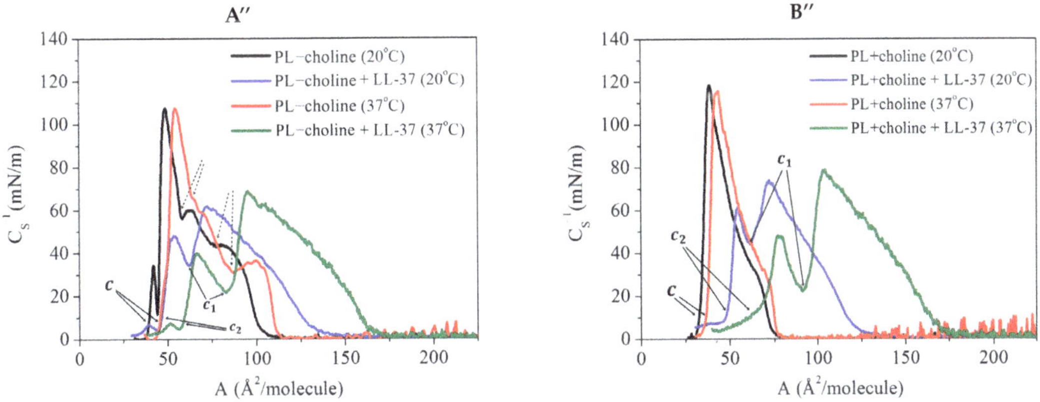

Figure 4. The (**A**,**B**) $\pi - A$ isotherms and (**A′**,**B′**) compression modulus (C_S^{-1}) as a function of surface pressure (π) or (**A″**,**B″**) mean molecular area (A) obtained for the monolayers composed of phospholipids isolated from the *L. gormanii* bacteria cultured on the (**A**,**A′**,**A″**) non-supplemented (PL−choline) or (**B**,**B′**,**B″**) choline-supplemented (PL+choline) medium, in the presence or absence of the LL-37 peptide at 20 °C or 37 °C. The dashed arrows (**A**,**A′**,**A″**) indicate phase transition locations.

Based on the $\pi - A$ compression isotherms, the lift-off area (A_0) and the collapse pressure (π_c), along with the corresponding area (A_c), were determined and are listed in Table 1. A_0 is ascribed to the gas–liquid expanded state transition of the monolayer occurring at π ~0.5 mN/m. The collapse pressure (π_c) is the pressure at which the rapid change in the slope of the isotherm takes place. A single collapse point (π_c) is typical of a monophasic or multiphasic mixed system in which all molecules remain at the interface below π_c. Meanwhile, two collapse points can be related to the expulsion of materials to the subphase at lower π (the first collapse pressure, π_{c1}) or to fracturing, followed by a loss of materials in the subphase or the formation of multilayered aggregates at the liquid–air interface at higher π (the second collapse pressure, π_{c2}).

Table 1. The lift-off area (A_0, expressed in Å^2/molecule), the collapse pressure (π_c, π_{c1}, π_{c2}, expressed in mN/m), and the collapse area (A_c, A_{c1}, A_{c2} expressed in Å^2/molecule) values obtained for the monolayers composed of phospholipids isolated from *L. gormanii* bacteria cultured on the non-supplemented (PL−choline) or choline-supplemented (PL+choline) medium, in the presence or absence of the LL-37 peptide at 20 °C or 37 °C.

	PL−choline (20 °C)	PL−choline + LL-37 (20 °C)	PL−choline (37 °C)	PL−choline + LL-37 (37 °C)	PL+choline (20 °C)	PL+choline + LL-37 (20 °C)	PL+choline (37 °C)	PL+choline + LL-37 (37 °C)
A_0	99.0	146.0	108.0	162.0	71.0	121.0	76.0	160.0
π_c	48.6	-	46.2	-	48.2	-	46.0	-
A_c	44.0	-	50.0	-	35.0	-	39.0	-
π_{c1}/π_{c2}	-	31.4/42.3	-	29.3/39.3	-	32.2/42.3	-	28.9/39.5
A_{c1}/A_{c2}	-	94.0/82.0	-	110.0/93.0	-	62.0/50.0	-	92.0/69.0

Moreover, on the basis of the $\pi - A$ isotherms, the compression modulus (C_S^{-1}) known as elasticity was determined for all analyzed monolayers according to the following Equation (1) [13]:

$$C_S^{-1} = -A\left(\frac{d\pi}{dA}\right)_{T,p} \quad (1)$$

where T means the temperature, and p is the atmospheric pressure.

The compression modulus provides information about the physical state of the monolayer and makes it possible to determine the degree of packing and ordering of molecules on the subphase surface. The $C_S^{-1} - \pi$ functions specified for the analyzed systems are presented in Figure 4A′,B′. According to the Davies and Rideal criterion, the C_S^{-1} parameter values in the range of 12.5–50 mN/m correspond to the liquid-expanded (LE) state, 100–250 mN/m indicates the liquid-condensed (LC) phase, and the in-between values (50–100 mN/m) point to the intermediate LE–LC state [13,14]. In general, the greater the C_S^{-1} value, the more the ordered monolayers with the molecules arranged in a more upright position to the surface plane, resulting in less elastic monolayer formation. To be able to specify the physical state of the *L. gormanii* model membranes, the maximal values of the compression modulus, along with the surface pressure at which they occur, were determined (Table 2). In addition, on the $C_S^{-1} - \pi$ plot, the minimum value more accurately identified the surface pressure at which the monolayer collapsed.

Table 2. Maximal ($C_{S,max1}^{-1}$, $C_{S,max2}^{-1}$, expressed in mN/m) and minimal ($C_{S,min}^{-1}$, expressed in mN/m) compression modulus values along with the corresponding surface pressure values (π_{max1}, π_{max2}, π_{min}, expressed in mN/m) for all obtained monolayers.

	PL−choline (20 °C)	PL−choline + LL-37 (20 °C)	PL−choline (37 °C)	PL−choline + LL-37 (37 °C)	PL+choline (20 °C)	PL+choline + LL-37 (20 °C)	PL+choline (37 °C)	PL+choline + LL-37 (37 °C)
$C_{S,max1}^{-1}$	107.0	62.0	109.0	69.0	119.0	74.0	116.0	79.0
π_{max1}	39.7	24.0	38.2	23.5	36.7	22.9	33.0	23.0
$C_{S,min}^{-1}$	-	34.0	-	22.0	-	44.0	-	22.0
π_{min}	-	31.4	-	29.3	-	32.2	-	28.9
$C_{S,max2}^{-1}$	-	48.0	-	40.0	-	62.0	-	48.0
π_{max2}	-	37.8	-	36.3	-	38.3	-	34.0

The changes in the phospholipid composition caused by the choline presence in the medium for bacteria growth and their influence on the behavior of molecules on the subphase were thoroughly discussed in our previous paper [8]. In summary, the smaller lift-off areas of the $\pi - A$ isotherms (Figure 4A,B, Table 1) for the PL+choline monolayers suggest that molecules are in closer proximity to each other in comparison to PL−choline. On the basis of the maximal C_S^{-1} values (Figure 4A′,B′, Table 2), it was possible to define that, without the peptide, both the PL−choline and PL+choline monolayers are in the liquid-condensed (LC) state, as the $C_{S,max}^{-1}$ at 20 °C and 37 °C exceeds 100 mN/m. For the monolayers without the peptide addition, only one collapse (at π_c) is observed on the $\pi - A$ isotherms (Figure 4A,B). However, in the case of PL−choline, subtle inflections are noted around 8 mN/m (at 37 °C) or 10 mN/m (at 20 °C) and 25 mN/m (both of) which correspond to two small minima in the course of the $C_S^{-1} = f(\pi)$ function (pointed out by the dashed arrows in Figure 4A,A′,A″). The first inflection can be described as the atypical first-order LE–LC phase transition (domains are visible on the BAM images, Figure S1), as the compression modulus does not decrease to 0 mN/m and the horizontal *plateau* on the $\pi - A$ isotherms is not noticeable (Figure 4A). The origin of the second inflection is difficult to interpret [8]. However, the further increase in the compression modulus to values greater than for the first local maximum can indicate that molecules are not expelled from the monolayer but that the second-order phase transition related to the changes in alkyl chain tilting takes place, and the monolayer becomes more condensed. The LC phase is revealed by the second local maximum at π~40 mN/m with $C_S^{-1} = 107$ mN/m.

As can be seen in Figure 4A,B and Table 1, the LL-37 peptide addition causes a shift of the $\pi - A$ isotherms to the right, towards the larger area per molecule at a given temperature, as well as a decrease in the collapse pressure value. Moreover, the temperature increase enhances the LL-37 effect on the phospholipid films. At a lower temperature, the LL-37 addition induces the A_0 increase by ~47 Å^2/molecule for the PL−choline and ~51 Å^2/molecule for PL+choline monolayers. At 37 °C the area changes are equal to

~55 Å^2/molecule and ~84 Å^2/molecule, respectively (Table 1). This indicates that the PL+choline monolayers are susceptible to the antimicrobial peptide action to a larger extent than the PL−choline model membranes at a higher temperature, while at 20 °C the effect is comparable. Furthermore, the LL-37 peptide addition to the subphase also results in a decrease in the compression modulus maximal values in all the analyzed systems, indicating that the antimicrobial agent causes the phospholipid molecules to form less condensed (more elastic) monolayers at the liquid–air interface (Figure 4A′,B′). The highest C_S^{-1} values (Table 2) indicate the intermediate LE–LC phase. The compression modulus dependencies obtained for both PL films with the addition of LL-37 (Figure 4A′,B′,A″,B″) are characterized by two maximal values separated by a minimum, which is in agreement with the inflection on the $\pi - A$ isotherms (Figure 4A,B) and corresponds to the first collapse (π_{c1}) of the monolayer. At the first maximum, the C_S^{-1} values are within 60–80 mN/m (depending on the monolayer type and temperature), while at the second maximum, the values are 14–29 mN/m smaller for PL-choline and 12–31 mN/m for PL+choline, respectively.

Thus LL-37, as the compound with amphipathic properties [15], accumulates at the liquid–air interface and disturbs the LC domain formation due to hindering the reorientation of PL molecules from a horizontal to a vertical position, which leads to elimination of the first-order LE–LC phase transition observed for PL−choline (no distinct domains on the BAM images, Figure S1). One inflection appears on the $\pi - A$ isotherms at the surface pressure of about 30 mN/m (Figure 4A,B) and correlates with the minimum point on the $C_S^{-1} - \pi$ (Figure 4A,B) and $C_S^{-1} - A$ curves (Figure 4A″,B″). The above value is close to the collapse pressure of the LL-37 monolayer described in the literature, which is in the range of 24–30 mN/m [16,17]. It is commonly known that when two components are immiscible, two collapse states occur which can be ascribed to the collapse surface pressures for the individual pure components [14,18,19]. When the components are partially miscible, some domains enriched mainly by either one or the other component can be formed. Once the surface pressure of collapse of the given component is reached, they collapse and are expelled from the monolayer [14,19]. In this study, the LL-37 or LL-37-enriched phase collapsing at a lower surface pressure should be removed from such PL/LL-37 mixed monolayer at the collapse pressure, which is indicated by Neville et al. as 24–30 mN/m [16,17]. The loss of molecules from the monolayer leads to the smaller degree of its packing revealed by smaller values of the second maximum on the $C_S^{-1} - \pi$ curve (Figure 4A′,B′, Table 2). This also correlates with a large reduction in the area per molecule after the first collapse indicating the material desorption from the interface. In contrast, for the monolayers without the peptide, the reduction in the mean molecular area during the compression is accompanied by an increase in the packing density of the monolayer as evidenced by the increase in the compression modulus values (Figure 4A″,B″).

Furthermore, in the presence of the peptide, the second collapse pressure (π_{c2}) values are smaller, which proves that the monolayers are less stable than the PL monolayers without the LL-37 in the subphase. The above statements and observations confirm that due to the complex interactions, the cathelicidin changes the PL molecules' behavior on the subphase surface and evokes the destabilization of both monolayers.

2.4. *Surface Potential Changes—Mean Molecular Area ($\Delta V - A$) Isotherms*

Along with the surface pressure, the surface potential changes (ΔV) were measured. Phospholipids, as amphiphilic compounds, contain two (or four in the cardiolipin molecule) hydrocarbon chains, orienting away from the polar subphase and hydrophilic head group, which interact strongly with the liquid phase [20]. When the phospholipids are far away from each other (at large values of the mean molecular area), the ΔV is fairly constant. Upon compression of the PL molecules at the liquid–air interface, their orientation changes, due to the increasingly stronger intermolecular interactions resulting from the molecular distance decrease. Formation of the more ordered and condensed structures evokes a change in the normal component of the dipole density, which is observed as the ΔV increase [20]. The maximal value corresponds to the most packed state of the monolayer in which molecules assume an orientation as vertical as possible [21]. Therefore, the measured values of the surface potential changes make it possible to analyze the orientation of molecules on the subphase surface [21]. The surface potential changes of the *L. gormanii* phospholipid monolayers with or without the peptide are presented as the $\Delta V - A$ isotherms in Figure 5.

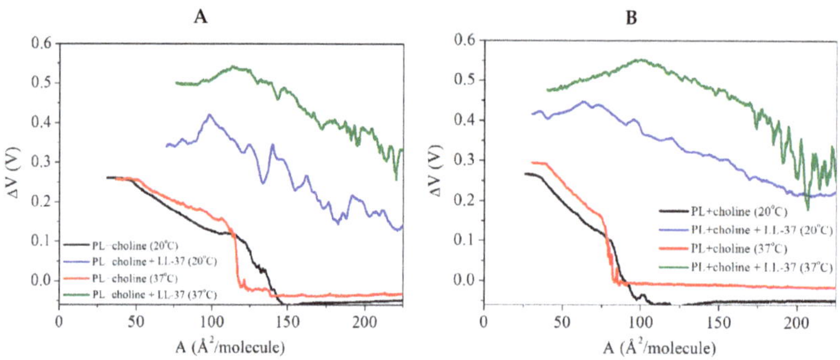

Figure 5. The surface potential changes (ΔV) as a function of mean molecular area (A) obtained for the monolayers composed of phospholipids isolated from *L. gormanii* bacteria cultured on the (**A**) non-supplemented (PL−choline) or (**B**) choline-supplemented (PL+choline) medium, in the presence or absence of the LL-37 peptide at 20 °C or 37 °C.

As follows from Figure 5, the maximal ΔV values are comparable and reach around 0.27–0.30 V for the pure PL−choline and PL+choline monolayers, both at 20 °C and 37 °C. The addition of the LL-37 peptide affects both films to a large extent, as the significant surface potential increase is observed for the PL+choline and PL−choline monolayers. The course of the $\Delta V - A$ dependencies is similar, and the maximal values are approximately the same at a given temperature. In the early stages of compression, the $\Delta V - A$ curve is characterized by many fluctuations that are more pronounced in comparison to the films without the peptide.

As the phospholipid molecules are applied to the surface with LL-37 already on it and the intermolecular interactions with PLs can already take place, the ΔV values are bigger from the early stages of compression. It is worth mentioning that the model membranes in the presence of the peptide at 20 °C and 37 °C obtain the greatest ΔV values near the monolayer's first collapse (Figure 4). Above that point, as the peptide-rich phase is pushed out of the monolayer, the molecules are less densely packed on the surface, which leads to a greater inclination of the molecules with respect to the subphase plane and, consequently, to a decrease in the surface potential (Figure 5).

2.5. Surface Morphology

The Brewster angle microscopy technique was used to observe the morphology of the model membrane surface. As aforementioned, the composition of the phospholipid mixtures and the specific interactions between the components can lead to formation of at least two-phase structures, referred to as domains [22]. They occur naturally in the bacterial membranes in order to minimize the system energy; however, the presence of antibiotics or other antimicrobial substances could also alter or induce the membrane components' sequestering [23]. The images of PL−choline and PL+choline surface morphology under the various conditions are presented in Figure 6. The images were taken at π~30 mN/m, as this value of surface pressure corresponds to the equilibrium lateral pressure of biomembranes [24]. The data obtained at the other values of surface pressure are included as the Supplementary Data (Figures S1 and S2).

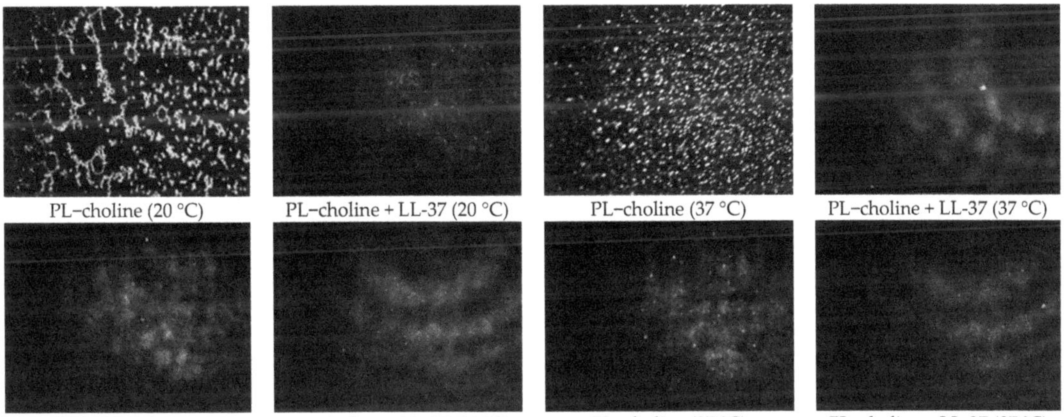

Figure 6. BAM images (720 μm × 400 μm) taken at 30 mN/m for the monolayers composed of phospholipids isolated from the *L. gormanii* bacteria cultured on the non-supplemented (PL−choline) or choline-supplemented (PL+choline) medium, in the presence or absence of the LL-37 peptide at 20 °C or 37 °C.

The morphology of the PL model membranes without the peptide addition was analyzed thoroughly in our previous paper [8]. The results showed greater homogeneity of PL+choline, as it comprises mostly zwitterionic phospholipids, in comparison to the PL−choline monolayers containing a larger amount of negatively charged molecules, promoting phase separation.

In all the analyzed monolayers, the LL-37 peptide addition results in more homogeneous monolayers. The greatest difference is visible for the PL−choline films, due to the high heterogeneity and large areas of condensed structures observed in the absence of the antimicrobial agent. The domains formed in the presence of LL-37 are small, circular, and occur in relatively minor quantities, even at great values of surface pressure.

Taking into account the LL-37's positive charge, it is justified to assume that the peptide interacts preferentially with the anionic PLs from the early stages of compression, thus modifying the surface morphology and limiting domain creation by decreasing the repulsive interactions among the PL molecules. The reduction in the phospholipid domain amount caused by the LL-37 peptide was also reported by Majewska et al. [25]. Moreover, at a higher temperature, the domains are less numerous compared to 20 °C, possibly as a result of the thermal motion of molecules. Disruption of the natural domain structure of the bacterial membranes can be responsible for the peptide antimicrobial effect.

2.6. Penetration Study

Monolayer penetration studies are widely used for investigating the interactions between model membranes and peptides, antibiotics, or other biomolecules. The experiments can be performed at constant area value or constant surface pressure value [26]. The peptide penetration of the phospholipid monolayers in the presented studies was investigated at the constant surface pressure of 30 mN/m, to better mimic the LL-37 influence on the bacterial membranes in human organism. This value is reported to be consistent with the molecular density occurring in the bilayers [27] and the equilibrium lateral pressure established for the biomembranes [24]. As the surface pressure is kept constant, the mean molecular area (A) changes describe the monolayer−peptide interactions. The increase in the area value indicates incorporation of the peptide into the PL film, resulting in an expansion of the monolayer. On the contrary, the A decrease follows from the desorption of molecules from the monolayer into the subphase [28]. This allows one to specify the effect of, e.g., antimicrobial agent on bacterial membranes.

It is important to point out that for obtaining the $\pi - A$ isotherms, the peptide was dropped on the subphase surface before the phospholipids were applied; thus, the LL-37 could interact with the unorganized PL molecules widespread throughout the surface. To the contrary, in the penetration studies, the peptide is injected under the compressed monolayer; therefore, the effect of LL-37 on the intermolecular interactions in the PL films can be significantly different. Nevertheless, both methods complement each other, allowing for the assessment of differences in the behavior of PL molecules in the monolayers due to the peptide presence. While all of the performed analyses provide a lot of information on the molecular behavior and lead to more detailed characteristics of the model membranes, the penetration studies seem to reflect better the bacterial cell response to the LL-37 peptide presence in the human organism. Therefore, the registered data provide adequate information considering the *L. gormanii* response to the peptide occurrence.

After compressing the monolayer to the surface pressure of 30 mN/m, the alterations of the mean molecular area were monitored for about 3 h. The obtained results were normalized (A/A_0) and expressed as a function of time in Figure 7. In the measurements with LL-37, the peptide was added after about 15 min of the monolayer equilibration at the given surface pressure. Thus, only the data registered after this time are considered in the further analysis.

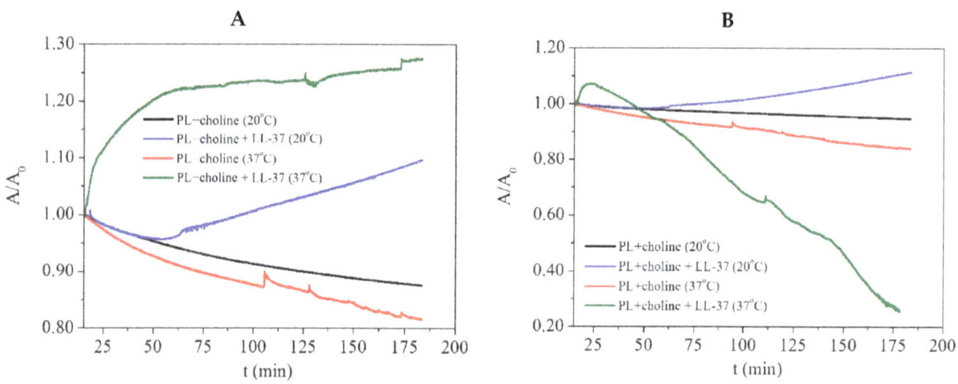

Figure 7. The normalized mean molecular area changes (A/A_0) in time (t) obtained for the monolayers composed of phospholipids isolated from the *L. gormanii* bacteria cultured on (**A**) the non-supplemented (PL−choline) or (**B**) the choline-supplemented (PL+choline) medium, in the presence or absence of the LL-37 peptide at 20 °C or 37 °C.

In general, the A/A_0 decrease can be noted in the pure PL monolayers, indicating instability of the obtained films at the given surface pressure due to the desorption of molecules. After the 3 h equilibration, the area reduction is 12% and 18% for PL−choline and 5% and 16% for PL+choline at 20 °C or 37 °C, respectively. On the other hand, the increase in the mean molecular area is caused by the penetration of the peptide into the monolayer. In order to determine the area changes induced by the addition of the peptide, the relative area change ($\Delta A/A$) was calculated as the difference between the mean molecular area recorded in the presence and absence of LL-37 at the given surface pressure value, in relation to that occupied by the molecule in the pure PL monolayers. These changes are expressed in percentages (Figure 8).

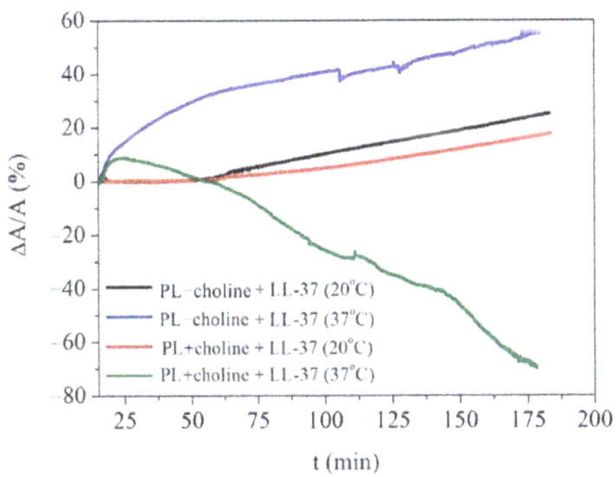

Figure 8. The relative area change ($\Delta A/A$) as a function of time (t) for the PL−choline and PL+choline monolayers at π = 30 mN/m caused by the addition of LL-37 peptide into the subphase at 20 °C or 37 °C.

For comparative purposes, the $\Delta A/A$ values obtained for the films after ~3 h from the peptide injection were taken into consideration. They are presented in Table 3. As already mentioned, the greater mean molecular area ($\Delta A/A > 0$) due to the barrier expansion at the constant surface pressure indicates the peptide's incorporation into the phospholipid monolayer [17], while the opposite phenomena ($\Delta A/A < 0$) can point out the molecules' expulsion into the subphase or molecular reorganization occurring on the subphase surface [16].

Table 3. The relative area change ($\Delta A/A$, expressed in %) for the PL−choline and PL+choline monolayers at π = 30 mN/m caused by the addition of LL-37 peptide into the subphase at 20 °C or 37 °C. The values given below correspond to the peptide effect at about 3 h after the injection.

	PL−choline (20 °C)	PL−choline (37 °C)	PL+choline (20 °C)	PL+choline (37 °C)
$\Delta A/A$	25	56	18	−70

At 20 °C, the effect of the LL-37 peptide on the PL−choline and PL+choline monolayers is similar, and no significant differences in the PL molecules' behavior are observed between the two. This is revealed in the comparable $\pi - A$ isotherm shifts (Figure 4A,B, Table 1) caused by the antimicrobial agent presence as well as the similar compression modulus decrease (Figure 4A′,B′,A″,B″, Table 2). In the penetration study, after the peptide addition, no effect is noted up until 50 min—the area decrease is identical to that of the pure phospholipid films (Figure 7). Subsequently, an increase is observed, and after 3 h from the injection, the molecular area still rises, and no *plateau* occurs. Therefore, it can be assumed that the LL-37 peptide incorporates gradually into the film and alters the PL molecules' organization on the subphase surface. No abrupt changes in the mean molecular area occur, and the effect of LL-37 (Figure 8) is only slightly greater for PL−choline ($\Delta A/A$ = 25%) than for PL+choline ($\Delta A/A$ = 18%).

It is of significant importance to note that measurements conducted at 37 °C better reflect the processes taking place in the human organism, as it is the physiological temperature of the body. For PL+choline, at first a 9% area increase is observed (Figure 8), which makes it possible to conclude that the peptide incorporates into the PL monolayer immediately after the injection. A similar conclusion was drawn by Sevcsik et al. [17]. Then, a significant area decrease is observed as the $\Delta A/A$ reaches −70%, suggesting that the so-called critical destabilization [16] occurs just 10 min after the LL-37 addition. On the contrary, the PL−choline analysis at this temperature showed that the area increases immediately after antimicrobial agent insertion, keeps rising, and then reaches a *plateau*.

Some fluctuations can be still noted; however, it can be due to the high kinetic energy of the molecules as well as the subphase evaporation. The $\Delta A/A$ is equal to 56%; thus, the peptide is built into the phospholipid film. However, it does not result in great destabilization, as was observed in the case of PL+choline. The penetration results indicate that the PL+choline films are more susceptible to the LL-37 action in comparison to PL−choline at a higher temperature, which is consistent with the $\pi - A$ isotherm courses (Figure 4A,B) and the compression modulus decrease (Figure 4A′,B′,A″,B″).

3. Discussion

The successful colonization of lung tissue by the *L. gormanii* bacteria can be attributed to their ability to produce enzymes that degrade a lung surfactant, an essential component in maintaining proper lung function and protecting against incoming pathogens. One of the major components of the lung surfactant is PC, which can serve as a substrate for the phospholipase PlaB produced by *L. gormanii* [29]. Utilizing extracellular choline to synthesize PC, *L. gormanii* modifies the quantitative composition of membrane lipids, resulting in the alteration of the physicochemical properties of the membrane and its interactions with antibacterial peptides. Our previous studies demonstrated that LL-37 possesses a bactericidal action against *L. micdadei*. The bacteria cultured on the medium supplemented with choline displayed increased susceptibility to this peptide compared to the bacteria cultured without the addition of exogenous choline. The peptide perturbed bacterial-derived phospholipid monolayers, and the interactions were dependent on the molar portion of PC [30]. LL-37 binds to the membrane of *L. gormanii* as demonstrated by fluorescence lifetime microscopy, and the interaction was stronger with the bacteria cultured on the medium supplemented with exogenous choline. Compared to *L. micdadei*, LL-37 exhibits a more potent bactericidal effect on *L. gormanii* as a result of its higher efficacy at a lower concentration. However, the concentration of 10 μM LL-37 resulted in an approximately 50% killing rate of *L. gormanii* in the live/dead assay, whereas in the plate assay, it showed a higher killing rate of 83% and 91% for the bacteria cultured on the medium without and with exogenous choline, respectively, surpassing the effect observed with the 20 μM concentration of LL-37. The probable cause of these differences is the transition of the *L. gormanii* bacteria into a viable but non-culturable state (VBNC) under the influence of 10 μM LL-37. The confirmation of this assumption is supported by the fact that after plating the LL-37-treated bacteria at the concentration of 10 μM onto the BCYE medium with sodium pyruvate, the number of grown colonies was larger compared to that of the colonies cultured on the standard medium. Despite sodium pyruvate (SP) being a well-known intermediate key metabolite in glycolysis and proven to facilitate the resuscitation of VBNC cells [31], the exact mechanism underlying the transition of *L. gormanii* bacteria into a cultivable state on the artificial medium remains to be elucidated.

To investigate the mechanism of the bactericidal action of the LL-37 peptide on *L. gormanii*, Langmuir monolayers composed of phospholipids isolated from the bacteria cultured on the medium without and with exogenous choline were exposed to this peptide. The analysis of changes in the membrane's physiochemical properties made it possible to propose the mechanism of antimicrobial agent action. Our previous study showed a substantial difference in the phospholipid composition of the *L. gormanii* membranes resulting from the presence of exogenous choline in the bacterial medium [7]. These various proportions of phospholipids influence not only the physicochemical properties of the model membranes composed of the four PL classes [8], but also the interactions with the human cathelicidin LL-37.

Considering all the results obtained for the PL−choline and PL+choline monolayers at 20 °C and 37 °C with or without the peptide addition, several conclusions can be drawn. The pure PL+choline films are more tightly packed and ordered than PL−choline, which is highlighted by the values of the compression modulus (Figure 4A′,B′,A″,B″, Table 2). The PL−choline films are characterized by a larger content of anionic PLs (24%), thus stronger repulsive forces occur between the monolayer components resulting in the more expanded monolayer [8]. At a higher temperature, due to the great kinetic energy of the molecules and thermal motions, these interactions become even stronger. On the contrary, PL+choline, comprising only 15% of negatively charged and 85% of zwitterionic lipids, are characterized by more attractive forces (reduced by a higher temperature), therefore the formed monolayers are more rigid. These differences are also reflected in the surface morphology (Figures S1 and S2). In all analyzed monolayers, the peptide presence leads to a decrease in the compression modulus (Figure 4A′,B′,A″,B″) as well as to a shift of the $\pi - A$ isotherms (Figure 4A,B) towards the greater values of the mean molecular area. These dependencies are more pronounced for the PL+choline films, indicating their greater susceptibility to the LL-37 action.

To develop a mechanism for the LL-37 peptide action on the model membranes, numerous factors need to be taken into account. Many authors describe the great affinity of cationic LL-37 for the anionic phospholipids in the monolayers and bilayers, which is believed to be crucial for the peptide activity towards the bacterial membranes [16,17,26,32]. In principle, a larger content of anionic lipids in the PL−choline model membranes would suggest that these films will be more susceptible to the LL-37 peptide. Meanwhile, our studies proved the opposite effect at 37 °C, where the PL+choline monolayers exhibit the largest destabilization in the presence of the peptide. Therefore, other possibilities should be considered (Figure 9).

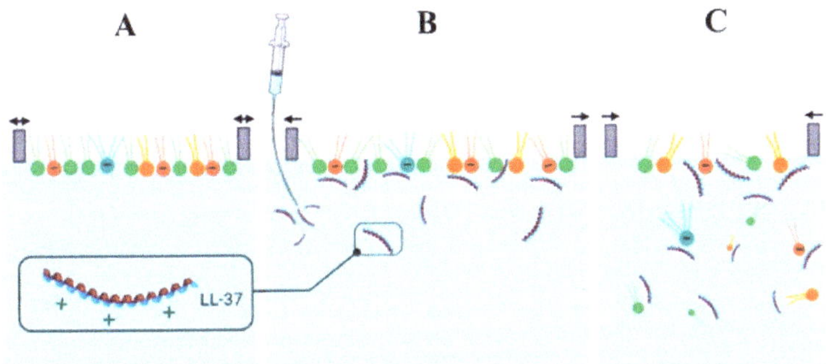

Figure 9. The possible mechanisms of LL-37 action on the *L. gormanii* model membranes: (**A**) equilibration of the PL monolayer—balance (↔); (**B**) changes in orientation of the PL molecules and incorporation of the peptide into the monolayer—expansion (←→); (**C**) membrane disruption and expulsion of the PL molecules into the subphase—compression (→←). The hydrophobic peptide residues are schematically designated in red.

The cationic peptide molecules are drawn towards the anionic phospholipids and interact strongly with the hydrophilic head groups. Furthermore, the charge of the monolayer defines the LL-37 organization in the subphase, as it can remain in the oligomeric or monomeric form [24,33,34]. These phenomena lead to the alterations of the PL molecules' orientation and overall monolayer ordering. At 20 °C, a slightly larger PL−choline area increases compared to the PL+choline (Figure 8), which most likely results from the higher percentage of negatively charged lipids present in the PL−choline film, as the affinity of the positively charged peptide for PG and CL promotes incorporation. The LL-37 penetration into the PL−choline monolayer is more pronounced at a higher temperature, in comparison to 20 °C, as the molecules have greater kinetic energy and are more mobile. Therefore, the monolayers are slightly disordered due to the constant movement of the molecules, and the LL-37 peptide can incorporate into the monolayer more readily and rapidly [35], which is observed as a sudden mean molecular area increase after the peptide injection (Figure 7).

As already mentioned, in the presented studies, the greatest monolayer destabilization is observed for PL+choline analyzed at 37 °C and, contrary to the less condensed PL−choline monolayer exposed to the LL-37 peptide, the drastic area decrease takes place

(Figures 7 and 8). Interestingly, PL+choline contains large amounts of PC [7], which in many studies did not show significant interactions with LL-37 [16,32]. These results indicate that not only electrostatic but Lifshitz–van der Waals interactions can define the peptide mechanism of action as well. Moreover, the presence of hydrogen bonds cannot be ruled out either [36]. Naturally, both types of interactions (polar and non-polar) can occur in the analyzed systems; nevertheless, their share and strength determine the monolayer's behavior under the LL-37 influence. The favorable electrostatic interactions drive the initial peptide anchoring [36]. Large affinity and strong attraction for the head groups keep the LL-37 peptide in the head group region, preventing the peptide from penetrating more deeply towards the hydrophobic chains. Hydrogen binding with the PL molecules allows aggregation on the membrane surface [36]. As a consequence, no critical destabilization occurs, and the peptide's incorporation into the PL−choline monolayer is observed (Figure 9B). If the cationic compound, in this case LL-37, is hydrophobic enough, it can penetrate more deeply into the monolayer and interact with the fatty acid chains [37,38], rather than just with the polar head groups. The stronger the electrostatic LL-37−PL head group attraction, the smaller depth the peptide can penetrate [35]. As the PL+choline monolayer does not contain as many anionic compounds, LL-37 passes through the zone of the polar groups. Then the peptide can interact with the fatty acid (FA) chains, causing severe alterations in the molecular orientation and resulting in critical destabilization as well as expulsion of the molecules into the subphase (Figure 9C).

The above mechanism indicates that, in fact, the LL-37 peptide is able to interact and penetrate the monolayers with a large content of zwitterionic PLs to a great extent. This is consistent with the studies by other researchers showing that the antimicrobial agent interactions with the FA chains are the factor responsible for the monolayers' disruption [35]. Therefore, the mechanism of the antimicrobial peptide action is dependent on the composition of the model membrane. Thus, it can be stated that the PC-enriched PL+choline monolayers are more susceptible to the LL-37 peptide, compared to the more anionic PL−choline, due to the larger share of Lifshitz–van der Waals interactions decisive for membrane disruption.

The effect of LL-37 on the *L. gormanii* model membranes is much stronger than on those of *L. micdadei* [30] and *L. dumoffii* [39] at the same peptide concentration in the bulk phase (0.08 μg/mL). This is revealed in the monolayer expansion confirmed by the shifts of the $\pi - A$ isotherms towards larger mean molecular areas. For instance, the changes in the lift-off areas determined for the PL−choline and PL+choline monolayers in the presence of LL-37 are 21% and 23% for *L. micdadei* [30], 18% and 28% for *L. dumoffii* [39], and 48% and 72% for *L. gormanii*. Thus, the observed changes in the area for the *L. gormanii* model membranes are two to three times greater than in the other membranes, indicating the strongest disruptive effect. This is also manifested in a significant decrease in the degree of packing and ordering of the membranes as indicated by the maximal compression modulus values. The percentage decreases are as follows: 0% and 21% for *L. micdadei* [30], 27% and 32% for *L. dumoffii* [39], and 42% and 38% for *L. gormanii*, in the absence or presence of exogenous choline, respectively. Additionally, the effect of the peptide on the PL+choline membranes was found to be greater than on PL−choline. This is due to the compositional differentiation in these three *Legionella* species caused by the choline addition to the growth medium [30,39]. It can be concluded that the negative charge of PL molecules does not play an exclusive role in the susceptibility of *Legionella* bacteria membranes to the positively charged peptide, but such a role can also be assigned to the Lifshitz–van der Waals interactions between the LL-37 and FA chains, which leads to a strong fluidization of the membranes. These studies demonstrated that the LL-37 peptide exhibited a bactericidal effect on *L. gormanii* cells by binding to their membrane and subsequently penetrating it. Moreover, the interaction between the LL-37 peptide and the *L. gormanii* membrane was found to be more pronounced at 37 °C compared to 20 °C.

The exploration of novel bactericidal compounds holds significant importance, especially considering the prevalence of antibiotic resistance genes like aminoglycoside phosphotransferase (APH(9)-Ia), Ambler class B metallo-β-lactamase (FEZ-1), and tetracycline-resistant ribosomal protection protein (tet56) in the genome of *L. gormanii* [4]. The phenotypic plasticity of bacterial membranes plays a crucial role in determining susceptibility to antimicrobial peptides, which are an integral part of host defenses. This feature holds significant implications for the design and evaluation of innovative therapeutic strategies aimed at treating and preventing bacterial infections. Antimicrobial peptides, initially derived from natural sources, serve as the foundation for developing synthetic analogs with enhanced efficacy, prolonged duration, and minimal adverse effects to address the potential issues encountered with drug candidates. Designing the sequence of these molecules and utilizing advanced bioinformatics methods and computational simulations, based on natural peptides like the LL-37 peptide, will enable effective combat against human pathogens such as *Legionella*.

4. Materials and Methods

4.1. Live/Dead Bacteria Staining Assay

L. gormanii (ATCC 33297) was cultured on the buffered charcoal yeast extract (BCYE) agar (Oxoid, Basingstoke, UK) and on this medium supplemented with 100 µg/mL choline chloride (Sigma-Aldrich, St. Louis, MO, USA) at 37 °C and 5% CO_2 in the humid atmosphere for 3 days. Next, 50 µL of the LL-37 peptide solution [LL-37 (human) trifluoroacetate salt, Merck, Darmstadt, Germany; purity $\geq$ 95%] at the final concentrations of 10 µM and 20 µM was added to 50 µL of the bacterial suspension in water with OD_{600} = 0.1, and incubated for 1 h at 37 °C. The suspension was transferred into a sterile Eppendorf tube and stained using the Live/Dead BacLight bacterial viability assay kit (ThermoFisher, Waltham, MA, USA). The bacterial suspension was stained in shielded tubes to minimize light exposure using 5 µM Syto9 and 30 µM propidium iodide in 0.3% DMSO for 15 min at room temperature. The live/dead stained images of the LL-37 peptide-treated and untreated bacteria (control) were collected using the Axiovert 200 M confocal microscope with the LSM 5 PASCAL scanning head (LSCM) (Carl Zeiss, Jena, Germany). The cells were counted from 20 microscopic images of each test sample. The images were captured with the AxioVision 4.8 software (Carl Zeiss) in the multichannel fluorescence technique, with the AxioCam HR3 camera (Carl Zeiss, Jena, Germany), using 470 nm and 546 nm filters for the green and red channels, respectively, with the same exposure time for each pair of images. The quantification of live/dead was performed with ImageJ 1.50i (Wayne Rasband, National Institutes of Health, Kensington, MD, USA). The sum of the lighting values was analyzed separately for each channel in the pair of images, which corresponds to the percentage of live and dead bacteria.

4.2. Colony Forming Unit (CFU)—Counting Assay

L. gormanii was cultured on BCYE agar (Oxoid, Basingstoke, UK) with and without 100 µg/mL choline chloride (Sigma-Aldrich, St. Louis, MO, USA) at 37 °C and 5% CO_2 in the humid atmosphere for 3 days. The bacteria were suspended in sterile MilliQ water (OD_{600} = 0.1), and the subsequent 10-fold dilutions were performed to 10^{-3}. Then, 5 µL of the last dilution of the bacterial suspension was transferred into a sterile Eppendorf tube, mixed with 5 µL of sterile MilliQ water (control) or the LL-37 peptide solution in water (10 µM and 20 µM). After 1 h of incubation at 37 °C, the bacterial suspensions were plated on the BCYE medium and incubated for 3 days at the same temperature. Subsequently, the number of CFUs was enumerated from the BCYE plates. The studies were carried out in three independent experiments, with each experiment including three replicates.

4.3. Fluorescence Lifetime Imaging Microscopy

The FRET (Förster resonance energy transfer) spectroscopy technique was employed to investigate the interactions between *L. gormanii* and the LL-37 peptide. The study utilized the MicroTime 200 microscopic confocal fluorescence system provided by PicoQuant GmbH (Berlin, Germany). In this experiment, the LL-37 peptide was labeled with FAM (carboxyfluorescein, ANASPEC, Seraing, Belgium, purity ≥ 95%), while the bacteria were labeled with Syto9 dye (Thermo Fisher Scientific, Waltham, MA, USA). A total of 50 µL of the 5-FAM-LL-37 peptide solution at the final concentration of 20 µM was added to 50 µL of the bacterial suspension in water with OD_{600} = 0.1, while 20 µL of the sample was applied on the microscope coverslips coated with polylysine, which aided in immobilizing the bacteria. To initiate the excitation, energy transfer process, and fluorescence, a pulsed laser beam operating at the wavelength of 440 nm was directed onto the sample through a dichroic mirror (ZT442rdc-UF3 Chroma) and a high numerical aperture objective (Olympus 100×, NA 1.3). The emitted fluorescence was collected by the same objective in the confocal mode using a 50 µm diameter pinhole. Only the donor fluorescence (Syto9) was collected, distinguished from the total emission signal through the combination of a dichroic beam splitter (FF495-Di03-25 × 36 Semrock) and a 465 long-pass filter (AT465lp Chroma) placed in front of the avalanche photodetector (Excelitas Technologies, Waltham, MA, USA). The required mirrors for the set-up were provided by Analysentechnik (Mainz, Germany).

The excitation laser was operated in the pulse mode with a repetition rate of 10 MHz, and the system was set to the single-channel lifetime resolution of 16 ps. Photon detection was accomplished using the time-correlated single photon counting mode facilitated by a TimeHarp 400 board. The acquired data were subsequently analyzed using the SymPhoTime software package (version 2.3) developed by PicoQuant in Germany. The intensity decays ($I(t)$) for each sample were deconvoluted using the multi-exponential model expressed as

$$I(t) = \sum_i \alpha_i exp\left(-\frac{t}{\tau_i}\right) \quad (2)$$

where τ_i represents the decay times, and α_i denotes the pre-exponential factors. Then, the average lifetime ($\langle\tau\rangle$) was calculated using the amplitude-weighted formula:

$$\langle\tau\rangle = \frac{\sum_i \alpha_i \tau_i}{\sum_i \alpha_i} \quad (3)$$

The Förster distance (R_0 = 57 Å) was calculated based on the spectral properties of the Syto9 and FAM dyes. The steady-state fluorescence measurements were performed utilizing a Shamrock 163 spectrograph, which was coupled to a MicroTime 200 microscope set-up, enabling direct spectral analysis of individual bacterial specimens. The detection system employed a Newton EMCCD DU970P BUF camera (manufactured by Andor Technology, Belfast, UK) that was cooled to the temperature of −50 °C. A Cary 60 UV-VIS spectrometer, developed by Agilent Technologies, was employed for the absorption spectra acquisition.

4.4. *Extraction of Lipids*

Lipids were extracted from the freeze-dried bacterial mass of *L. gormanii* grown on the medium with and without the addition of exogenous choline using the Bligh and Dyer (1959) method [40]. Briefly, the bacterial mass (450 mg) was suspended in 34 mL chloroform/methanol (1/2; *v*/*v*) and vigorously mixed using a magnetic stirrer for 4 h at room temperature. Next, the suspension was centrifuged for 30 min, 6000× *g*, 4 °C. The organic phase was collected, and a new portion of chloroform/methanol (1/2; *v*/*v*) was added to the bacterial pellet. Lipid extraction was continued for 3 h. After the re-centrifugation with the parameters as laid out above, organic phases were pooled, and chloroform and water were added to the final methanol/chloroform/water ratio of 2/2/1.8 (*v*/*v*/*v*). After thorough mixing of the reagents and centrifugation for 30 min, 6000× *g*, 4 °C, the lower organic phase was collected and dried using a rotary evaporator (Rotavapor R-100). Next, the samples were suspended in a mixture of hexane/isopropanol (3/2; *v*/*v*), mixed thoroughly, and centrifuged for 20 min, 6000× *g*, 4 °C. The lipid-containing organic phase was dried under a nitrogen stream and stored at −80 °C for further analysis. *L. gormanii* cultured without exogenous choline yielded 27 mg lipids (6% dried biomass). The bacteria cultured with the exogenous choline supplementation produced 31.5 mg lipids (7% dried biomass).

4.5. Separation of Phospholipids into Classes by Thin-Layer Chromatography

Before the chromatographic separation of phospholipids into classes, one-dimensional thin-layer chromatography (TLC) was performed to purify the extracted lipids from the pigments, mainly legioliulin, which exhibits the blue-white autofluorescence under long-wavelength UV [41]. Chromatographic separations were carried out on the silica gel 60 F254 plates of the size 10 cm × 10 cm (Merck, Darmstadt, Germany). The plates were washed with chloroform/methanol (1/1; v/v) prior to use in order to remove all contaminants and air-dried at room temperature. Lipids (about 2 mg) resolved in 20 μL of chloroform/methanol (1/2; v/v) were applied to the silica gel as a narrow band and developed with the solvent system chloroform/methanol/acetic acid (98/2/1; $v/v/v$). The lipids were visualized with iodine vapor, and pigments were detected under long-wavelength UV (Transiluminator UV-953). The lipid-containing band was scraped off, transferred to the screw-capped tubes, and extracted from the silica gel with a mixture of chloroform/methanol (1/1; v/v). PLs purified from the pigments were intended for the Langmuir monolayer study.

4.6. Langmuir Monolayer Study

The phospholipids isolated from the *L. gormanii* bacteria cultured on the non-supplemented (PL−choline) or choline-supplemented (PL+choline) medium were dissolved in chloroform (Avantor Performance Materials Poland S.A., Gliwice, Poland; purity > 99.9%) and methanol (ROMIL Chemicals Ltd., Cambridge, UK; purity > 99.9%) at the ratio of 4/1 (v/v). 1 mg/mL solutions were obtained. Langmuir–Blodgett (KSV 2000 Standard, KSV Instruments, Helsinki, Finland) and Langmuir (KSV Nima, Biolin Scientific, Stockholm, Sweden) troughs equipped with symmetric barriers and a Wilhelmy plate for the surface tension determination (with the 0.1 mN/m accuracy) were employed for the surface pressure–molecular area ($\pi - A$) isotherm measurements. The 0.01% acetic acid solution used as a subphase was prepared by diluting the concentrated one (Avantor Performance Materials Poland S.A., 99.7%) with water purified by the Milli-Q system (resistivity 18.2 MΩcm). An external water circulating system (Lauda, Schwechat, Austria) was used to keep a constant temperature (20 °C or 37 °C; ± 0.1 °C) during the measurements.

The prepared phospholipid solutions were applied to the subphase surface by means of a Hamilton microsyringe. To ensure that the solvents were fully evaporated, the trough was left for 10 min. Then, the monolayer was compressed with the barrier rate of 10 mm/min until the film collapsed. Each experiment was repeated 2–3 times, giving the mean error of ±2 Å^2/molecule. The $\pi - A$ isotherms for the PL monolayers with the addition of the LL-37 peptide (LL-37 (human cathelicidin) trifluoroacetate salt, Sigma-Aldrich, St. Louis, MO, USA; purity ≥ 95%) to the subphase were obtained by dropping 50 μL of the peptide solution into 0.01% acetic acid (1 mg/mL) onto the subphase surface, giving a final concentration of 0.08 μg/mL in the bulk phase, leaving it for two hours and then applying the phospholipid solution and carrying on as described above. Acetic acid was employed to keep the peptide's structure while in the solution. Simultaneously, the surface potential changes were recorded with 1 mV accuracy as a function of area (the $\Delta V - A$ isotherms) by means of a surface potential sensor (SPOT, Biolin Scientific, Stockholm, Sweden). Moreover, to directly visualize the monolayer's morphology, a Brewster angle microscope (nanofilm_ultrabam, Accurion, Göttingen, Germany) was used. A solid-state 50 mW laser, emitting p-polarized light (wavelength = 658 nm), was employed, and the incident angle was 53.2°.

To be able to determine the monolayer penetration by the peptide, the experiments were conducted as follows. The phospholipid solution was applied to the subphase, and after the solvent's evaporation, the obtained monolayer was compressed up to the surface pressure of 30 mN/m. Then, the compression was stopped, the monolayer was equilibrated at the constant surface pressure, and, after 15 min, the peptide solution was injected into the subphase, underneath the monolayer and the barrier level. After that, the changes in the mean molecular area were observed for about 3 h. As a control, the measurement without the peptide addition was performed analogously for the same period of time.

5. Conclusions

The human cathelicidin LL-37 exhibited a dose-dependent bactericidal effect on *L. gormanii* cells. At a concentration of 10 μM, the LL-37 peptide rendered the bacteria viable but non-culturable. The antibacterial activity of the LL-37 peptide was attributed to its strong binding to the bacterial membrane, as evidenced by the fluorescence lifetime imaging microscopy. The changes in the phospholipid composition of *L. gormanii* cells, caused by the exogenous choline presence, altered the model bacterial membrane response to the LL-37 peptide significantly. Depending on the zwitterionic and anionic phospholipid contents, two mechanisms of the LL-37 action were proposed. Both the polar (electrostatic and hydrogen bonding) and non-polar (Lifshitz–van der Waals) interactions were found to play a pivotal role in the molecular mechanism of the peptide operation at the interface. PL+choline was more susceptible to the LL-37 antimicrobial action than PL−choline, since, at 37 °C, the intermolecular LL-37−PL interactions led to extensive monolayer disruption. A larger content of negatively charged PLs in PL−choline provoked the peptide binding into the film's head group region, due to the strong polar interactions. To the contrary, large amounts of zwitterionic compounds (PL+choline) resulted in critical membrane destabilization, as a result of dominant Lifshitz–van der Waals forces between the peptide hydrophobic parts and the acyl chains of phospholipids.

Supplementary Materials: The following supporting information can be downloaded at https://www.mdpi.com/article/10.3390/ijms241512039/s1.

Author Contributions: Conceptualization, M.J. and M.P.-S.; investigation, K.P., B.K., J.T., R.L., M.J. and M.P.-S.; resources, B.K., J.T. and M.P.-S.; writing—original draft preparation, K.P., B.K., W.I.G., M.J. and M.P.-S.; writing—review and editing, R.L.,W.I.G., M.J. and M.P.-S.; visualization, K.P. and J.T.; supervision, R.L., W.I.G., M.J. and M.P.-S. All authors have read and agreed to the published version of the manuscript.

Funding: B. Kowalczyk—Individual project for a young scientist at the Institute of Biological Sciences of the Maria Curie-Skłodowska University, MN/2021/7.

Institutional Review Board Statement: Not applicable.

Informed Consent Statement: Not applicable.

Data Availability Statement: Not applicable.

Conflicts of Interest: The authors declare no conflict of interest. The funders had no role in the design of the study; in the collection, analyses, or interpretation of data; in the writing of the manuscript; or in the decision to publish the results.

References

1. Ephros, M.; Engelhard, D.; Maayan, S.; Bercovier, H.; Avital, A.; Yatsiv, I. *Legionella gormanii* pneumonia in a child with chronic granulomatous disease. *Pediatr. Infect. Dis. J.* **1989**, *8*, 726–727.
2. Han, X.Y.; Ihegword, A.; Evans, S.E.; Zhang, J.; Li, L.; Cao, H.; Tarrand, J.J.; El-Kweifi, O. Microbiological and clinical studies of legionellosis in 33 patients with cancer. *J. Clin. Microbiol.* **2015**, *53*, 2180–2187. [CrossRef]
3. Lei, C.; Zhou, X.; Ding, S.; Xu, Y.; Yang, B.; Guo, W.; Song, M.; Yang, M.; Jia, Y.; Luo, H. Case report: Community-acquired *Legionella gormanii* pneumonia in an immunocompetent patient detected by metagenomic next-generation sequencing. *Front. Med.* **2022**, *9*, 819425.
4. Svetlicic, E.; Jaén-Luchoro, D.; Klobucar, R.S.; Jers, C.; Kazazic, S.; Franjevic, D.; Klobucar, G.; Shelton, B.G.; Mijakovic, I. Genomic characterization and assessment of pathogenic potential of *Legionella* spp. Isolates from environmental monitoring. *Front. Microbiol.* **2023**, *13*, 1091964. [CrossRef] [PubMed]
5. Qin, T.; Zhou, H.; Ren, H.; Liu, W. Distribution of secretion systems in the genus *Legionella* and its correlation with pathogenicity. *Front. Microbiol.* **2017**, *8*, 388. [CrossRef] [PubMed]

6. Chmiel, E.; Galuska, C.E.; Koper, P.; Kowalczyk, B.; Urbanik-Sypniewska, T.; Palusińska-Szysz, M.; Fuchs, B. Unusual lipid components of *Legionella gormanii* membranes. *Metabolites* **2022**, *12*, 418. [CrossRef]
7. Palusińska-Szysz, M.; Szuster-Ciesielska, A.; Janczarek, M.; Wdowiak-Wróbel, S.; Schiller, J.; Reszczyńska, E.; Gruszecki, W.I.; Fuchs, B. Genetic diversity of *Legionella pcs* and *PmtA* genes and the effect of utilization of choline by *Legionella* spp. on induction of proinflammatory cytokines. *Pathog. Dis.* **2019**, *77*, ftz065. [CrossRef]
8. Pastuszak, K.; Chmiel, E.; Kowalczyk, B.; Tarasiuk, J.; Jurak, M.; Palusińska-Szysz, M. Physicochemical characteristics of model membranes composed of *Legionella gormanii* lipids. *Membranes* **2023**, *13*, 356. [CrossRef]
9. Scharf, S.; Vardarova, K.; Lang, F.; Schmeck, B.; Opitz, B.; Flieger, A.; Heuner, K.; Hippenstiel, S.; Suttorp, N.; N'Guessan, P.D. *Legionella pneumophila* induces human beta defensin-3 in pulmonary cells. *Respir. Res.* **2010**, *11*, 93. [CrossRef]
10. Yang, B.; Good, D.; Mosaiab, T.; Liu, W.; Ni, G.; Kaur, J.; Liu, X.; Jessop, C.; Yang, L.; Fadhil, R.; et al. Significance of LL-37 on immunomodulation and disease outcome. *BioMed Res. Int.* **2020**, *2020*, 8349712. [CrossRef]
11. Zeth, K.; Sancho-Vaello, E. Structural plasticity of LL-37 indicates elaborate functional adaptation mechanisms to bacterial target structures. *Int. J. Mol. Sci.* **2021**, *22*, 5200. [CrossRef] [PubMed]
12. Ridyard, K.E.; Overhage, J. The potential of human peptide LL-37 as an antimicrobial and anti-biofilm agent. *Antibiotics* **2021**, *10*, 650. [CrossRef]
13. Davies, J.T.; Rideal, E.K. *Interfacial Phenomena*; Academic Press: New York, NY, USA, 1963.
14. Jurak, M.; Szafran, K.; Cea, P.; Martín, S. Analysis of molecular interactions between components in phospholipid-immunosuppressant-antioxidant mixed Langmuir films. *Langmuir* **2021**, *37*, 5601–5616. [CrossRef]
15. Kościuczuk, E.M.; Lisowski, P.; Jarczak, J.; Strzałkowska, N.; Jóźwik, A.; Horbańczuk, J.; Krzyżewski, J.; Zwierzchowski, L.; Bagnicka, E. Cathelicidins: Family of antimicrobial peptides. A review. *Mol. Biol. Rep.* **2012**, *39*, 10957–10970. [CrossRef] [PubMed]
16. Neville, F.; Cahuzac, M.; Konovalov, O.; Ishitsuka, Y.; Lee, K.Y.C.; Kuzmenko, I.; Kale, G.M.; Gidalevitz, D. Lipid headgroup discrimination by antimicrobial peptide LL-37: Insight into mechanism of action. *Biophys. J.* **2006**, *90*, 1275–1287. [CrossRef] [PubMed]
17. Sevcsik, E.; Pabst, G.; Richter, W.; Danner, S.; Amenitsch, H.; Lohner, K. Interaction of LL-37 with model membrane systems of different complexity: Influence of the lipid matrix. *Biophys. J.* **2008**, *94*, 4688–4699. [CrossRef] [PubMed]
18. Dynarowicz-Łątka, P.; Hąc-Wydro, K. Interactions between phosphatidylcholines and cholesterol in monolayers at the air/water interface. *Colloids Surf. B* **2004**, *37*, 21–25. [CrossRef]
19. Jurak, M.; Mroczka, R.; Łopucki, R. Properties of artificial phospholipid membranes containing lauryl gallate or cholesterol. *J. Membr. Biol.* **2013**, *251*, 277–294. [CrossRef]
20. Vogel, V.; Möbius, D. Local surface potentials and electric dipole moments of lipid monolayers: Contributions of the water/lipid and the lipid/air interfaces. *J. Colloid Interface Sci.* **1988**, *126*, 408–420. [CrossRef]
21. Chachaj-Brekiesz, A.; Kobierski, J.; Wnętrzak, A.; Dynarowicz-Latka, P. Electrical properties of membrane phospholipids in Langmuir monolayers. *Membranes* **2021**, *11*, 53. [CrossRef]
22. Vanounou, S.; Parola, A.H.; Fishov, I. Phosphatidylethanolamine and phosphatidylglycerol are segregated into different domains in bacterial membrane. A study with pyrene-labelled phospholipids. *Mol. Microbiol.* **2003**, *49*, 1067–1079. [CrossRef] [PubMed]
23. Epand, R.M.; Epand, R.F. Domains in bacterial membranes and the action of antimicrobial agents. *Mol. BioSyst.* **2009**, *5*, 580–587. [CrossRef]
24. Sood, R.; Domanov, Y.; Pietiäinen, M.; Kontinen, V.P.; Kinnunen, P.K.J. Binding of LL-37 to model biomembranes: Insight into target vs host cell recognition. *Biochim. Biophys. Acta* **2008**, *1778*, 983–996. [CrossRef]
25. Majewska, M.; Zamlynny, V.; Pieta, I.S.; Nowakowski, R.; Pieta, P. Interaction of LL-37 human cathelicidin peptide with a model microbial-like lipid membrane. *Bioelectrochemistry* **2021**, *141*, 107842. [CrossRef]
26. Maget-Dana, R. The monolayer technique: A potent tool for studying the interfacial properties of antimicrobial and membrane-lytic peptides and their interactions with lipid membranes. *Biochim. Biophys. Acta* **1999**, *1462*, 109–140. [CrossRef] [PubMed]
27. Crosio, M.A.; Via, M.A.; Camara, C.I.; Mangiarotti, A.; Pópolo, M.G.D.; Wilke, N. Interaction of polyarginine peptide with membranes of different mechanical properties. *Biomolecules* **2019**, *9*, 625. [CrossRef] [PubMed]
28. Rojewska, M.; Smułek, W.; Kaczorek, E.; Prochaska, K. Langmuir monolayer techniques for the investigation of model bacterial membranes and antibiotic biodegradation mechanisms. *Membranes* **2021**, *11*, 707. [CrossRef] [PubMed]
29. Bender, J.; Rydzewski, K.; Broich, M.; Schunder, E.; Heuner, K.; Flieger, A. Phospholipase PlaB of *Legionella pneumophila* represents a novel lipase family: Protein residues essential for lipolytic activity, substrate specificity, and hemolysis. *J. Biol. Chem.* **2009**, *284*, 27185–27194. [CrossRef] [PubMed]
30. Palusińska-Szysz, M.; Jurak, M.; Gisch, N.; Waldow, F.; Zehethofer, N.; Nehls, C.; Schwudke, D.; Koper, P.; Mazur, A. The human LL-37 peptide exerts antimicrobial activity against *Legionella micdadei* interacting with membrane phospholipids. *Biochim. Biophys. Acta Mol. Cell Biol. Lipids* **2022**, *1867*, 159138. [CrossRef]
31. Ducret, A.; Chabalier, M.; Dukan, S. Characterization and resuscitation of 'non-culturable' cells of *Legionella pneumophila*. *BMC Microbiol.* **2014**, *14*, 3. [CrossRef]
32. Barzyk, W.; Rogalska, E.; Więcław-Czapla, K. Penetration of milk-derived antimicrobial peptides into phospholipid monolayers as model biomembranes. *Biochem. Res. Int.* **2013**, *2013*, 914540. [CrossRef]

33. Oren, Z.; Lerman, J.C.; Gudmundsson, G.H.; Agerberth, B.; Shai, Y. Structure and organization of the human antimicrobial peptide LL-37 in phospholipid membranes: Relevance to the molecular basis for its non-cell-selective activity. *Biochem. J.* **1999**, *341*, 501–513. [CrossRef]
34. Lu, F.; Zhu, Y.; Zhang, G.; Liu, Z. Renovation as innovation: Repurposing human antibacterial peptide LL-37 for cancer therapy. *Front. Pharmacol.* **2022**, *13*, 944147. [CrossRef] [PubMed]
35. Henzler-Wildman, K.A.; Martinez, G.V.; Brown, M.F.; Ramamoorthy, A. Perturbation of the hydrophobic core of lipid bilayers by the human antimicrobial peptide LL-37. *Biochemistry* **2004**, *43*, 8459–8469. [CrossRef]
36. Martynowicz, M.W.; Rice, A.; Andreev, K.; Nobre, T.M.; Kuzmenko, I.; Wereszczynski, J.; Gidalevitz, D. Salmonella membrane structural remodeling increases resistance to antimicrobial peptide LL-37. *ACS Infect. Dis.* **2019**, *5*, 1214–1222. [CrossRef]
37. Sevcsik, E.; Pabst, G.; Jilek, A.; Lohner, K. How lipids influence the mode of action of membrane-active peptides. *Biochim. Biophys. Acta* **2007**, *1768*, 2586–2595. [CrossRef]
38. Takahashi, H.; Caputo, G.A.; Vemperala, S.; Kuroda, K. Synthetic random copolymers as a molecular platform to mimic host-defense antimicrobial peptides. *Bioconjug. Chem.* **2017**, *28*, 1340–1350. [CrossRef] [PubMed]
39. Pastuszak, K.; Kowalczyk, B.; Tarasiuk, J.; Jurak, M.; Palusińska-Szysz, M. Influence of the antimicrobial LL-37 peptide on *Legionella dumoffii* phospholipids adsorbed at the air–liquid interface. *Sustainability* **2023**, *15*, 6670. [CrossRef]
40. Bligh, E.G.; Dyer, J.W. A rapid method of total lipid extraction and purification. Can. *J. Biochem. Physiol.* **1959**, *37*, 911–917.
41. Amemura-Maekawa, J.; Hayakawa, Y.; Sugie, H.; Moribayashi, A.; Kura, F.; Chang, B.; Wada, A.; Watanabe, H. Legioliulin, a new isocoumarin compound responsible for blue-white autofluorescence in *Legionella (Fluoribacter) dumoffi* under long-wavelength UV light. *Biochem. Biophys. Res. Commun.* **2004**, *323*, 954–959. [CrossRef]

International Journal of
Molecular Sciences

MDPI

Article

QbD-Optimized, Phospholipid-Based Elastic Nanovesicles for the Effective Delivery of 6-Gingerol: A Promising Topical Option for Pain-Related Disorders

Mohammed Ghazwani [1,*], Mohammed H. Alqarni [2], Umme Hani [1] and Aftab Alam [2]

[1] Department of Pharmaceutics, College of Pharmacy, King Khalid University, P.O. Box 1882, Abha 61441, Saudi Arabia; uahmed@kku.edu.sa

[2] Department of Pharmacognosy, College of Pharmacy, Prince Sattam Bin Abdulaziz University, Al Kharj 11942, Saudi Arabia; m.alqarni@psau.edu.sa (M.H.A.); a.alam@psau.edu.sa (A.A.)

* Correspondence: myghazwani@kku.edu.sa

Abstract: In this study, elastic nanovesicles, constructed of phospholipids optimized by Quality by Design (QbD), release 6-gingerol (6-G), a natural chemical that may alleviate osteoporosis and musculoskeletal-related pain. A 6-gingerol-loaded transfersome (6-GTF) formulation was developed using a thin film and sonication approach. 6-GTFs were optimized using BBD. Vesicle size, PDI, zeta potential, TEM, *in vitro* drug release, and antioxidant activity were evaluated for the 6-GTF formulation. The optimized 6-GTF formulation had a 160.42 nm vesicle size, a 0.259 PDI, and a −32.12 mV zeta potential. TEM showed sphericity. The 6-GTF formulation's *in vitro* drug release was 69.21%, compared to 47.71% for the pure drug suspension. The Higuchi model best described 6-G release from transfersomes, while the Korsmeyer–Peppas model supported non-Fickian diffusion. 6-GTF had more antioxidant activity than the pure 6-G suspension. The optimized transfersome formulation was converted into a gel to improve skin retention and efficacy. The optimized gel had a spreadability of 13.46 ± 4.42 g·cm/s and an extrudability of 15.19 ± 2.01 g/cm^2. The suspension gel had a 1.5 μg/cm^2/h ex vivo skin penetration flux, while the 6-GTF gel had 2.71 μg/cm^2/h. Rhodamine B-loaded TF gel reached deeper skin layers (25 μm) compared to the control solution in the CLSM study. The gel formulation's pH, drug concentration, and texture were assessed. This study developed QbD-optimized 6-gingerol-loaded transfersomes. 6-GTF gel improved skin absorption, drug release, and antioxidant activity. These results show that the 6-GTF gel formulation has the ability to treat pain-related illnesses effectively. Hence, this study offers a possible topical treatment for conditions connected to pain.

Keywords: 6-gingerol; osteoporosis; musculoskeletal disorder; antioxidant; confocal laser scanning microscopy; transfersomes; Box–Behnken design

1. Introduction

The global prevalence of musculoskeletal problems is estimated at millions of people of all ages [1]. Musculoskeletal disorders are the second leading cause of disability worldwide, according to Global Burden of Disease Research 2016. Painful musculoskeletal disorders afflict 20–33% of the global population [2], depending on age and diagnosis. Musculoskeletal disorders include skeletal-joint, musculotendinous, and ligamentous disorders [1,3]. This includes osteoarthritis, rheumatoid arthritis, back and neck pain, soft tissue rheumatism (tendinitis, bursitis, and myofascial pain), and sports injuries. Chronic pain hinders work, sleep, and social activities, lowering the quality of life [4]. Musculoskeletal diseases, including pathological ones, cause pain and disability, which raise healthcare costs [5]. Bone remodeling helps maintain the skeleton's mechanical integrity. Poor diet, inactivity, and low estrogen and testosterone levels induce osteoporosis and reduced bone density [6,7]. This requires careful bone growth and resorption [8]. Natural substances are becoming more popular for lowering osteoporosis and other pain-related risks [9].

Zingiber officinale (ZO) rhizome-isolated polyphenols (gingerols) have oesteo-protective properties [10]. Much like other medicinal plants, ginger displays a wide range of therapeutic qualities owing to a variety of phytoconstituents [11,12]. Its extracts have been reported to exhibit an anti-inflammatory impact [13], a protective effect on the gastrointestinal mucosa, antioxidant effects [14], antipyretic, antibacterial, antiviral, and hypoglycemic actions [15]. Gingerols and gingerol-related compounds are two of the most active phenolic compounds in terms of their pharmacological effects [16]. The secondary metabolite 6-gingerol (6-G) is the most abundant phytochemical in ginger oleoresin [17]. In the treatment of osteoarthritis and other chronic inflammatory illnesses, 6-G has proven to be a beneficial bioactive [18]. Gingerol is a powerful antioxidant that reduces the formation of reactive oxygen species (ROS) and manages oxidative stress [19,20]. Gingerol-rich compounds reduce inflammation and pain [21]. Zerumbone, a gingerol-related polyphenol, prevents osteoclastogenesis caused by receptor activators of nuclear factor-κB (NF-κB) ligand (RANKL) [22]. This suggests ginger may help alleviate pain. Gingerols have low bioavailability due to limited water solubility and severe phase II hepatic metabolism, even though 6-gingerol is the primary bioactive component [23]. Several strategies have been used to increase the bioavailability of these compounds by various researchers [24,25]. For skin delivery of bioactive compounds, lipid-based nanocarriers such as liposomes [26], nanoemulsions [27], microemulsions [28], ethosomes [29], and solid lipid nanoparticles [30] have been frequently reported. The delivery of active chemical compounds through the barriers of the skin, namely the stratum corneum, is the most difficult aspect of developing drug delivery [31]. Transfersomes (TF), or deformable liposomes, are among the nanocarriers that are regarded as ideal for administering drugs via the skin [32]. Drug penetration is enhanced by these modified liposomes made of phospholipids and edge activators. These ultra-flexible particles can carry drugs and squeeze themselves to penetrate tissue [33]. Transfersomes improve drug penetration through the stratum corneum and deliver the drug into the epidermis [34]. Transfersomes can overcome vesicular system issues like restricted skin permeability, vesicle splitting, and drug leakage, aggregation, and fusion [35]. Transfersomes penetrate a small opening by acting as deformers and changing their shape as they go [36]. Surfactants, also known as edge activators, generate flexible liposomes and increase drug absorption through the skin. Drug concentrations increase in the deeper skin layers. These substances can be delivered by transfersomes due to their high skin penetration or poor water solubility [37].

Topical administration may avoid first-pass hepatic metabolism, increase drug duration, reduce adverse effects, and improve patient convenience [38]. Topical bioactive compounds are limited because the stratum corneum blocks the penetration of these molecules [39]. However, transfersomes, a promising lipid-based vesicular mechanism, are widely used in topical medication delivery [40].

Response Surface Methodology (RSM) studies explanatory variables that affect response variables. BBD can be used to minimize the number of experimental runs while still achieving the desired results [41]. We created 6-gingerol-loaded transfersomes (6-GTF) for osteoporosis treatment using thin-film hydration and sonication. A Box–Behnken design (BBD) optimized them. Viscosity controls medication absorption and skin residence time. Carboxy vinyl polymer (Carbopol 934) is beneficial in topical gels due to its non-appreciable rheological change in the pH range of 5–8 [42]. Vesicle shape, *in vitro* and ex vivo, CLSM, antioxidant activity, and dermatokinetic investigations were assessed for the optimized 6-gingerol-loaded transfersomes and gel formulation.

This study offers significant information on the development and optimization of transfersomes as a potential topical delivery strategy for 6-gingerol in pain-related illnesses. The clinical effectiveness and safety of these transfersomes in treating pain and enhancing patient outcomes call for further research. The study's findings add to the body of evidence on nanocarrier-based pain management drug delivery systems, providing new directions for future study and development in this area.

2. Results and Discussions

2.1. Optimization of 6-Gingerol-Loaded Transfersomes

The Box–Behnken Design, using Design Expert® software (version 13, Minneapolis, MN, USA), assessed the relationships between independent variables, i.e., phospholipid (X_1), sodium cholate (X_2), and sonication time (X_3), at three levels (−1, 0, +1), and dependent responses, i.e., vesicle size (Y_1), PDI (Y_2), and encapsulation efficiency percentage (Y_3). Quadratic models were found to be the best fit for both the independent and dependent responses. According to Box–Behnken Design, the compositions of various trial runs of phospholipid, sodium cholate, and sonication time are tabulated in Tables 1 and 2. These findings confirm this study's independent variables. ANOVA evaluated model significance. The quadratic model describes the link between independent factors and dependent responses.

Table 1. Formulation runs and responses of various compositions as per BBD.

Formulations	Independent Variables			Dependent Variables		
	Phospholipid (mg) (X_1)	Sodium Cholate (mg) (X_2)	Sonication Time (min) (X_3)	Vesicle Size (nm) (Y_1)	PDI (Y_2)	Entrapment Efficiency (%) (Y_3)
1	70	10	2	173.15	0.272	84.22
2	90	15	6	169.05	0.267	85.97
3	90	5	2	166.87	0.269	84.04
4	90	10	4	161.05	0.268	88.67
5	70	15	4	179.25	0.288	80.06
6	110	5	4	178.11	0.286	83.36
7	110	10	2	176.04	0.281	86.25
8	70	5	4	174.98	0.276	80.25
9	70	10	6	164.18	0.268	85.26
10	90	10	4	160.97	0.266	88.98
11	90	5	6	166.08	0.269	87.85
12	90	10	4	160.42	0.259	89.43
13	110	10	6	169.92	0.271	87.94
14	90	15	2	179.38	0.289	84.17
15	90	10	4	160.88	0.267	88.48
16	90	10	4	161.04	0.268	88.24
17	110	15	4	188.19	0.297	83.18

X_1 = Phospholipid (mg), X_2 = Sodium cholate (mg), X_3 = Sonication time (min), Y_1 = Vesicle size (nm), Y_2 = PDI, and Y_3 = EE (%).

Table 2. Results of regression analysis of Y_1, Y_2, and Y_3 responses.

Quadratic Model	R^2	Adjusted R^2	Predicted R^2	SD	%CV	p-Values		
						X_1	X_2	X_3
Response (Y_1)	0.9965	0.9921	0.9480	0.7448	0.4382	<0.0001	<0.0001	<0.0001
Response (Y_2)	0.9845	0.9647	0.8158	0.0019	0.6752	0.0121	0.0030	0.0059
Response (Y_3)	0.9825	0.9601	0.8097	0.5877	0.6860	<0.0001	0.003	0.0019

2.2. Independent Factors' Impact on Vesicle Size (Y_1)

According to the experimental results, the independent variables, on vesicle size as indicated by the equation below:

$$\text{vesicle size} = +160.87 + 2.59\ X_1 + 3.73\ X_2 - 3.28\ X_3 + 1.45\ X_1 \times 2 + 0.7125\ X_1X_3 - 2.38\ X_2X_3 + 9.87\ X_{12} + 9.39\ X_{22} + 0.0815\ X_{32}$$

As per the above-mentioned equation and results indicated in Table 1, the 6-Gloaded transfersome vesicle sizes ranged from 160.42 nm to 188.19 nm. Figure 1A–D depict 3D and contour plots that show how influences affect size. Transfersome size marginally increases with increasing phospholipid concentration (X_1). At high concentrations and greater availability of lipid, higher amounts of drug can be entrapped, which increases the size of the particles. To achieve the ideal vesicle size, it is crucial to have the ideal lipid content, so the minimum vesicle size is achievable by having a moderate phospholipid concentration. According to the results of our study, which are in line with earlier studies, the size of the effect increases as the phospholipid concentration in the formulation rises [43]. The vesicle size increases with the increase in sodium cholate concentration (X_2) from 5 mg to 15 mg, as indicated in Table 1. Formulation 11 has a vesicle size of 166.08 nm (5 mg sodium cholate), formulation 7 has a vesicle size of 176.04 nm (10 mg sodium cholate), and formulation 5 has a vesicle size of 179.25 nm (15 mg sodium cholate). This is due to the edge activator's ability to cause aggregates in excess, increasing the vesicle's size [44]. Additionally, sonication time (X_3) had a significant impact on vesicle size, as it reduced the size of the vesicles that may have resulted from the dispersion of agglomerated vesicles into smaller vesicles as the sonication time increased. These findings are in line with earlier studies [43,45].

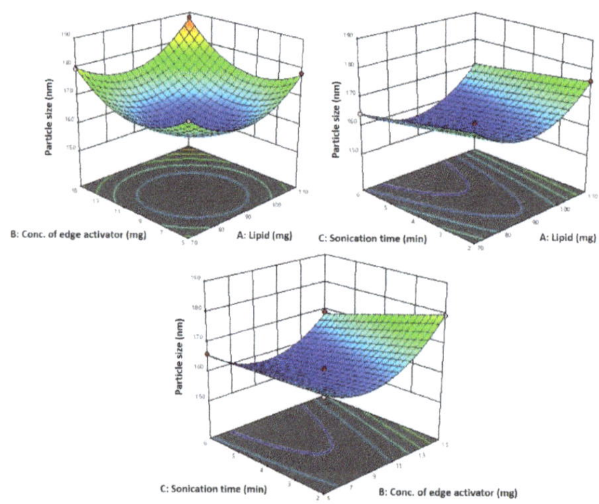

(A)

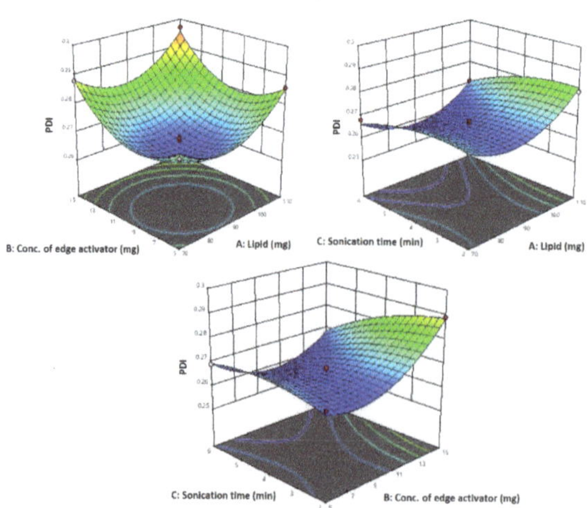

(B)

Figure 1. *Cont.*

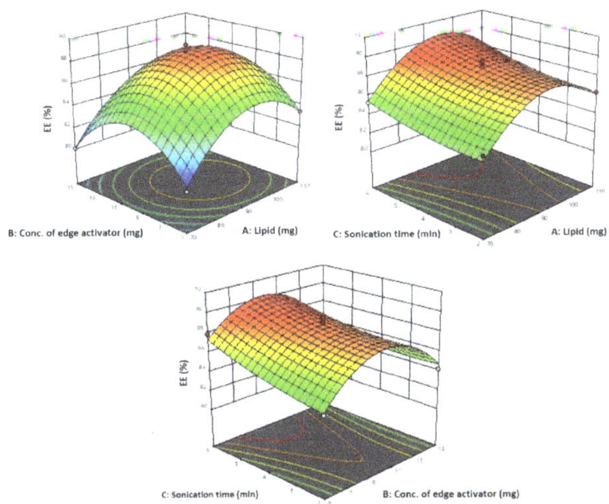

(C)

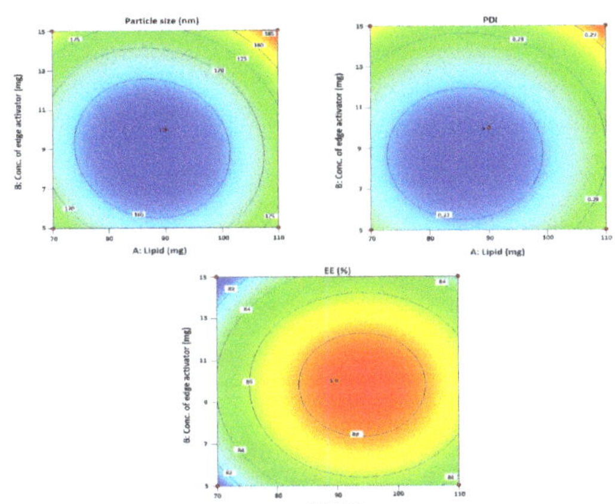

(D)

Figure 1. (**A**) A 3D surface plot of the effect of independent variables on vesicle size. (**B**) A 3D surface plot of the effect of independent variables on PDI. (**C**) A 3D surface plot of the effect of independent variables on entrapment efficiency. (**D**) Counterplots for response vesicle size, PDI, and entrapment efficiency of 6-GTF.

2.3. Independent Factors' Impact on PDI (Y_2)

The PDI for each of the 17 runs ranged from 0.259 to 0.302, between the lower and upper limits.

$$PDI = +0.2602 - 0.0001\,X_1 + 0.0020\,X_2 - 0.0036\,X_3 + 0.0003\,X_1X_2 - 0.0015\,X_1X_3 - 0.0013\,X_2X_3 + 0.0101\,X_{12} + 0.0049\,X_{22} + 0.0182\,X_{32}$$

The PDI decreases with an increase in phospholipid concentration (X_1) from 90 mg to 110 mg and sonication time from 2 to 6 min (Figure 1B). This may be due to the vesicle size reduction caused by the sonication (X_3), which leads to less agglomeration of vesicles by breaking them into smaller sizes with a longer sonication period, thereby increasing the homogeneity [43]. As per the above-mentioned equation, PDI increases with an increase in sodium cholate (X_2). This is due to the edge activator's ability to cause aggregates in excess, increasing the vesicle's size. These findings are in line with previous reports [40].

2.4. Independent Factors' Impact on Entrapment Efficiency (Y_3)

The 6-G transfersome produced had an entrapment efficiency that varied from 80.06 to 89.43%, in accordance with the equation given below and Table 1.

$$\text{Entrapment efficiency} = +88.76 + 1.37\ X_1 - 0.2650\ X_2 + 1.04\ X_3 + 0.0025\ X_1X_2 + 0.1625\ X_1X_3 - 0.5025\ X_2X_3 - 3.32\ X_{12} - 3.73\ X_{22} + 0.4762\ X_{32}$$

Since a lipophilic drug would be drawn to the lipophilic phase and be deposited there, 6-G intrinsic lipophilicity may be the reason for the steady increase in entrapment efficiency with an increase in phospholipid concentration. In the current investigation, the effectiveness of drug entrapment in formulation was found to positively correlate with the concentration of phospholipid (X^1) and the length of the sonication (X^3) process (Figure 1C). A high concentration of lipophilic drugs can exist in the context of high phospholipid concentrations. Because the linear increase in phospholipid concentration also influences encapsulation effectiveness, finding the ideal phospholipid concentration is crucial [43]. It was discovered that a decrease in entrapment effectiveness occurs with an increase in edge activator concentration. The encapsulation decreased as the concentration of the edge activator increased gradually. At a high concentration of edge activator, a greater amount of drug leaches out of the vesicles. With regard to the third factor, the encapsulation efficiency rises as sonication time increases, but at longer sonication times, phospholipids are degraded and vesicle structures are disrupted, leading to drug leakage and a lower EE [44].

2.5. Validation of Experimental Design

In order to create the ideal formulation with the lowest vesicle size, lowest PDI, and highest entrapment efficiency, BBD was used. Table 1 lists the expected values for every response and factors derived from the design. Therefore, 6-GTF formulations based on the runs, process variables, and experimental results attained using the formulations were compared to expected responses. The fact that the experimental and projected values were nearly identical supports the validity of the optimization procedure, as shown in Figure S1 (Supplementary File). The desirability values for the responses were determined to be 1, indicating the response that had the lowest prediction error and was most desired.

2.6. Optimized Formulation

The confirmation of the experimental design for the optimized formulation showed that the ideal formulation contained 95.16 mg of lipid, 8.91 mg of sodium cholate, and 5.94 min of sonication time. The results of the experiments were discovered to be comparable to one another, indicating the validity and dependability of the optimized formulation, as shown in Figure S1 (Supplementary File).

2.7. Development of Optimized 6-G-Loaded Transfersomes

The method of thin film followed by sonication was used to develop 6-G-loaded transfersomes. The detailed procedure for synthesizing 6-gingerol-loaded transfersomes can be found in the method section. After preparing 6-gingerol-loaded transfersomes, the formulation was kept in a cold, dry environment. By altering the composition and process variables, the formulation was optimized.

2.8. Characterization of 6-Gingerol-Loaded Transfersomes

2.8.1. Vesicle Size, Zeta Potential, and Morphology

The 6-GTF that had been prepared had an opaque look and a faintly yellowish opalescence. The PS of 6-GTF was 160.42 ± 0.25 nm on average, as shown in Figure 2A. The polydispersity index (PDI), which expresses the degree of non-uniformity in the distribution of vesicle size, is a measure of vesicle size distribution. For stability and effectiveness, nanovesicles should be monodispersed and have a PDI value of 0.3 or lower. The lower PDI value for 6-GTF in this study (0.259) denotes a monodisperse size distribution. Similar to above, the value of ZP shown in Figure 2B was frequently employed to assess the physical stability of nanoformulation during storage. The zeta potential value of the 6-GTF in this investigation was −32.12 mV, which is consistent with a stable system. Figure 2C displays the transmission electron micrographs of transfersomes. The image showed a spherical shape of vesicles without aggregation or crystal formation within the formulation [45].

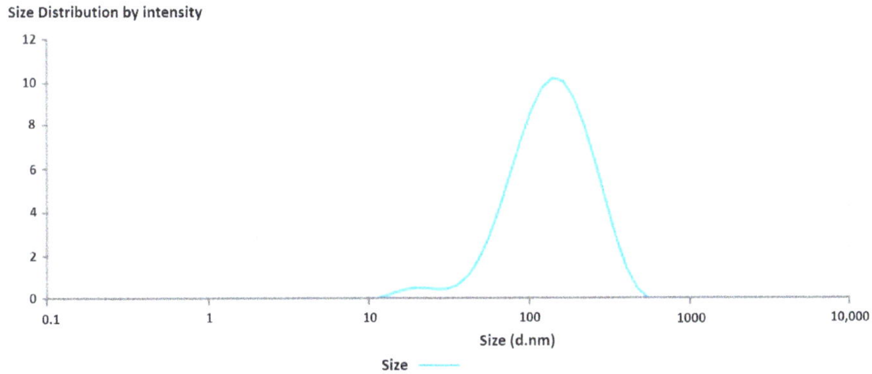

Statistics Table

Name	Mean	Standard Deviation	RSD	Minimum	Maximum
Z-Average (nm)	160.42	-	-	160.42	160.42
Polydispersity Index (PI)	0.259	-	-	0.259	0.259
Intercept	0.986	-	-	0.986	0.986
Peak 1 Mean by Intensity ordered by area (nm)	160.42	-	-	160.42	160.42
Peak 1 Area by Intensity ordered by area (%)	100	-	-	100	100

(A)

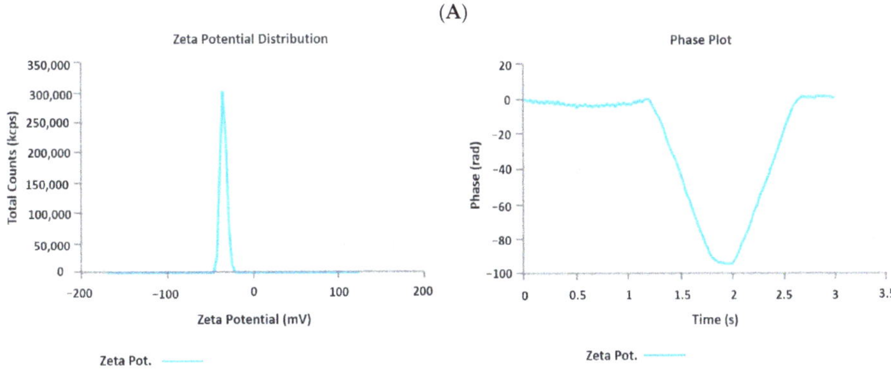

Statistics Table

Name	Mean	Standard Deviation	RSD	Minimum	Maximum
Zeta Potential (mV)	-32.12	-	-	-32.12	-32.12
Zeta Peak 1 Mean (mV)	-32.12	-	-	-32.12	-32.12
Conductivity (mS/cm)	0.03546	-	-	0.03546	0.03546
Wall Zeta Potential (mV)	-25.57	-	-	-25.57	-25.57
Zeta Deviation (mV)	3.649	-	-	3.649	3.649

(B)

Figure 2. *Cont.*

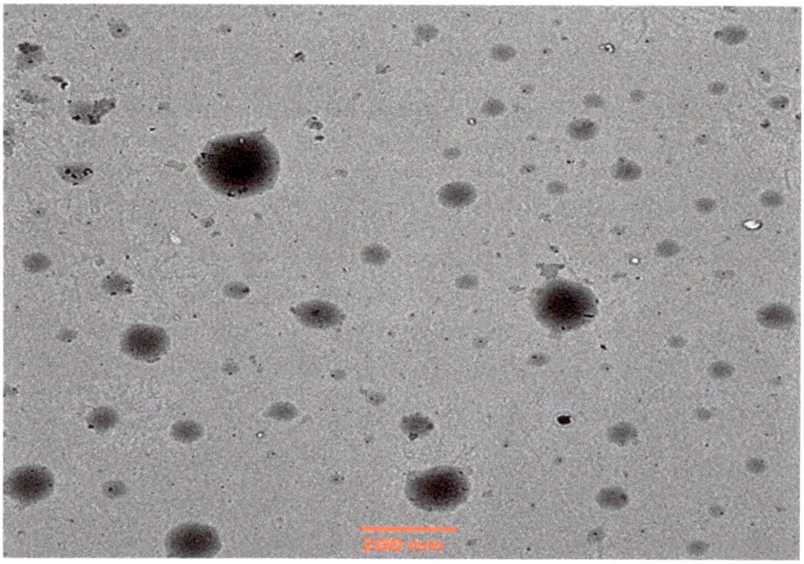

(C)

Figure 2. (**A**) Vesicle size of transfersomes; (**B**) Zeta potential of transfersomes; (**C**) Morphology of transfersomes.

2.8.2. Estimation of % Entrapment Efficiency and Drug Loading

The percentage entrapment efficiency and drug loading of 6-gingerol-loaded transfersomes were found to be 89.43 ± 1.01% and 10.57%, respectively.

2.9. Assessment of the Transfersome Gel

2.9.1. Drug Content

The most crucial aspect in the formulation of transfersomes is the drug content, and the data obtained is acceptable. It was discovered to be between 81.26 and 86.14%, demonstrating the formulation's strong ability to hold the drug. The optimized 6-G-loaded transfersome gel formulation (1% Carbopol) contained the highest amount of drug (86.14 ± 3.14%). Dermal drug delivery techniques depend significantly on the pH of the skin, and the transfersome formulation results show that all formulations are suitable for skin delivery. The produced transfersome gels' pH value was discovered to be between 6.98 and 7.14. The pH of the optimized 6-G-loaded transfersome gel formulation was discovered to be 7.14, as stated in Table 3.

Table 3. Results of 6-G-loaded transfersome gel formulations.

Formulation Gel Code with Carbopol Conc.	Homogeneity	pH	Drug Content (%)	Spreadability (g·cm/s)	Extrudability (g/cm^2)
TF1 (0.5%)	Homogeneous	6.98	81.26 ± 2.14	10.26 ± 1.29	10.36 ± 1.11
TF2 (1%)	Homogeneous	7.14	86.14 ± 3.14	13.46 ± 4.42	15.19 ± 2.01
TF3 (2%)	Homogeneous	7.09	85.14 ± 2.30	11.12 ± 2.64	13.18 ± 0.78

2.9.2. Spreadability and Extrudability

The most crucial characteristics for ensuring patient compliance and a uniform distribution of topical formulations are extrudability and spreadability. The extrudability and spreadability of the developed transfersome gel were discovered to range from 10.36 to 15.19 g/cm^2 and 10.26 to 13.46 g·cm/s, respectively, as shown in Table 3. According to these results, the generated formulation spreads very well across the skin's surface and can be easily extruded by pressing down with the thumb [46].

2.10. Texture Analysis

For 6-GTFG, a texture analysis was conducted, and many elements like cohesion, consistency, stiffness, and index of viscosity were evaluated. The outcomes of the texture analysis of the formulation's (6-GTFG) software output have been given, and the texture image is shown in Figure 3. The optimized 6-GTFG was found to exhibit cohesiveness (−93.55 g), consistency (783.94 g s), firmness (191.86 g), and work of cohesion (−653.11 g s), respectively.

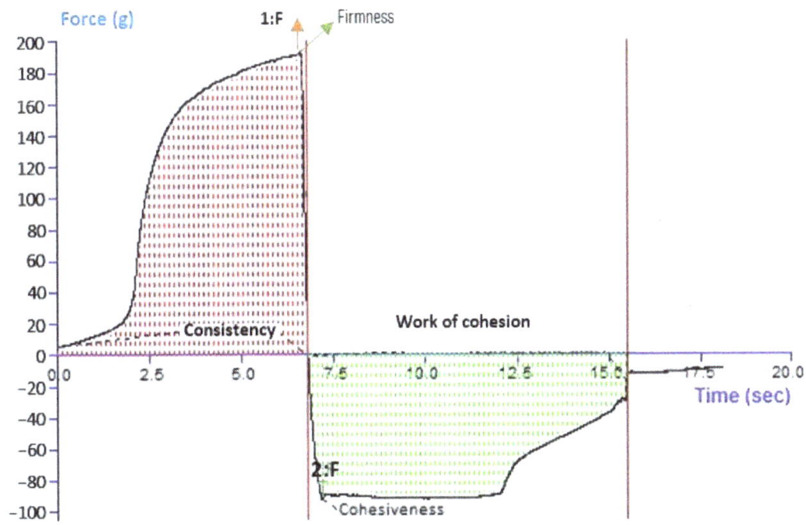

Figure 3. Analysis of texture of optimized transfersome gel.

2.10.1. *In Vitro* Drug Release

Our studies found that after 24 h, the optimized transfersomes released 69.21 percent of the drug, while the pure drug suspension released 47.71 percent, as shown in Figure 4. The *in vitro* data revealed a biphasic release process, with a quick release from the surface-adsorbed drug in the first 2 h, followed by a gradual release profile for up to 24 h [47]. The Korsmeyer–Peppas model exhibited the highest R^2 value of 0.9901 and an n value of 0.697, indicating the pattern of non-Fickian diffusion release of drug, and it follows the Higuchi model release pattern with an R^2 value of 0.995, followed by zero- and first-order R^2 values of 0.9291 and 0.9893, respectively. According to the data of the *in vitro* drug release investigation, the optimized TF formulation has a suitable structure capable of entrapping 6-G to ensure their gradual release as the drug was successfully entrapped by the transfersomes, enabling a sustained release that could be effective in a drug delivery system that demands prolonged release profiles.

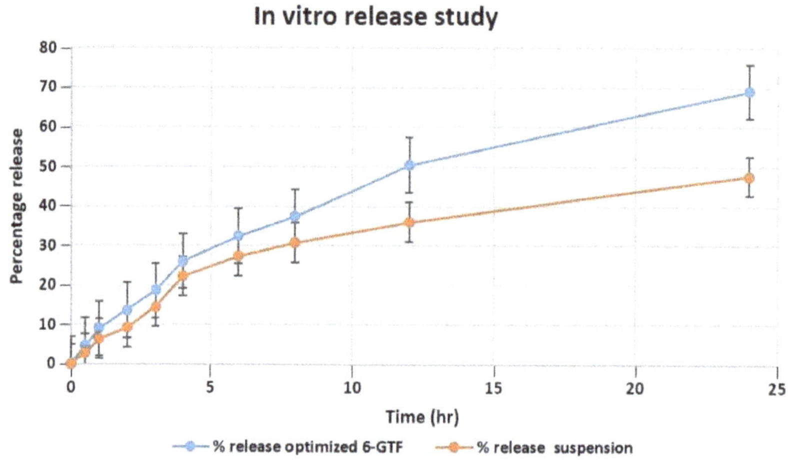

Figure 4. Comparative *in vitro* release profile of 6-GTF through a dialysis membrane.

2.10.2. Determination of Antioxidant activity

Using the ABTS assay, the antioxidant capability of 6-gingerol and its developed formulation was assessed. The ability of 6-gingerol to operate as reducing agents is based on their redox capabilities; this ability is typically linked to the presence of reductants, which break the free radical chain by donating a hydrogen atom or prevent the development of peroxide to produce antioxidant action [48]. With experiments proving the drug's effectiveness in scavenging free radicals and reducing lipid peroxidation, 6-gingerol has been found to have considerable antioxidant potential. Both the pure 6-G and the optimized 6-GTF formulations had their antioxidant properties evaluated and contrasted with a standard solution. Antioxidant activity ranged from 91.14 $\pm$ 0.24 percent in a standard solution, 67.01 $\pm$ 1.11 percent in free 6-G, and 76.89 $\pm$ 3.19 percent in a 6-GTF formulation [49].

2.10.3. Ex Vivo Study of the Permeation of Skin

According to an ex vivo skin permeation investigation, 40.63% of 6-G from 6-G suspension gel permeated the skin. While the optimized 6-GTF gel formulation demonstrated 77.01% 6-G permeation across rat skin, as demonstrated in Figure 5. The 6-GTF gel showed initial burst release (36.45%), which is higher than that of the 6-G suspension gel (18.45) after 4 h. The drug transport profile from the 6-GTFG formulation across rat skin was found to have a considerably higher flux value, i.e., 2.71 $\mu g/cm^2/h$ compared to the 6-G suspension gel, which showed a flux of 1.5 $\mu g/cm^2/h$. The considerable distinction in the percentage transport of drugs might be due to the incorporation of 6-G into the optimized TF gel system [46], which could have been due to the nanosized TF vesicles. However, due to transfersomes' flexibility, they are able to penetrate the skin and overcome the barrier function by squeezing themselves through the pores of the stratum corneum.

2.10.4. CLSM

Confocal microscopic images showed that the rhodamine B hydroalcoholic solution only contacted the topmost layers of skin and that the dye's fluorescence permeated as deep as 5.0 μm (Figure S2B; Supplementary File). Rhodamine B-loaded TF gels were penetrated 25.0 μm into the skin (Figure S2A; Supplementary File). The finding of higher fluorescence intensity indicates that the formulation has a sustained presence and greater levels in the deeper layers of the skin. Using the newly developed TF formulation, rhodamine B was successfully delivered into the dermal layers of rat skin.

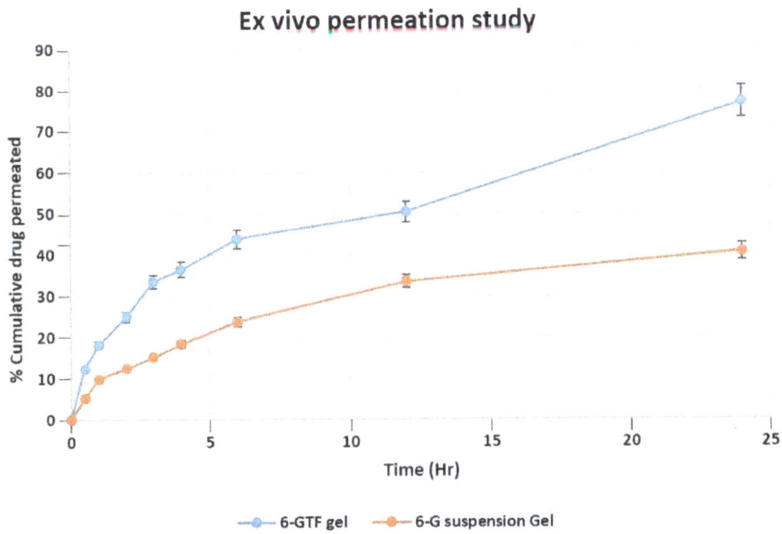

Figure 5. Permeation of 6-G from 6-GTF gel and 6-G suspension gel over the rat excised skin in a unit of time.

2.10.5. Dermatokinetics for Optimized Formulation

The relative concentration of 6-G in the layers of rat skin's dermis and epidermis following treatment with gel formulations of 6-GTFG and 6-G-suspension gel at different time intervals is displayed in Figure 6A,B. The dermatokinetic parameter values are presented in Table 4. The excised rat skin treated with 6-GTF gel and 6-G-suspension gel revealed that the 6-GTF gel produced much larger concentrations of 6-G in the epidermis, with values of Cskin max of 294.61 ± 2.14 μg/cm^2 and 198.49 ± 1.11 μg/cm^2, respectively, and in the dermis, the values of Cskin max for 6-GTFG and 6-G suspension gel were found to be 274.92 ± 1.49 μg/cm^2 and 170.70 ± 0.99 μg/cm^2, respectively. The 6-GTFG delivery demonstrated quick drug absorption when compared to the suspension gel formulation, reaching the Cmax in the dermis and epidermis after topical administration within 1.5 and 2 h, respectively. The $AUC_{0–8h}$ of 6-G in 6-GTFG was much higher than the 6-G suspension gel in the epidermis, and it was found to be 1286.16 ± 2.49 μg/cm^2 h and 762.99 ± 0.99 μg/cm^2 h, respectively. In dermis, the $AUC_{0\text{-}8}$ of 6-G concentration in 6-GTFG was considerably greater when compared to 6-G suspension gel; it was found to be 1213.16 ± 1.01 μg/cm^2 h and 648.34 ± 0.11 μg/cm^2 h, respectively. Vesicles in 6-GTF gel improve drug absorption and retention in skin layers. It was discovered that the 6-G concentration fell over the course of the experiment until 8 h had passed and the drug's concentration was detectable. The above results of the TF delivery suggest increased penetration of the delivery tool when applied topically, which may be related to the earlier findings of the confocal laser microscopic study and skin permeation study, where the increased penetration of the formulation was obvious due to its elasticity and the presence of an edge activator.

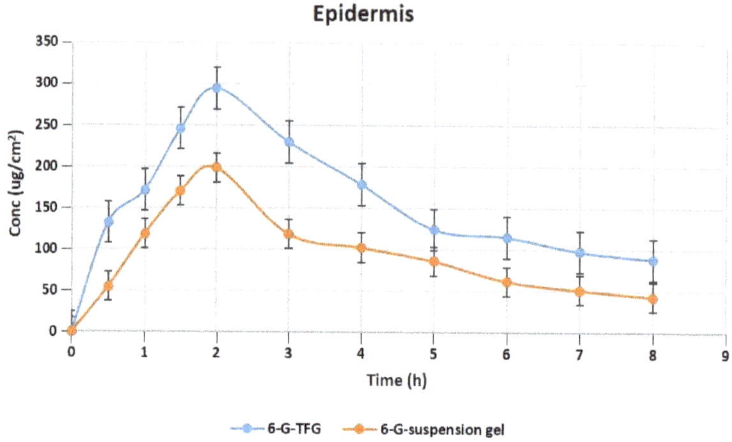

(A)

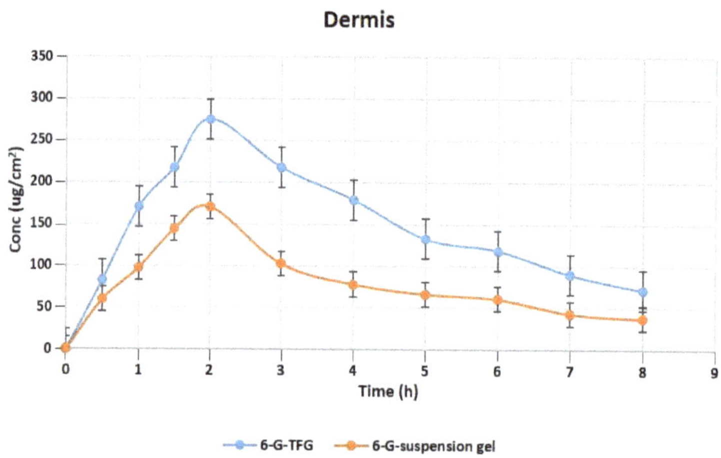

(B)

Figure 6. 6-G concentration on rat skin (**A**) epidermis (**B**) dermis after application to the skin of 6-GTF gel and 6-G-suspension gel.

Table 4. Dermatokinetic parameters (Mean ± SD) of 6-GTF-gel and 6-G suspension-gel.

Dermatokinetics Parameters	6-GTF-Gel		6-G Suspension Gel	
	Epidermis Mean ± SD	Dermis Mean ± SD	Epidermis Mean ± SD	Dermis Mean ± SD
$T_{skin\ max}$ (h)	2	2	2	2
$C_{skin\ max}$ (μg/cm^2)	294.61 ± 2.14	274.92 ± 1.49	198.49 ± 1.11	170.70 ± 0.99
$AUC_{0\text{–}8}$ (μg/cm^2 h)	1286.16 ± 2.49	1213.16 ± 1.01	762.99 ± 0.99	648.34 ± 0.11
Ke (h^{-1})	0.113 ± 0.14	0.113 ± 0.88	0.128 ± 0.46	0.136 ± 0.89

Tskin max = Time to maximum concentration, Cskin max = Maximum concentration, AUC = Area under curve, Ke = Elimination rate constant.

2.11. Stability Studies

Physiochemical stability tests were performed to determine the period of time the 6-GTF gel may be expected to remain stable after it has been prepared. Based on the findings, 6-GTF gel is the best option. The product met all required specifications for color, clarity, pH, and other parameters shown in Table 5.

Table 5. Stability studies for optimized 6-GTF gel stored at 25 ± 2 °C/60 ± 5% and 4 ± 2 °C.

Evaluation Parameters	Initial	1 Month		3 Months		6 Months	
		4 ± 2 °C	25 ± 2 °C/ 60 ± 5% RH	4 ± 2 °C	25 ± 2 °C/ 60 ± 5% RH	4 ± 2 °C	25 ± 2 °C/ 60 ± 5% RH
Color	Slightly yellowish	Slightly yellowish	Slightly yellowish	Slightly yellowish	Slightly yellowish	Slightly yellowish	Slightly yellowish
Appearance	Translucent	Translucent	Translucent	Translucent	Translucent	Translucent	Translucent
Homogeneity	***	***	***	***	**	***	**
Clarity	√	√	√	√	√	√	√
pH	7.14	7.14	7.09	7.07	7.04	7.06	7.03
Washability	Washable	Washable	Washable	Washable	Washable	Washable	Washable

** Good, *** Excellent, √ Yes.

3. Materials and Method

3.1. Material

6-Gingerol was procured from Sigma Aldrich (Mumbai, India); sodium cholate and methanol were obtained from Merck Pvt. Ltd. (Mumbai, India). Phospholipid, Carbopol 934, and triethanolamine were purchased from SD Fine Chemicals Ltd. (Mumbai, India). Dialysis membrane bags were procured from Sigma-Aldrich Co., St. Louis, MO, USA. All other substances were of an experimental grade, such as sodium dihydrogen phosphate and sodium chloride, which were provided by S D Fine Chemicals Limited, Mumbai, India.

3.2. Methods

3.2.1. Developing Transfersomes Loaded with 6-Gingerol (6-GTF)

The 6-GTF were prepared by the traditional method of thin film hydration followed by sonication. Initially, to produce vesicle suspension, a weighed amount of phospholipid, drug (6-G), and edge activator (sodium cholate) was dissolved in a chloroform: methanol (5:1 v/v) mixture. The dry lipid film was prepared by evaporating the solvent in the rotary evaporator. The lipid layer was vacuum-dried to remove the solvents. The thin dried layer was hydrated with distilled water (10 mL), and it was then allowed to rotate for 1 h at 90 °C at 60 rpm to swell. Further, the formulation was sonicated using a probe sonicator (UP100H, Hielscher GmbH, Teltow, Germany) for 5.94 min, and then extruded through membrane filters (0.25 μm). After the preparation, the formulation was stored in a cool, dry place. The formulation was optimized by changing the formulation and process variables [50].

3.2.2. Transfersome Optimization Using Box–Behnken Design (BBD)

Using response surface methodology (RSM), a three-factor, three-level Box–Behnken design (BBD) was utilized to optimize the transfersomes. Many researchers found that phospholipid (X_1), edge activator (X_2), and sonication time (X_3) affect formulation stability and entrapment efficiency; hence, they were chosen as independent variables. Vesicle size (Y_1), polydispersity index (PDI) (Y_2), and entrapment efficiency (Y_3) were chosen as dependent variables for optimization. After many preliminary experiments, phospholipid, sodium cholate, and probe sonication time were selected in the ranges of 70–110 mg, 5–15 mg, and 2–6 min, respectively, as independent variables for 6-gingerol-loaded transfersome optimization (Table 6). Table 1 shows 17 BBD-prepared trial formulations of various factor combinations, including five center points. The information was put into BBD, and mathematical modeling was used to analyze the results. ANOVA, in addition to the coefficient of correlation (R_2), adjusted R_2, anticipated R_2, and predicted residual sum of squares, was used to analyze the data fitting the selected quadratic second-order model.

The optimum conditions for the development of transfersomes were determined with the use of the numerical desirability function and the graphical optimization technique. The study maintained all other formulation and processing characteristics as constants.

Table 6. Independent and dependent variables used in BBD for the development of 6-GTF.

Variables	Used Levels		
	Low (−1)	**Medium (0)**	**High (+1)**
	Independent variables		
X_1 = Phospholipid (mg)	70	90	110
X_2 = Sodium cholate (mg)	5	10	15
X_3 = Sonication time (min)	2	4	6
	Dependent variables		
Y_1 = Vesicle size (nm)		Minimize	
Y_2 = PDI		Minimize	
Y_3 = EE (%)		Maximize	

3.2.3. Measurement of Vesicle Size, PDI and Zeta Potential for 6-GTF

Using dynamic light scattering, the vesicles' size, size distribution curve, and electric potential were determined using a Malvern Zetasizer, ver.7.11, serial number MAL1121994 (Malvern Panalytical Ltd., Malvern, United Kingdom). Prior to the study of the aforementioned parameters, the optimized 6-GTF was diluted roughly 100 times. The temperature was maintained at about 25 ± 2 °C, while the scattering angle was kept at 90° [51]. The same procedure was followed for the blank formulation.

3.2.4. Determination of 6-G Encapsulation Efficiency (%EE) and % Drug Loading (DL)

The ultracentrifugation technique (Beckman Coulter India Pvt. Ltd., New Delhi, India) was used to determine % EE of 6-G-loaded in the transferosomes. The obtained formulations were centrifuged for one hour at 6000 rpm. The supernatant was withdrawn, diluted, and filtered through a membrane filter (0.25 μm), and the drug concentration was determined at 278 nm using a UV spectrophotometer (Shimadzu, Tokyo, Japan). Drug entrapment effectiveness is an important consideration when assessing transfersomes ability to load drugs. This metric is influenced by the drug's physicochemical characteristics, development process, and formulation considerations. The blank sample was prepared using the same method. The drug was not incorporated into the formulation's development. The following equation was used to determine the percentage of the drug encapsulated and loaded [35,52].

$$Encapuslation\ efficiency\ (\%) = \frac{(Total\ drug - Free\ drug)}{Total\ drug} \times 100$$

$$Drug\ loading\ (\%) = \frac{Amount\ of\ drug\ entraped}{Total\ amount\ of\ drug\ and\ lipid} \times 100$$

3.2.5. Morphological Study

Transmission electron microscopy (TEM) was used to study the surface morphology of the optimized transfersomes utilizing a microscope (FEI Company, Eindhoven, Netherlands). A diluted drop of TF was applied to a copper grid that had been covered with carbon. The sample was dyed with 1% phosphotungstic acid for 10 s, and then it was set to air-dry at room temperature before being examined by a TEM at 70 kV [53].

3.2.6. Preparation of 6-Gingerol-Based TF Gel (6-GTFG)

The optimized 6-G transfersomes (6-GTF) were incorporated into a gel (6-GTFG). Carbopol 934 was employed as the gelling agent. In order to create the dispersion, distilled water was first gently added to the Carbopol 934. After that, the mixture was allowed to sit overnight in the dark to fully swell. Triethanolamine (0.5% w/v) was gradually added to the dispersion to further neutralize it, enabling the development of a clear, viscous gel. The Carbopol gels at 0.5%, 1%, and 2% were made. Finally, the optimized 6-GTF formulation was gently added to the gel while being continuously stirred [54].

3.2.7. Evaluation of 6-G Based TF Gel (6-GTFG)

Organoleptic Assessment and pH Determination

The resulting formulations' appearance or homogeneity were visually inspected for any grittiness or agglomerates. Additionally, a precise weight of the 6-GTF gel formulation (1 g) was used to dilute it with double-distilled water. The pH value of the developed formulation was evaluated at room temperature using a pH meter (Mettler Toledo, Ohio, United States).

Drug Content

A weighed amount (100 mg) of the produced gel was dispersed in phosphate-buffered saline (PBS) (pH 7.4) (100 mL). To ensure complete drug solubility, the volumetric flask holding the gel solution was agitated on a mechanical shaker for 2 h. After this, the solution was filtered using a membrane (pore size 0.45 mm). Using a UV–visible spectrophotometer and solvent as a blank (followed by the same procedure without the incorporation of drug in a gel formulation), the absorbance of the sample was calculated at 278 nm [55].

Spreadability and Extrudability Determination

The spreadability and extrudability of the developed formulation are critical factors in determining uniform distribution and spreading. The spreadability of the formulations describes the area where they can easily spread following skin application. A reference load (500 g) was administered to the top slide for five minutes while the experimental formulations (1 g) were sandwiched in between two glass slides [56]. The time (in seconds) required to separate the two slides served as a gauge of their spreadability. The formula used to calculate the spreadability, as follows:

$$S = M \cdot L/T$$

S stands for spreadability, M for the weight supported by the higher slide, L for the glass slide's length, and T for the processing time.

Extrudability is defined as the quantity of material in g required to extrude no less than a half-centimeter ribbon of the formulation from a breakable tube within ten seconds. The 6-G-loaded transfersome gel (20 g) was packed in a collapsible tube, and the crimped end of the tube was pushed to force the gel out when the cap was removed [57]. The following equation was used to determine the extrudability (g/cm^2):

$$Extrudability = \frac{Applied\ weight\ to\ the\ tube\ (gm)}{Area\ (cm^2)}$$

Gel Texture Evaluation

The texture of the generated 6-GTFG was inspected using a texture analyzer (TA.XT Plus Texture Analyzer, Stable Micro Systems Ltd., Surrey, UK). A 100 mL glass beaker was filled with 100 g of the optimized gel formulation (6-GTFG), and the surface was maintained as smooth as possible to avoid air bubble entrapment. The analytical probe was inserted twice into the gel-filled beaker at a depth of 15 mm, moving slowly (2 mm per s) and allowing 20 s between compressions. A force curve that includes mechanical traits like cohesiveness, consistency, cohesiveness index, and hardness was used to describe the results [58].

In Vitro 6-G Release and Release Kinetic Modeling

- Cellophane membrane activation

To remove glycerin and sulfur compounds and to open pores, the dialysis membrane is always activated prior to *in vitro* experiments. By washing the dialysis membrane under running water for three to four hours, the glycerin was removed. Using a 0.3% *w/v* sodium sulfide solution at 80 °C for 1 min, the sulfur compounds were removed. The tube was first rinsed in hot water (60 °C) for 2 min, then acidified with sulfuric acid (0.2% *v/v*), and finally treated with hot water to remove the acid. Membranes preserve the majority of proteins with molecular weights of 12,000 or more after this treatment [59].

- Conducting *in vitro* release studies

The *in vitro* drug release from the optimized 6-GTF formulation under standard circumstances was examined using the cellophane dialysis membrane bag technique (molecular weight cutoff 12,400 Da, Sigma-Aldrich Co., St. Louis, MO, USA). In the dialysis bags, 2 mL of the 6-GTF formulation and 6-G suspension (10 mg of drug was dissolved in methanol and made up to 10 mL with a suitable solvent) were loaded before being bound at both ends. The dialysis bags were submerged in a beaker filled with 25 mL of PBS (pH 7.4) that was magnetically stirred at 100 rpm and 37 ± 0.2 °C throughout the experiment. The samples were taken at various intervals (0, 0.5, 1, 2, 4, 6, 8, 12, and 24 h) from the beaker while sink conditions were maintained, and they were UV spectrophotometrically examined for the drug content against the PBS pH 7.4 as a blank [60].

Each measurement was made in triplicate. By fitting the results into various models, including zero-order, first-order, Korsmeyer–Peppas, and Higuchi mathematical models, the drug release mechanism was further evaluated. In accordance with the great value of the coefficient of correlation (R2), the model that best describes the 6-G release order was chosen [58].

$$\% \, Drug \ release = \frac{Conc. \ (\mu g/mL) \times DF \times Volume \ of \ the \ release \ medium \ (mL)}{Initial \ dose \ (\mu g)} \times 100$$

where *DF* stands for dilution factor.

Ex Vivo Skin Permeation Study

- Preparation of skin

The skins of healthy Wistar rats were removed for the ex vivo skin permeation study and dermatokinetic investigation. The skin permeation experiments and dermatokinetics on rodent skin were properly approved by the Standing Committee of Bioethics Research at Prince Sattam bin Abdulaziz University in Al-Kharj, Saudi Arabia (SCBR-020-2023). The 1 cm^2 of abdomen skin of the rat was cut, and the hair was removed by the use of hair removal cream. Manual removal of the subcutaneous tissue was performed, and any remaining clinging fat was removed by wiping the dermis side with isopropyl alcohol. The skin's entire thickness was cleaned with PBS 7.4. Later, the skin was covered in aluminum foil and kept at −20 °C until use (used within 2 weeks of preparation) [61]. In the Franz diffusion cell assembly, the skin was placed in such a way that the stratum corneum side faces the donor compartment and the dermis side faces the receiver compartment.

3.2.8. Permeation Study

"Franz diffusion cells", having a diffusion surface area of 0.785 cm^2 over rat skin, were used to conduct a skin permeation experiment. In the donor compartment, 1 gm of the 6-G-loaded transfersome gel formulation was placed. Then, 10 mL of phosphate-buffered saline (pH 7.4) was placed in the receiver cell, which was maintained at 37 ± 1 °C during the experiment and stirred with a magnetic bead rotating at 600 rpm [62]. Aliquots (1 mL) of samples were taken out and replaced with the dissolving media at predetermined intervals of time, as performed in *in vitro* studies. The withdrawn samples underwent additional filtering, were diluted with PBS 7.4, and were measured spectrophotometrically using UV spectrophotometry for their drug content by taking PBS as a blank [63]. The slope of the linear component of the plot was used as the steady-state flux (*Jss*) to calculate the cumulative amount of 6-G that penetrates per unit area over time. The following equation was used to measure the permeability coefficient (*Kp*):

$$Kp = \frac{Jss}{Cv}$$

where *Cv* is the formulation's overall donor concentration [64].

3.2.9. Confocal Laser Scanning Microscopy (CLSM)

The CLSM (Leica Microsystems GmbH, Wetzlar, Germany) investigation measured the penetration depth of the produced TF vesicles. Rhodamine B dye was added to the optimized TF gel formulation, and CLSM measurements of the rat skin's penetration depth were made in comparison to the control (solution of rhodamine B-hydroethanolic). Franz diffusion cells were used to conduct a study of skin permeation for 8 h using rhodamine B-loaded TF gel and a hydroethanolic solution of rhodamine B for the CLSM experiment. Glass slides of skin samples were made and viewed under a CLMS microscope, and probe dye penetration depth was measured [65].

3.2.10. 2,2′-Azino-bis(3-Ethylbenzothiazoline-6-sulphonic) Acid (ABTS) Radical Scavenging Activity

The ability of the nanocarriers to scavenge the free radical ABTS was used to measure their antioxidant activity. To create an ABTS free radical, a stock solution of 7 mM ABTS was combined with 2.45 mM potassium persulfate and agitated in the dark for 12 h. The solution was diluted with ethanol prior to use until it had an absorbance of 0.7. Then, a standard solution (ascorbic acid) was mixed with the ABTS+ solution and incubated for 10 min in a 37 °C water bath. After that, various concentrations (10–100 µg/mL) of all samples were prepared in ethanol, 3 mL of the solution and 1 mL of the samples were combined, and the absorbance was measured at 734 nm using a UV spectrophotometer after 6 min [66]. Finally, the following equation was used to compute ABTS scavenging activity:

$$Scavenging\ (\%) = \frac{(Absorbance\ blank - Absorbance\ sample)}{Absorbance\ blank} \times 100$$

3.2.11. Dermatokinetics

Wistar rats' skins that had been removed were employed in the study. As mentioned in the section on ex vivo skin permeation, the skin tissue was prepared. For the experiments, Franz cell diffusion assembly (PermeGear, Inc., Hellertown, PA, USA) was used, as discussed under permeation studies. In this instance, the entire skin was peeled off of the Franz cell at the appropriate sampling intervals. In order to separate the epidermis from the dermis, the skin was cleansed three times to eliminate any remaining adhesive formulation. For complete drug extraction, each of the sections was cut into small pieces and macerated in methanol (5 mL) for 24 h. The filtrate was examined using the UV method after the solution was filtered through a membrane (0.45 µm). The concentration of 6-gingerol per square centimeter of skin versus time was plotted separately for the epidermis and dermis. The dermatokinetic parameters such as *Tskin max*, *Cskin max*, $AUC_{0\text{–}8h}$, and *Ke* were also calculated [67]. An equation was used to fit the collected data into a single compartment open model.

$$Cskin = \frac{Kp \cdot Cskin\ max}{Kp - Ke}\left(e^{-Kpt} - e^{-Ket}\right)$$

where *Cskin* is the drug's concentration in the skin at time *t*, *Kp* is the constant for dermal penetration, *Cmax* skin is the highest concentration that can be found in the skin, and *Ke* is the constant for skin elimination.

3.2.12. Stability Studies

Short-term accelerated stability studies were conducted according to ICH Q1 A (R2) requirements, whereby the 6-GTF gel was subjected to testing for a total of six months. In order to test the stability of the optimized 6-GTF gel, it was kept at a temperature of 25 ± 2 °C and a relative humidity of $60 \pm 5\%$ RH in an aluminum-covered container. Throughout the process of developing and using the gel, measurements were made of its appearance, clarity, homogeneity, pH, spreadability, extrudability, and 6-G content [58].

4. Conclusions

The thin film approach followed by sonication was utilized to develop the 6-gingerol-loaded transfersomes. Quality-by-Design (QbD) was used to optimize 6-gingerol-loaded transfersomes. The 6-GTF gel formulation showed improved skin absorption, sustained drug release, and increased antioxidant activity. These advances lay the groundwork for a future topical treatment for pain linked to osteoporosis and musculoskeletal diseases, giving tailored relief and improving patient outcomes. To verify the effectiveness and safety of this formulation, future research should concentrate on clinical studies and additional optimization. If it is successful, this strategy could offer a useful substitute for pain management, bringing more effective and tailored relief to people with these illnesses.

Supplementary Materials: The supporting information can be downloaded at: https://www.mdpi.com/article/10.3390/ijms24129983/s1.

Author Contributions: Conceptualization, A.A. and U.H.; methodology, M.G., A.A. and U.H.; software, M.H.A.; validation, MG.; formal analysis, A.A.; investigation, A.A., U.H. and M.G.; resources, U.H.; data curation, M.H.A.; writing—original draft preparation, A.A. and U.H.; writing—review and editing, M.H.A. and M.G.; visualization, M.H.A.; supervision, M.G.; project administration, M.H.A. and M.G.; funding acquisition, M.G. All authors have read and agreed to the published version of the manuscript.

Funding: The authors extend their appreciation to the Deanship of Scientific Research at King Khalid University for funding this work through the Small Groups Project under grant number (RGP1/399/44).

Institutional Review Board Statement: The Standing Committee of Bioethics Research (SCBR-020-2023), Prince Sattam bin Abdulaziz University Al-Kharj, Saudi Arabia, had approved the animal study prior to the study.

Informed Consent Statement: Not applicable.

Data Availability Statement: Not applicable.

Conflicts of Interest: The authors declare no conflict of interest.

References

1. Atchison, J.W.; Herndon, C.M.; Rusie, E. NSAIDs for Musculoskeletal Pain Management: Current Perspectives and Novel Strategies to Improve Safety. *J. Manag. Care Pharm.* **2015**, *19*, S3–S19. [CrossRef]
2. Vos, T.; Abajobir, A.A.; Abbafati, C.; Abbas, K.M.; Abate, K.H.; Abd-Allah, F.; Abdulle, A.M.; Abebo, T.A.; Abera, S.F.; Aboyans, V.; et al. Global, Regional, and National Incidence, Prevalence, and Years Lived with Disability for 328 Diseases and Injuries for 195 Countries, 1990–2016: A Systematic Analysis for the Global Burden of Disease Study 2016. *Lancet* **2017**, *390*, 1211–1259. [CrossRef]
3. World Health Organization. Musculoskeletal Conditions: Fact Sheet. Available online: https://www.who.int/mediacentre/factsheets/musculoskeletal/en/ (accessed on 11 May 2023).
4. Marinho, O.D.S.; Pereira, A. Clinical Efficacy and Safety Profile of Topical Etofenamate in the Treatment of Patients with Musculoskeletal Disorders: A Systematic Review. *Pain Ther.* **2020**, *9*, 393–410. [CrossRef] [PubMed]

5. De Sire, A.; Agostini, F.; Lippi, L.; Mangone, M.; Marchese, S.; Cisari, C.; Bernetti, A.; Invernizzi, M. Oxygen–Ozone Therapy in the Rehabilitation Field: State of the Art on Mechanisms of Action, Safety AndEffectiveness in Patients with Musculoskeletal Disorders. *Biomolecules* **2021**, *11*, 356. [CrossRef] [PubMed]
6. Karsenty, G.; Wagner, E.F. Reaching a Genetic and Molecular Understanding of Skeletal Development. *Dev. Cell* **2002**, *2*, 389–406. [CrossRef]
7. Boyle, W.J.; Simonet, W.S.; Lacey, D.L. Osteoclast Differentiation and Activation. *Nature* **2003**, *423*, 337–342. [CrossRef]
8. Desoutter, J.; Mentaverri, R.; Brazier, M.; Kamel, S. Le Remodelage Osseux Normal et Pathologique. *Rev. Francoph. des Lab.* **2012**, *2012*, 33–42. [CrossRef]
9. Kim, I.G.; Kang, S.C.; Kim, K.C.; Choung, E.S.; Zee, O.P. Screening of Estrogenic and Antiestrogenic Activities from Medicinal Plants. *Environ. Toxicol. Pharmacol.* **2008**, *25*, 75–82. [CrossRef]
10. Shubhashree, M.N.; Naik, R.; Doddamani, S.H.; Bhat, S. An Updated Review of Single Herbal Drugs in the Management of Osteoporosis. *Int. J. Complement. Altern. Med.* **2018**, *11*, 82–86. [CrossRef]
11. Alqasoumi, S.; Yusufoglu, H.; Farraj, A.; Alam, A. Effect of 6-shogaol and 6-gingerol on diclofenac sodium induced liver injury. *Int. J. Pharmacol.* **2011**, *7*, 868–873. [CrossRef]
12. Bakht, M.A.; Alajmi, M.F.; Alam, P.; Alam, A.; Alam, P.; Aljarba, T.M. Theoretical and experimental study on lipophilicity and wound healing activity of ginger compounds. *Asian Pac. J. Trop. Biomed.* **2014**, *4*, 329–333. [CrossRef] [PubMed]
13. Thomson, M.; Al-Qattan, K.K.; Al-Sawan, S.M.; Alnaqeeb, M.A.; Khan, I.; Ali, M. The Use of Ginger (Zingiber Officinale Rosc.) as a Potential Anti-Inflammatory and Antithrombotic Agent. *Prostaglandins Leukot. Essent. Fat. Acids* **2002**, *67*, 475–478. [CrossRef]
14. Haksar, A.; Sharma, A.; Chawla, R.; Kumar, R.; Arora, R.; Singh, S.; Prasad, J.; Gupta, M.; Tripathi, R.P.; Arora, M.P.; et al. Zingiber Officinale Exhibits Behavioral Radioprotection against Radiation-Induced CTA in a Gender-Specific Manner. *Pharmacol. Biochem. Behav.* **2006**, *84*, 179–188. [CrossRef]
15. Grzanna, R.; Lindmark, L.; Frondoza, C.G. Ginger—An Herbal Medicinal Product with Broad Anti-Inflammatory Actions. *J. Med. Food* **2005**, *8*, 125–132. [CrossRef]
16. Dugasani, S.; Pichika, M.R.; Nadarajah, V.D.; Balijepalli, M.K.; Tandra, S.; Korlakunta, J.N. Comparative Antioxidant and Anti-Inflammatory Effects of [6]-Gingerol, [8]-Gingerol, [10]-Gingerol and [6]-Shogaol. *J. Ethnopharmacol.* **2010**, *127*, 515–520. [CrossRef] [PubMed]
17. Kim, Y.H.; Nguyen, H.T.; Ding, Y.; Park, S.-H.; Choi, E.-M. Effect of [6]-Gingerol, a Pungent Ingredient of Ginger, on Osteoblast Response to Extracellular Reducing Sugar. *Food Sci. Biotechnol.* **2007**, *18*, 807–811.
18. Abusarah, J.; Benabdoune, H.; Shi, Q.; Lussier, B.; Martel-Pelletier, J.; Malo, M.; Fernandes, J.C.; de Souza, F.P.; Fahmi, H.; Benderdour, M. Elucidating the Role of Protandim and 6-Gingerol in Protection against Osteoarthritis. *J. Cell. Biochem.* **2017**, *118*, 1003–1013. [CrossRef] [PubMed]
19. Rondanelli, M.; Fossari, F.; Vecchio, V.; Gasparri, C.; Peroni, G.; Spadaccini, D.; Riva, A.; Petrangolini, G.; Iannello, G.; Nichetti, M.; et al. Clinical Trials on Pain Lowering Effect of Ginger: A Narrative Review. *Phyther. Res.* **2020**, *34*, 2843–2856. [CrossRef]
20. Mohd Yusof, Y.A. Gingerol and Its Role in Chronic Diseases. *Adv. Exp. Med. Biol.* **2016**, *929*, 177–207. [CrossRef]
21. Funk, J.L.; Frye, J.B.; Oyarzo, J.N.; Timmermann, B.N. Comparative Effects of Two Gingerol-Containing Zingiber Officinale Extracts on Experimental Rheumatoid Arthritis. *J. Nat. Prod.* **2009**, *72*, 403–407. [CrossRef]
22. Sung, B.; Murakami, A.; Oyajobi, B.O.; Aggarwal, B.B. Zerumbone Abolishes RaNKL-Induced NF-KB Activation, Inhibits Osteoclastogenesis, and Suppresses Human Breast Cancer-Induced Bone Loss in Athymic Nude Mice. *Cancer Res.* **2009**, *69*, 1477–1484. [CrossRef]
23. Yagihashi, S.; Miura, Y.; Yagasaki, K. Inhibitory Effect of Gingerol on the Proliferation and Invasion of Hepatoma Cells in Culture. *Cytotechnology* **2008**, *57*, 129–136. [CrossRef] [PubMed]
24. Giri, P.M.; Banerjee, A.; Layek, B. A Recent Review on Cancer Nanomedicine. *Cancers* **2023**, *15*, 2256. [CrossRef]
25. Mitchell, M.J.; Billingsley, M.M.; Haley, R.M.; Wechsler, M.E.; Peppas, N.A.; Langer, R. Engineering precision nanoparticles for drug delivery. *Nat. Rev. Drug Discov.* **2021**, *20*, 101–124. [CrossRef] [PubMed]
26. Thangavelu, P.; Sundaram, V.; Gunasekaran, K.; Mujyambere, B.; Raju, S.; Kannan, A.; Arasu, A.; Krishna, K.; Ramamoorthi, J.; Ramasamy, S.; et al. Development of Optimized Novel Liposome Loaded with 6-Gingerol and Assessment of Its Therapeutic Activity Against NSCLC In Vitro and In Vivo Experimental Models. *Chem. Phys. Lipids* **2022**, *245*, 105206. [CrossRef]
27. Ahmad, N.; Khalid, M.S.; Khan, M.F.; Ullah, Z. Beneficial Effects of Topical 6-Gingerol Loaded Nanoemulsion Gel for Wound and Inflammation Management with Their Comparative Dermatokinetic. *J. Drug Deliv. Sci. Technol.* **2023**, *80*, 104094. [CrossRef]
28. Akram, A.; Akhtar, N.; Waqas, M.K.; Rasul, A.; Rehman, K.U.; Khan, J.; Iqbal, M.; Khan, B.A. Development, Characterization and Evaluation of Ginger Extract Loaded Microemulsion: In Vitro and Ex Vivo Release Studies. *Pak. J. Pharm. Sci.* **2019**, *32*, 1873–1877. [PubMed]
29. Mazyed, E.A.; Badria, F.A.; Elnaggar, M.H.; El-Masry, S.M.; Helmy, S.A. Development of Cyclodextrin-Functionalized Transethoniosomes of 6-Gingerol: Statistical Optimization, In Vitro Characterization and Assessment of Cytotoxic and Anti-Inflammatory Effects. *Pharmaceutics* **2022**, *14*, 1170. [CrossRef] [PubMed]
30. Bhalekar, M.R.; Madgulkar, A.J. Demonstration of Lymphatic Uptake of (6)-Gingerol Solid Lipid Nanoparticles. *J. Drug Deliv. Ther.* **2019**, *9*, 461–469.

31. Ruela, A.L.M.; Perissinato, A.; Lino, M.E.D.S.; Mudrik, P.S.; Pereira, G.R. Evaluation of Skin Absorption of Drugs from Topical and Transdermal Formulations. *Brazilian J. Pharm. Sci.* **2016**, *52*, 527–544. [CrossRef]
32. Cevc, G.; Blume, G. New, Highly Efficient Formulation of Diclofenac for the Topical, Transdermal Administration in Ultradeformable Drug Carriers, Transfersomes. *Biochim. Biophys. Acta- Biomembr.* **2001**, *1514*, 191–205. [CrossRef] [PubMed]
33. Cevc, G.; Gebauer, D.; Stieber, J.; Schätzlein, A.; Blume, G. Ultraflexible Vesicles, Transfersomes, Have an Extremely Low Pore Penetration Resistance and Transport Therapeutic Amounts of Insulin across the Intact Mammalian Skin. *Biochim. Biophys. Acta-Biomembr.* **1998**, *1368*, 201–215. [CrossRef] [PubMed]
34. Gupta, P.N.; Mishra, V.; Rawat, A.; Dubey, P.; Mahor, S.; Jain, S.; Chatterji, D.P.; Vyas, S.P. Non-Invasive Vaccine Delivery in Transfersomes, Niosomes and Liposomes: A Comparative Study. *Int. J. Pharm.* **2005**, *293*, 73–82. [CrossRef] [PubMed]
35. Zeb, A.; Qureshi, O.S.; Kim, H.S.; Cha, J.H.; Kim, H.S.; Kim, J.K. Improved Skin Permeation of Methotrexate via Nanosized Ultradeformable Liposomes. *Int. J. Nanomed.* **2016**, *11*, 3813–3824. [CrossRef]
36. Shamma, R.N.; Elsayed, I. Transfersomal Lyophilized Gel of Buspirone HCl: Formulation, Evaluation and Statistical Optimization. *J. Liposome Res.* **2013**, *23*, 244–254. [CrossRef]
37. Ramezani, V.; Honarvar, M.; Seyedabadi, M.; Karimollah, A.; Ranjbar, A.M.; Hashemi, M. Formulation and Optimization of Transfersome Containing Minoxidil and Caffeine. *J. Drug Deliv. Sci. Technol.* **2018**, *44*, 129–135. [CrossRef]
38. Chen, M.; Shamim, M.A.; Shahid, A.; Yeung, S.; Andresen, B.T.; Wang, J.; Nekkanti, V.; Meyskens, F.L.; Kelly, K.M.; Huang, Y. Topical Delivery of Carvedilol Loaded Nano-Transfersomes for Skin Cancer Chemoprevention. *Pharmaceutics* **2020**, *12*, 1151. [CrossRef]
39. Paudel, K.S.; Milewski, M.; Swadley, C.L.; Brogden, N.K.; Ghosh, P.; Stinchcomb, A.L. Challenges and Opportunities in Dermal/Transdermal Delivery. *Ther. Deliv.* **2010**, *1*, 109–131. [CrossRef]
40. Opatha, S.A.T.; Titapiwatanakun, V.; Chutoprapat, R. Transfersomes: A Promising Nanoencapsulation Technique for Transdermal Drug Delivery. *Pharmaceutics* **2020**, *12*, 855. [CrossRef]
41. Abedullahh, M.H. Box-Behnken Design for Development and Optimization of Acetazolamide Microspheres. *Int. J. Pharm. Sci. Res.* **2014**, *5*, 1228–1239.
42. Zheng, Y.; Ouyang, W.Q.; Wei, Y.P.; Syed, S.F.; Hao, C.S.; Wang, B.Z.; Shang, Y.H. Effects of Carbopol®934 Proportion on Nanoemulsion Gel for Topical and Transdermal Drug Delivery: A Skin Permeation Study. *Int. J. Nanomed.* **2016**, *11*, 5971. [CrossRef] [PubMed]
43. Kazmi, I.; Al-Abbasi, F.A.; Nadeem, M.S.; Altayb, H.N.; Alshehri, S.; Imam, S.S. Formulation, Optimization and Evaluation of Luteolin-Loaded Topical Nanoparticulate Delivery System for the Skin Cancer. *Pharmaceutics* **2021**, *13*, 1749. [CrossRef]
44. He, Y.; Luo, L.; Liang, S.; Long, M.; Xu, H. Influence of Probe-Sonication Process on Drug Entrapment Efficiency of Liposomes Loaded with a Hydrophobic Drug. *Int. J. Polym. Mater. Polym. Biomater.* **2018**, *68*, 193–197. [CrossRef]
45. Shah, N.V.; Seth, A.K.; Balaraman, R.; Aundhia, C.J.; Maheshwari, R.A.; Parmar, G.R. Nanostructured lipid carriers for oral bioavailability enhancement of raloxifene: Design and in vivo study. *J. Adv. Res.* **2016**, *7*, 423–434. [CrossRef] [PubMed]
46. Qadir, A.; Aqil, M.; Ali, A.; Warsi, M.H.; Mujeeb, M.; Ahmad, F.J.; Ahmad, S.; Beg, S. Nanostructured lipidic carriers for dual drug delivery in the management of psoriasis: Systematic optimization, dermatokinetic and preclinical evaluation. *J. Drug Deliv. Sci. Tech.* **2020**, *57*, 101775. [CrossRef]
47. Abdallah, M.H. Transfersomes as a transdermal drug delivery system for enhancement the antifungal activity of nystatin. *Int. J. Pharm. Pharm. Sci.* **2013**, *5*, 560–567.
48. Kumaran, A. Antioxidant and free radical scavenging activity of an aqueous extract of Coleus aromaticus. *Food Chem.* **2006**, *97*, 109–114. [CrossRef]
49. Ghasemzadeh, A.; Jaafar, H.Z.; Rahmat, A. Optimization protocol for the extraction of 6-gingerol and 6-shogaol from Zingiber officinale var. rubrum Theilade and improving antioxidant and anticancer activity using response surface methodology. *BMC Complement. Med.* **2015**, *15*, 258. [CrossRef]
50. Moqejwa, T.; Marimuthu, T.; Kondiah, P.P.D.; Choonara, Y.E. Development of Stable Nano-Sized Transfersomes as a Rectal Colloid for Enhanced Delivery of Cannabidiol. *Pharmaceutics* **2022**, *14*, 703. [CrossRef]
51. Mitri, K.; Shegokar, R.; Gohla, S.; Anselmi, C.; Müller, R.H. Lipid Nanocarriers for Dermal Delivery of Lutein: Preparation, Characterization, Stability and Performance. *Int. J. Pharm.* **2011**, *414*, 267–275. [CrossRef]
52. Dey, S.; Pramanik, S.; Malgope, A. Formulation and Optimization of Sustained Release Stavudine Microspheres Using Response Surface Methodology. *ISRN Pharm.* **2011**, *2011*, 627623. [CrossRef]
53. Hanna, P.A.; Ghorab, M.M.; Gad, S. Development of Betamethasone Dipropionate-Loaded Nanostructured Lipid Carriers for Topical and Transdermal Delivery. *Antiinflamm. Antiallergy Agents Med. Chem.* **2018**, *18*, 26–44. [CrossRef]
54. Alam, A.; Foudah, A.I.; Alqarni, M.H.; Yusufoglu, H.S. Microwave-Assisted and Chemically Tailored Chlorogenic Acid-Functionalized Silver Nanoparticles of Citrus Sinensis in Gel Matrix Aiding QbD Design for the Treatment of Acne. *J. Cosmet. Dermatol.* **2023**, *22*, 1613–1627. [CrossRef]
55. Nawaz, A.; Farid, A.; Safdar, M.; Latif, M.S.; Ghazanfar, S.; Akhtar, N.; Al Jaouni, S.K.; Selim, S.; Khan, M.W. Formulation Development and Ex-Vivo Permeability of Curcumin Hydrogels under the Influence of Natural Chemical Enhancers. *Gels* **2022**, *8*, 384. [CrossRef] [PubMed]

56. Abdallah, M.H.; Abu Lila, A.S.; Anwer, K.; Khafagy, E.-S.; Mohammad, M.; Soliman, M.S. Formulation, Development and Evaluation of Ibuprofen Loaded Nano-Transferosomal Gel for the Treatment of Psoriasis. *J. Pharm. Res. Int.* **2019**, *31*, 1–8. [CrossRef]
57. Rajan, R.; Vasudevan, D.T. Effect of permeation enhancers on the penetration mechanism of transfersomal gel of ketoconazole. *J. Adv. Pharm. Technol. Res.* **2012**, *3*, 112. [PubMed]
58. Aodah, A.H.; Hashmi, S.; Akhtar, N.; Ullah, Z.; Zafar, A.; Zaki, R.M.; Khan, S.; Ansari, M.J.; Jawaid, T.; Alam, A.; et al. Formulation Development, Optimization by Box–Behnken Design, and In Vitro and Ex Vivo Characterization of Hexatriacontane-Loaded Transethosomal Gel for Antimicrobial Treatment for Skin Infections. *Gels* **2023**, *9*, 322. [CrossRef]
59. Merck India Life Science Products & Service Solutions. Available online: https://www.sigmaaldrich.com/IN/en?&msclkid=a4eaa568a753188f0b7dc135abc10c7a&utm_source=bing&utm_medium=cpc&utm_campaign=sigmaaldrich_countries_NA_(bingebizbrand)&utm_term=sigmaaldrichcanadaltd&utm_content=CanadaSIALBranded&gclid=a4eaa568a753188f0b7dc135abc10c7a&gclsrc=3p.ds (accessed on 11 May 2023).
60. Jahan, S.; Aqil, M.; Ahad, A.; Imam, S.S.; Waheed, A.; Qadir, A.; Ali, A. Nanostructured Lipid Carrier for Transdermal Gliclazide Delivery: Development and Optimization by Box-Behnken Design. *Inorg. Nano-Metal Chem.* **2022**, 1–4. [CrossRef]
61. Dudhipala, N.; Phasha Mohammed, R.; Adel Ali Youssef, A.; Banala, N. Effect of Lipid and Edge Activator Concentration on Development of Aceclofenac-Loaded Transfersomes Gel for Transdermal Application: In Vitro and Ex Vivo Skin Permeation. *Drug Dev. Ind. Pharm.* **2020**, *46*, 1334–1344. [CrossRef] [PubMed]
62. Mura, S.; Manconi, M.; Sinico, C.; Valenti, D.; Fadda, A.M. Penetration Enhancer-Containing Vesicles (PEVs) as Carriers for Cutaneous Delivery of Minoxidil. *Int. J. Pharm.* **2009**, *380*, 72–79. [CrossRef] [PubMed]
63. Mashru, R. Development and validation of analytical method of berberine and its quantification in multicomponent homeopathic formulations. *World J. Pharm. Res.* **2017**, *6*, 793–802. [CrossRef]
64. Tas, C.; Ozkan, Y.; Okyar, A.; Savaser, A. In Vitro and Ex Vivo Permeation Studies of Etodolac from Hydrophilic Gels and Effect of Terpenes as Enhancers. *Drug Deliv.* **2008**, *14*, 453–459. [CrossRef] [PubMed]
65. Moolakkadath, T.; Aqil, M.; Ahad, A.; Imam, S.S.; Praveen, A.; Sultana, Y.; Mujeeb, M. Preparation and Optimization of Fisetin Loaded Glycerol Based Soft Nanovesicles by Box-Behnken Design. *Int. J. Pharm.* **2020**, *578*, 119125. [CrossRef] [PubMed]
66. Mazzotta, E.; Orlando, C.; Muzzalupo, R. New Nanomaterials with Intrinsic Antioxidant Activity by Surface Functionalization of Niosomes with Natural Phenolic Acids. *Pharmaceutics* **2021**, *13*, 766. [CrossRef]
67. Raza, K.; Singh, B.; Singla, S.; Wadhwa, S.; Garg, B.; Chhibber, S.; Katare, O.P. Nanocolloidal Carriers of Isotretinoin: Antimicrobial Activity against Propionibacterium Acnes and Dermatokinetic Modeling. *Mol. Pharm.* **2013**, *10*, 1958–1963. [CrossRef] [PubMed]

MDPI AG
Grosspeteranlage 5
4052 Basel
Switzerland
Tel.: +41 61 683 77 34

International Journal of Molecular Sciences Editorial Office
E-mail: ijms@mdpi.com
www.mdpi.com/journal/ijms

www.ingramcontent.com/pod-product-compliance
Lightning Source LLC
Chambersburg PA
CBHW040800250926
51850CB00007BA/1047
9783725849413